A Modern Course on

Statistical Distributions in Scientific Work

Volume 1 – Models and Structures

NATO ADVANCED STUDY INSTITUTES SERIES

Proceedings of the Advanced Study Institute Programme, which aims at the dissemination of advanced knowledge and the formation of contacts among scientists from different countries

The series is published by an international board of publishers in conjunction with NATO Scientific Affairs Division

A	Life Sciences	Plenum Publishing Corporation London and New York
B	Physics	
C	Mathematical and Physical Sciences	D. Reidel Publishing Company Dordrecht and Boston
D	Behavioral and Social Sciences	Sijthoff International Publishing Company Leiden
E	Applied Sciences	Noordhoff International Publishing Leiden

Series C – Mathematical and Physical Sciences

Volume 17 – Statistical Distributions in Scientific Work
Volume 1 – Models and Structures

A Modern Course on

Statistical Distributions in Scientific Work

Volume 1 – Models and Structures

Proceedings of the NATO Advanced Study Institute held at the University of Calgary, Calgary, Alberta, Canada July 29 – August 10, 1974

edited by

G. P. PATIL, *The Pennsylvania State University, University Park, Pa., U.S.A.*

S. KOTZ, *Temple University, Philadelphia, Pa., U.S.A.*

J. K. ORD, *University of Warwick, Coventry, England*

D. Reidel Publishing Company

Dordrecht-Holland / Boston-U.S.A.

Published in cooperation with NATO Scientific Affairs Division

Library of Congress Cataloging in Publication Data

NATO Advanced Study Institute, University of Calgary, 1974.
A modern course on statistical distributions in scientific work.

(NATO Advanced Study Institutes series : Series C, mathematical and physical sciences ; v. 17)
Includes bibliographies and indexes.
CONTENTS: v. 1. Models and structures.—v. 2. Model building and model selection.—v. 3. Characterizations and applications.
1. Distribution (Probability theory)—Congresses.
I. Patil, Ganapati P. II. Kotz, Samuel. III. Ord, J. K. IV. Title. V. Series.
QA273.6.N37 1974 519.5′3 75–11704
ISBN 90–277–0609–3

The set of three volumes: ISBN 90 277 0609 3
Volume 1: 90 277 0606 9
Volume 2: 90 277 0607 7
Volume 3: 90 277 0608 5

Published by D. Reidel Publishing Company
P.O. Box 17, Dordrecht, Holland

Sold and distributed in the U.S.A., Canada, and Mexico
by D. Reidel Publishing Company, Inc.
306 Dartmouth Street, Boston, Mass. 02116, U.S.A.

Printed in the Netherlands by D. Reidel, Dordrecht

CONTENTS

Contents of Volumes 2 and 3 IX

Preface XIII

Introduction to Volume 1 XVII

1. INAUGURAL ADDRESS

1.1 Concept and Conduct of Calgary Course and Conference: Some Thoughts...G. P. Patil 1

2. POWER SERIES AND RELATED FAMILIES

2.1 Some Recent Advances With Power Series Distributions...S. W. Joshi 9

2.2 Multiparameter Stirling and C-Type Distributions T. Cacoullos 19

2.3 Models for Gaussian Hypergeometric Distributions Adrienne W. Kemp and C. D. Kemp 31

2.4 On the Probabilistic Structure and Properties of Discrete Lagrangian Distributions...P. C. Consul and L. R. Shenton 41

2.5 Estimation of Parameters on Some Extensions of the Katz Family of Discrete Distributions Involving Hypergeometric Functions John Gurland and Ram Tripathi 59

2.6 A Characteristic Property of Certain Generalized Power Series Distributions...G. P. Patil and V. Seshadri 83

3. RECENT TRENDS IN UNIVARIATE MODELS

3.1 Stable Distributions: Probability, Inference, and Applications in Finance--A Survey, and a Review of Recent Results...S. J. Press 87
3.2 Structural Properties and Statistics of Finite Mixtures...Javad Behboodian 103
3.3 Distribution Theory for the von Mises-Fisher Distribution and Its Application...K. V. Mardia 113
3.4 Certain Statistical Distributions Involving Special Functions and Their Applications Frank McNolty, J. Richard Huynen and Eldon Hansen 131
3.5 Tailweight, Statistical Inference and Families of Distributions - A Brief Survey Thomas P. Hettmansperger and Michael A. Keenan 161
3.6 The Families With a "Universal" Location Estimator...A. L. Rukhin 173

4. MOMENTS-RELATED PROBLEMS

4.1 Approximation Theory, Moment Problems and Distribution Functions...M. S. Ramanujan 185
4.2 Kurtosis and Departure From Normality C. C. Heyde 193
4.3 Convergence of Sequences of Transformations of Distribution Functions and Some Moment Problems W. L. Harkness 203

5. LIMIT DISTRIBUTIONS AND PROCESSES

5.1 Weak Convergence for Exponential and Monotone Likelihood Ratio Families and the Convergence of Confidence Limits...Bernard Harris and Andrew P. Soms 213
5.2 On Efficiency and Exponential Families in Stochastic Process Estimation...C. C. Heyde and P. D. Feigin 227
5.3 A Lagrangian Gamma Distribution...D. L. Nelson and P. C. Consul 241

6. MULTIVARIATE CONCEPTS AND MODELS

6.1 Multivariate Distributions at a Cross Road Samuel Kotz 247
6.2 Dependence Concepts and Probability Inequalities Kumar Jogdeo 271
6.3 New Families of Multivariate Distributions J. J. J. Roux 281

6.4 Asymptotic Expansions for the Nonnull Distributions of the Multivariate Test Statistics
Minoru Siotani 299

7. CERTAIN MULTIVARIATE DISTRIBUTIONS

7.1 A Multivariate Gamma Type Distribution Whose Marginal Laws Are Gamma, and Which Has a Property Similar to a Characteristic Property of the Normal Case...A. Dussauchoy and R. Berland 319
7.2 The Bivariate Burr Distribution
Frederick C. Durling 329
7.3 Multivariate Beta Distribution...R. P. Gupta 337
7.4 Distribution of a Quadratic Form in Normal Vectors (Multivariate Non-Central Case)
C. G. Khatri 345
7.5 Bivariate and Multivariate Extreme Distributions
J. Tiago de Oliveira 355

8. SAMPLING DISTRIBUTIONS AND TRANSFORMATIONS

8.1 On the Distribution of the Minimum and of the Maximum of a Random Number of I.I.D. Random Variables...Moshe Shaked 363
8.2 Transformation of the Pearson System With Special Reference to Type IV...K. O. Bowman and W. E. Dusenberry 381
8.3 Distributions of Characteristic Roots of Random Matrices...V. B. Waikar 391
8.4 On the Arithmetic Means and Variances of Products and Ratios of Random Variables
Fred Frishman 401
8.5 Exact and Approximate Sampling Distribution of the F-Statistic Under the Randomization Procedure...Junjiro Ogawa 407

SUBJECT INDEX 419

CONTENTS OF VOLUMES 2 AND 3

Volume 2: Model Building and Model Selection

Preface. Introduction to Volume 2.

Modelling and Simulation: J. K. ORD and G. P. PATIL, Statistical Modelling: An Alternative View. G. P. PATIL and M. BOSWELL, Chance Mechanisms for Discrete Distributions in Scientific Modelling. E. J. DUDEWICZ, Random Numbers: The Need, the History, the Generators. G. P. PATIL, M. BOSWELL, and D. FRIDAY, Chance Mechanisms in the Computer Generation of Random Variables. J. S. RAMBERG, A Probability Distribution With Applications to Monte Carlo Simulation Studies. *Model Identification and Discrimination*: R. SRINIVASAN and C. E. ANTLE, Discrimination Between Two Location and Scale Parameter Models. M. CSORGÖ, V. SESHADRI, and M. YALOVSKY, Applications of Characterizations in the Area of Goodness of Fit. S. K. KATTI and K. MC DONALD, Tests for Normality Using a Characterization. J. GURLAND and R. C. DAHIYA, Tests for Normality Using Minimum Chi-Square. M. L. TIKU, A New Statistic for Testing an Assumed Distribution. J. J. GART, The Poisson Distribution: The Theory and Application of Some Conditional Tests. *Models in the Social Sciences and Management*: J. K. ORD, The Size of Human Settlements. J. K. ORD, Statistical Models for Personal Income Distributions. R. W. RESEK, Symmetric Distributions With Fat Tails: Interrelated Compound Distributions Estimated by Box-Jenkins Methods. C. CHATFIELD, A Marketing Application of a Characterization Theorem. *Models in the Physical and Biomedical Sciences*: J. E. MOSIMANN, Statistical Problems of Size and Shape I. J. E. MOSIMANN, Statistical Problems of Size and Shape II. M. E. WISE, Skew Distributions in Biomedicine Including Some With Negative Powers of Time. S. TALWALKER, Certain Models in Medicine and Toxicology. D. M. SCHULTZ, Mass-Size Distributions - A Review and a Proposed New Model. S. S. SHAPIRO, Application of Statistical Distributions to Engineering Problems. A. G. LAURENT, Failure and Mortality From Wear and Ageing. The Teissier Model. E. ELVERS, Some Statistical Models for Seismological Discrimination. *Models in the Environmental Sciences*: M. F. DACEY, Probability Laws for Topological Properties of Drainage Basins. D. V. GOKHALE, Indices and Models for Aggregation in Spatial Patterns. G. RAMACHANDRAN, Extreme Order

Statistics in Large Samples From Exponential Type Distributions and Their Applications to Fire Loss. W. G. WARREN, Statistical Distributions in Forestry and Forest Products Research. A Modern Course on Statistical Distributions: G. P. PATIL, S. KOTZ, and J. K. ORD, Introduction and an Outline. PARTICIPANTS, Discussion.

Subject Index.

Volume 3: Characterizations and Applications

Preface. Introduction to Volume 3.

Linnik Memorial Inaugural Lecture: C. RADHAKRISHNA RAO, Some Problems in the Characterization of the Multivariate Normal Distribution. Mathematical Tools for Characterization Problems: EUGENE LUKACS, Properties of Non-Negative Definite Solutions of Certain Ordinary Differential Equations. J. ACZÉL, General Solution of a Functional Equation Connected With a Characterization of Statistical Distributions. H. J. ROSSBERG, An Extension of the Phragmén-Lindelöf Theory Which is Relevant for Characterization Theory. Characterizations Using Order Statistics: J. GALAMBOS, Characterizations of Probability Distributions by Properties of Order Statistics I. J. GALAMBOS, Characterizations of Probability Distributions by Properties of Order Statistics II. PREM S. PURI, Characterization of Distributions Based on the "Scores" Arising in the Theory of Rank Tests. KENNETH S. KAMINSKY and PAUL I. NELSON, Characterization of Distributions by the Form of Predictors of Order Statistics. Z. GOVINDARAJULU, Characterization of the Exponential Distribution Using Lower Moments of Order Statistics. M. AHSANULLAH, A Characterization of the Exponential Distribution. J. S. HUANG and J. S. HWANG, L_1-Completeness of a Class of Beta Densities. Z. GOVINDARAJULU, J. S. HUANG, and A. K. MD. EHSANES SALEH, Expected Value of the Spacings Between Order Statistics. Characterizations by Other Statistical Properties: A. L. RUKHIN, Characterizations of Distributions by Statistical Properties on Groups. F. S. GORDON and S. P. GORDON, Transcendental Functions of a Vector Variable and a Characterization of a Multivariate Poisson Distribution. A. M. KAGAN and L. B. KLEBANOV, On the Conditions of Asymptotic ε-Admissibility of the Polynomial Pitman Estimators of a Location Parameter and Certain Properties of Information Measures of Closeness. LENNART BONDESSON, Characterizations of the Gamma Distribution and Related Laws. C. G. KHATRI, Characterization of Normal Law by Constancy of Regression. BÉLA GYIRES, A Characterization of the Normal Distribution. Characterizations on Spaces and Processes: IGNACY I. KOTLARSKI, On Characterization of Probability Distributions by Conditional Expectations. B. L. S. PRAKASA RAO, On a Characterization of Probability Distributions on Locally Compact Abelian Groups-II. RASHID AHMAD, Some

Characterizations of the Exchangeable Processes and Distribution-Free Tests. E. SENETA, Characterization by Functional Equations of Branching Process Limit Laws. Characterization Problems for Damaged Observations: G. P. PATIL and M. V. RATNAPARKHI, Problems of Damaged Random Variables and Related Characterizations. R. C. SRIVASTAVA and JAGBIR SINGH, On Some Characterizations of the Binomial and Poisson Distributions Based on a Damage Model. P. C. CONSUL, Some New Characterizations of Discrete Lagrangian Distributions. KEITH ORD, A Characterization of a Dependent Bivariate Poisson Distribution. Characterizations Using Entropy Measures and Related Problems: D. V. GOKHALE, Maximum Entropy Characterizations of Some Distributions. RYOICHI SHIMIZU, On Fisher's Measure of Information for Location Family. ADRIENNE W. KEMP, Characterizations Based on Second-Order Entropy. J. ACZÉL, Some Recent Applications of Functional Equations and Inequalities to Characterizations of Probability Distributions, Combinatorics, Information Theory and Mathematical Economics. Characterizations for Discrete Distributions and Families: A. V. GODAMBE and G. P. PATIL, Some Characterizations Involving Additivity and Infinite Divisibility and Their Applications to Poisson Mixtures and Poisson Sums. A. M. NEVILL and C. D. KEMP, On Characterizing the Hypergeometric and Multivariate Hypergeometric Distributions. K. G. JANARDAN, Characterizations of Certain Discrete Distributions. Characterizations for Continuous Distributions and Families: K. V. MARDIA, Characterizations of Directional Distributions. M. S. BINGHAM and K. V. MARDIA, Maximum Likelihood Characterization of the von Mises Distribution. A. P. BASU and HENRY W. BLOCK, On Characterizing Univariate and Multivariate Exponential Distributions With Applications. J. K. WANI and G. P. PATIL, Characterizations of Linear Exponential Families.

Subject Index.

STATISTICAL DISTRIBUTIONS IN SCIENTIFIC WORK

Based on the Nato Advanced Study Institute

A Modern Course on Statistical Distributions in Scientific Work and The International Conference on Characterizations of Statistical Distributions With Applications

Held at

The University of Calgary, Calgary, Alberta, Canada
July 29-August 10, 1974

Sponsored by

International Statistical Institute
The Pennsylvania State University
The University of Calgary
Indian Statistical Institute

With the Support of

North Atlantic Treaty Organization
National Research Council of Canada
United States Army Research Office

DIRECTOR

G. P. Patil

SCIENTIFIC DIRECTORS

G. P. Patil, S. Kotz, J. K. Ord

JOINT DIRECTORS

E. G. Enns (Local Chairman), J. K. Wani, P. C. Consul

ADVISORS

T. Cacoullos
J. B. Douglas
A. Hald
W. L. Harkness
N. L. Johnson
A. M. Kagan
C. D. Kemp
I. Kotlarski
E. Lukacs
L. J. Martin
W. Molenaar
P. A. P. Moran
J. E. Mosimann
I. Olkin
C. R. Rao
L. R. Shenton
D. A. Sprott
H. Teicher
J. Tiago de Oliveira

PREFACE

These three volumes constitute the edited Proceedings of the NATO Advanced Study Institute on Statistical Distributions in Scientific Work held at the University of Calgary from July 29 to August 10, 1974. The general title of the volumes is "Statistical Distributions in Scientific Work". The individual volumes are: Volume 1 - Models and Structures; Volume 2 - Model Building and Model Selection; and Volume 3 - Characterizations and Applications. These correspond to the three advanced seminars of the Institute devoted to the respective subject areas.

The planned activities of the Institute consisted of main lectures and expositions, seminar lectures and study group discussions, tutorials and individual study. The activities included meetings of editorial committees to discuss editorial matters for these proceedings which consist of contributions that have gone through the usual refereeing process. A special session was organized to consider the potential of introducing a course on statistical distributions in scientific modeling in the curriculum of statistics and quantitative studies. This session is reported in Volume 2. The overall perspective for the Institute is provided by the Institute Director, Professor G. P. Patil, in his inaugural address which appears in Volume 1. The Linnik Memorial Inaugural Lecture given by Professor C. R. Rao for the Characterizations Seminar is included in Volume 3.

As discussed in the Institute inaugural address, not much systematic attention has been paid to the study of statistical distributions with the modern emphasis of families, models, structures and characterizations using relevant inspiration from both statistical methodology and scientific method. The purpose of the Advanced Study Institute program was to provide an open forum with focus on recent, current and forward-looking communications on different aspects of statistical distributions arising in scientific or statistical work. The program was geared for statisticians, scientists and engineers interested in the varied work relating to statistical distributions so that they might come and learn from each other in a stimulating environment. The purpose of the Characterizations Conference was to bring together research workers investigating characterizations problems that have motivation in scientific concepts and formulations or that have application or potential use for statistical theory.

For purposes of convenience, the Proceedings are being published in the three volumes as stated above. All together, they consist of 23 topical sections of 99 contributions of 1219 pages of research, review and exposition, in addition to a common preface in each followed by individual volume introductions. Subject indexes are also prepared at the end. Every effort has been made to keep the coverage of the volumes close to their individual titles. However, a quick glance of the three volumes will reveal that the volumewise overlaps of the topics as well as contents are not quite void! For example, several contributions appearing in Volume 2 do properly belong in Volume 3 as well. May this three volume set in its own modest way provide an example of synergism.

In order to appreciate the development and maturity of the field of statistical distributions during the last decade, these volumes may be compared with the Proceedings of the International Symposium on Classical and Contagious Discrete Distributions held at McGill in 1963--a milestone in the recognition and development of the theory and application of statistical distributions. The period between the McGill and Calgary meetings is marked by the appearance of Patil and Joshi's Dictionary and Bibliography of Discrete Distributions, Ord's and Mardia's monographs on Families of Distributions, Random Counts in Scientific Work edited by Patil in a three-volume set, and the encyclopedic four volumes on Distributions in Statistics by Johnson and Kotz.

Several participants from the McGill Symposium continued to work in the distributions area and were able to participate in the Calgary program. Moreover, topics such as the Gurland-Tripathi contribution in Volume 1 and the contributions in Volume 3 on the damage model introduced by C. R. Rao at McGill are a direct outgrowth of the pioneering results initiated at that Symposium. A number of new faces and topics were also visible at the Calgary Institute.

The realization of any institute or conference often fails to match the initial expectations and objectives of the organizers. Factors that are both logistic and psychological in nature tend to contribute to this discrepancy. Logistic difficulties include optimality problems for time and location. Other difficulties which must be attended to involve conflicting attitudes towards the importance of contributions to the proceedings.

We tried to cope with these problems by seeking active advice from a number of special advisors. The advice we received was immensely helpful in guiding our selection of the best experts in the field to achieve as representative and balanced a picture as possible. Simultaneously, the editors together with the referees and editorial collaborators took a rather critical and constructive attitude from initial to final stages of preparation of papers by offering specific suggestions concerning the structure, content and size. These efforts of coordination and revision were intensified through editorial sessions at the Institute itself as a necessary step for the benefit of both the general readership and the participants. It is our pleasure to record with appreciation the spontaneous cooperation of the participants. Everyone went by scientific interests often at the expense of personal preferences. The Institute atmosphere became truly creative and friendly, and this remarkable development contributed to the maximal cohesion of these volumes within the limited time period available.

Clearly the volumes attempt to cover a wide spectrum of topics in the mainstream of contemporary statistical distributions. Hopefully, an alert reader will find abundant information about the present state of art and will also be able to detect prevailing trends. These studies are important in view of the ever-increasing modeling-consciousness and sophistication of the real-world studies. The availability of the computer devices is also a significant factor. One can't help wondering what course statistical distributions (and statistics in general) would have taken, had the early giants of statistical methodology had access to the modern computer! Very likely the development of statistical distributions would have speeded up and this would have provided a direction towards better collating and interweaving of statistical theory and practice. A purpose of the Institute will be served if these proceedings help provide that direction.

In retrospect, our goals were perhaps ambitious! We had close to 100 lectures and discussions during 12 days in the middle of the summer season. For several reasons, we decided that an overworked Advanced Study Institute was to be preferred to a leisurely one. First of all, gatherings of such dimension are possible only every 5-10 years. Secondly, the previous meeting of this nature occurred some 11 years back, and the subject area of statistical distributions has witnessed an unprecedented growth in this time. Thirdly, but most importantly, was the overwhelming response from the potential participants, many of whom were to come across the continents!

Some of the above remarks, which might initially appear as irrelevant as Monday morning quarterbacking, should help the reader to appreciate what one can find in these volumes and to understand (if not to wholly forgive!) any inevitable rough edges, minor duplications and inconsistencies. We very much hope that these three volumes together will provide comprehensive and convenient reading and reference material to interested researchers, instructors and students. Each volume carries an introduction for its contents and it is hoped that these introductions will have also served a useful purpose.

In any collaborative exercise of this magnitude and nature, the unstinting support of a large number of individuals and institutions is a prerequisite for success. We are particularly grateful to the Scientific Affairs Division of NATO whose grant made the whole project feasible. Also, to the National Research Council of Canada and to the United States Army Research Office for their support, and to the various other governments and institutions whose assistance to individual members of the Institute made it a truly international gathering representing all the continents! The International Statistical Institute and the Indian Statistical Institute cosponsored the program. Our thanks go also to The Pennsylvania State University and to the University of Calgary for providing facilities. We must thank our advisors at this time too.

The success of the Institute was due, in no small measure, to the endeavors of the Local Arrangements Committee: E. G. Enns (Chairman), J. K. Wani, and P. C. Consul. We thank them for their hospitality and support. We also wish to express our sincere appreciation to Mr. M. V. Ratnaparkhi for his varied assistance during the entire project.

Miss Karen McDermid was an ever cheerful and industrious Institute secretary in the face of every adversity. Our thanks also go to the secretaries who prepared the final versions of the manuscripts. Mrs. Bonnie Henninger very ably looked after Volumes 1 and 3 in addition to the continuing chores of the Institute Director's office. Mrs. Anne Kempson took care of Volume 2 in England.

All these three volumes have been included in the ongoing NATO Advanced Study Institutes Series. They are published by the D. Reidel Publishing Company, a member of the Board of Publishers of the NATO ASI Series. It is only proper that we conclude here with our sincere thanks to both the publisher and the NATO Scientific Affairs Division for these cooperative arrangements.

January 31, 1975

G. P. Patil
S. Kotz
J. K. Ord

INTRODUCTION TO VOLUME 1

This first volume of the Proceedings is concerned with the study of theoretical models and structures. It consists of eight sections covering thirty-three contributions. The first section of the inaugural address by Institute Director, Professor G. P. Patil, opens the volume and describes the background and aims of the Institute. The inaugural also attempts to develop a general perspective for the subject area of statistical distributions and, at the end, provides a bibliography of books which dwell on statistical distributions as tools or topics for study.

The second section is devoted to an important class of non-negative lattice distributions, called power series distributions, introduced around the middle of this century by Kosambi, Noack, Tweedie and Khatri, and generalized and extensively studied by Patil in a number of publications in the early sixties. This family includes major discrete distributions, enjoys truncation and additive closures, and has become an attractive tool for various models. The advances in the past ten years are surveyed by Joshi, one of the original contributors to this topic. Whereas Patil and Seshadri announce a small elegant result for this class, Cacoullos takes us to an area which has been a cornerstone of combinatorial applications to distribution theory.

While Kemp and Kemp introduce us to various uses of hyper-geometric functions in their development of discrete models covering Kemp and Kemp family, Lagrangian inversion formula has been exploited by Consul and Shenton as a starting base for theirs. Gurland and Tripathi discuss certain discrete distributions involving hypergeometric functions and provide some estimation results. A little reflection on these papers leads one to wonder how deeply rooted the late twentieth century discrete distribution theory is in the classical theory of special functions!

The contribution of McNolty, et al. from the next section would again bring the above point home for continuous distributions also! The authors report their ongoing work of great interest which shows that a certain generalized Bessel distribution is a useful theoretical tool because it unifies the theory of a broad class of special distributions; and that it is a useful tool because it has applications in radio communication. The random sine wave problem and distributions for fluctuating radar targets are studied in terms of this distribution with particular emphasis on amplitude, phase and component distributions.

The day is hopefully not too far off when these "special functions" models will be a standard tool and part of the kit of a statistician!

The third section of recent trends in univariate models opens with a comprehensive paper by Press, who examines the theory and practice of stable distributions--a challenging topic that has received prominent attention both in the Western World and in the Soviet Union.

In his paper related to directional distributions, Mardia draws attention to a class of models that have come to the limelight only recently. Their treatment reveals results not always parallel to the conventional normal theory. The interested reader would greatly benefit further from Mardia's well-known book on "Statistics of Directional Data" along with its review by Watson in Technometrics, November, 1973.

Hettmansperger and Keenan discuss certain theoretical aspects of long-tailed distributions and provide a comparative study of various measures of tail-weight, another topic of great contemporary interest.

The vast field of mixtures is represented by Behboodian. His lucid paper covers a wide range of topics, and yet general results and unsolved problems remain to be covered. A recent bibliography compiled by Brisebois of the University of Sherbrooke should be a useful reference supplement to Behboodian's brief review. The third section concludes with a theoretical discussion of families of universal location parameters by Rukhin. This type of investigation is in the mainstream of the Russian school.

The fourth section dealing with the moments-related problems is prefaced by Ramanujan by connecting approximation theory, moment problems and distribution functions. While Heyde illuminates on the role of kurtosis in measuring non-normality, Harkness develops a pretty problem using moment sequences to generate new families of distributions. Exponential distributions turn out to be pivotal in these studies.

The fifth section arose from our concern for somewhat narrowing the gap between distribution theory and the theory of stochastic processes and limits. The sooner that the close link between these major subjects is recognized, the better it will serve the purpose of intelligent model-building. The papers by

Harris and Soms, Heyde and Feigin, and Nelson and Consul appear here in these proceedings to be indicative of this view. A much more well-concerted effort in this direction should prove fruitful.

Sections six and seven are devoted to multivariate concepts and distributions. Many of these models have been neglected thus far in applied statistical work. The orthodox reliance on multivariate normal theory is at times simply puzzling since in several cases the basic empirical situation clearly contradicts the rigid normal assumption of linear dependence and spherical invariance. Kotz in his, admittedly selective, survey of multivariate continuous distributions tries to focus on the main stages in the development of this theory and suggests some alternatives to remedy the weakness of the accepted approach. His paper may be profitably read in conjunction with the Annotated Bibliography of some 100 recent papers on multivariate distributions compiled by him for the purposes of the Study Sessions at the Institute. The bibliography is available from the author upon request.

Jogdeo presents a brief and lucid review of several concepts of dependence in multivariate distribution theory. The subject of concepts of dependence has received renewed interest and merits more attention, especially because of prevalent inadequacies resulting from the absence of fully satisfactory and sufficiently general definitions of dependence. Roux gives an exposition of generalizations of the classical Wishart and related distributions to the matrix case, whereas the work of Siotani relates to the asymptotic expansions concerning multivariate test statistics.

In the seventh section, Dussauchoy and Berland present multivariate gamma distributions based on a recursive iteration procedure. The model seems to have some useful properties and may be studied further along with the available models in this field. Durling generalizes the analytically flexible Burr distribution to the bivariate case. His work may be supplemented with the investigations of Srivastava and Mielke reported in the Bulletin of the Institute of Mathematical Statistics, July 1973, pp. 140-141. Gupta gives a useful survey for multivariate beta distribution. It may be of particular interest to many readers who are not familiar with this rather recent development.

Khatri's contribution extends the results of Kotz, et al. on the expansion of quadratic forms in normal variables in terms of Laguerre polynomials to the matrix variate case. Tiago de Oliveira, who has contributed prominently to extreme value theory in the last decade, discusses the bivariate and multivariate problems that need further attention. Some of the readers may take the author up on his invitation to this field!

The last and eighth section on sampling distributions and transformations opens with Shaked, in which the distribution of the minimum and the maximum of a random number of i.i.d. variables is investigated on the background of reliability theory. Also, the methods of construction of multivariate distributions with desired marginals are discussed. The reader would surely benefit in several ways from further study of life distributions discussed in "Reliability and Biometry" edited by Proschan and Serfling, and very recently published by SIAM.

Using sophisticated numerical techniques, Bowman and Dusenberry obtain useful conclusions concerning the classical characteristics such as moment-ratios for Pearsonian curves. Waikar surveys the distribution of characteristic roots of random matrices--a topic of his expertise joint with Krishnaiah, whereas Frishman presents some useful approximate formulae for means and variances of products and ratios of random variables.

In the last paper, Ogawa reminds us that such classical topics as Analysis of Variance are alive and well, but it may be about time for some reexamination. He leads us to modifications of the classical F-distribution, and the experts in approximations may have gained an additional challenging problem to tackle!

Let us hope that the volume as a whole will serve as a catalyst in speeding up the coordination between various (often unrelated) directions in the development of statistical distributions.

January 31, 1975

G. P. Patil
S. Kotz
J. K. Ord

INAUGURAL ADDRESS - CONCEPT AND CONDUCT OF CALGARY COURSE AND CONFERENCE: SOME THOUGHTS

G. P. Patil

Department of Statistics, The Pennsylvania State University, University Park, Pennsylvania, U.S.A.

Friends:

For a long time now, we have looked forward to this day--the day we were to meet in Calgary for the two week get together for a favorite pursuit of ours. And that is the program on statistical distributions in scientific work. It is a great pleasure for me to see that most of us have been able to travel to Calgary in spite of difficulties of sorts. Let us very much hope that, through our constant awareness and tireless effort, our communication in Calgary will prove fruitful for everyone and that each one will return home with true satisfaction and lasting inspiration. It will surely be what we want to make of it and I very fondly hope that we will make our stay in Calgary most enjoyable and profitable.

1. INTRODUCTORY REMARKS. My mind goes back eleven years when the international symposium on classical and contagious discrete distributions was held. It was the first attempt to identify the area of discrete distributions as a subject area by itself. The symposium was a great success in that it stimulated growth in the field and more importantly provided a certain direction to it. Inspired by this symposium, a number of specialized symposia have been held and comprehensive books in the field of distributions and related areas have appeared. These events have naturally prompted efforts to cover the wider ground of all possible distributions.

My mind also goes back six years when the Biometric Society symposium at AAAS was held on random counts in scientific work.

G. P. Patil et al. (eds.), Statistical Distributions in Scientific Work, Vol. 1, 1-7.

The first symposium had emphasized models and structures; the second one focused its attention on the useful role of discrete distributions in applied work.

The published proceedings and subsequent monographs impress the fact that most of the books having special emphasis on discussions and applications of statistical distributions have appeared in the last decade, particularly in the last five years. Applied aspects for distributions as a whole have begun to receive a renewed interest again partly because of the modern computer and partly because of the contemporary cost-efficiency-mindedness. The overall thrust of this is clear. We need to respond.

2. STUDY INSTITUTE PURPOSE. We need to respond on all fronts--research, training, writing, consulting, and collaborating--and with awareness for quality; and with effort for balance. The subject area is vast. The diversity is great. Within the vastness and the diversity, we however get to see overwhelming, and at times bewildering, streams of unity. They call for systematic attention.

Not much systematic attention has been paid so far to the study of statistical distributions with an emphasis on families, models, structures and characterizations. These studies need to be carried out with relevant inspiration from and side lights on both statistical methodology and scientific method. The purpose of the present advanced study institute program is naturally to provide an open forum with focus on recent, current and forward-looking communcations on various aspects of statistical distributions arising in scientific and statistical work.

We can consider ourselves fortunate indeed in having this timely opportunity under the primary sponsorship of the well-established Nato Advanced Study Institutes Program, known for years for its recognition and support of important subject areas needing a new identity or renewed thrust. The National Research Council of Canada and the United States Army Research Office have also provided support. Individual governments, institutions and colleagues have extended support and have made representation from all continents possible. Recognizing the importance of the subject area, the International Statistical Institute and the Indian Statistical Institute have co-sponsored the program. The Pennsylvania State University and the University of Calgary have provided the necessary facilities.

Let us take full advantage of this unusual opportunity and get together in every way. This two week get together constitutes a course--an advanced course--a super special advanced course! It is neither a symposium nor a conference. It is a timely occasion to learn from each other. Let each one of us be both a professor and a student.

3. PLANNED PROGRAM. You may have noticed that our activities here, throughout the two week period, will be in the form of main lectures and expositions, seminar lectures and study group discussions, tutorials and individual study. The activities will also consist of discussions and deliberations on matters relating to the edited Proceedings to be published. We will also consider seriously the potential and promise of introducing an overdue course on statistical distributions in scientific modeling in the curriculum for statistics and quantitative studies.

We have a great deal of structured program. And with our contagious enthusiasm we have scope for spontaneous activities also. Let us feel informal and do our best to meet the desired goals. Our good hosts have arranged for our comfort and cheer. The scientific directors have formulated plans for comprehensive interaction and constructive exchange. Let us contribute our individual enthusiasm, experience and expertise to help make the planning fruitful and rewarding for all. The afternoon planning session today has been specially organized to develop a total working perspective and a sound working plan to fulfill the desired concept of the long awaited advanced course. I am confident that in the pleasant and scientifically inspiring atmosphere of the Calgary Campus we will feel at home with both the technical and social activities that we are anxious to commence.

4. FUTURE NEEDS AND DIRECTIONS. In view of the ever expanding needs of society for quantification, random quantities arising in modeling, in simulation and in decision making will increasingly lead to various kinds of distributional problems and requests for solutions. Thus, statistical distributions will remain an important and focal area of study. Its base, however, needs to be strengthened and broadened well above and beyond the subject area traditionally known as distribution theory. In view of its vastness and diversity, we need to take stock periodically. Otherwise, we may lose the perspective and direction that are essential for scientific relevance. There is an urgent need for systematization, unification and consolidation. Directions for research, for education and for dissemination need to be identified and assessed for the entire field of distributions. I hope that the present Study Institute will investigate these issues and assume responsibility in suggesting some working solutions.

In relation to directions in research, a careful look at the publications on distributions in statistical and scientific courses shows the dangers of isolation and duplication resulting from the multidisciplinary nature of the subject. The distributional area draws upon both the abstract and the applied. Yet the gap between theory and practice remains. Thus, there is an urgent need to achieve the delicate balance and harmony at both individual and professional levels related to problems of current concern. Among these are problems of models and structures, of model building and selection, of characterizations and their applications. A glance of the unit titles reveals many more important areas. Further, within statistics, the modeling and sampling aspects of the subject have tended to draw apart. These interrelations also need to be strengthened. While one can have an historical appreciation for empirical modeling that involves trial and error methods of fitting data to some mathematically or otherwise inspired curves, a definite need exists for conceptual modeling that involves consideration of the chance mechanisms underlying the observed real world phenomenon. Further, we must be concerned with the statistical validation of derived models, and be able to cope with the distortions prevalent in real data, and other data-oriented problems. It is hoped that the present institute will help to make some headway in developing the use of statistical distributions in scientific modeling.

In relation to directions for training, it would be a great contribution if a proper course sequence on statistical distributions in scientific modeling could be developed soon. The thrust of such a course on statistical distributions should emphasize their development and identification through chance mechanisms relevant to scientific modeling. You are all urged to contribute to the two units set up to study this need. I hope that the concluding session of the Institute on this training problem will provide an impressive highlight of our present two week get together.

In relation to directions for dissemination, improved communication at all levels is urgently needed. There seems to be a definite need for computerized bibliographies with classifications and users' listings, for dictionaries of distributions and their interrelations, and for monographs on individual distributions and families of importance utilizing the available and powerful modern computer technology. The subject is growing fast, and more and more applications are developing rapidly. Members of the Institute may wish to consider possibilities such as a newsletter or any other effective forms of communication whether periodic or occasional.

As many of you are aware, plans have been made for the publication of the proceedings in three volumes consisting of reviewed and revised articles that are edited, indexed and interlinked. The volumes will be published with the general title of (a modern course on) Statistical Distributions in Scientific Work with tentative titles for individual volumes as: Volume I: Models and Structures; Volume II: Model Building and Model Selection; Volume III: Characterizations and Their Applications. Let us all strive to make these volumes a monumental mark of our mutual effort and of our mutual concern for proper research training in statistical distributions. I invite you and urge you to mutually interact to inject the necessary homogeneity and continuity into the published proceedings. Let us hope that these volumes will provide a comprehensive, yet high quality, exposition useful to the student, instructor and researcher.

5. CONCLUDING REMARKS. I am indeed very pleased to see that the plans for the program have come to fruition today. The plans originated in the labor of love and have climaxed in the love of labor! I saw the need to organize a program that covered all distributions at this time and was not restricted to discrete distributions as before. I also saw the need for a program concerned with characterizations of distributions and their applications. I could see that both of these activities should preferably be together. The plans for the characterizations activity emerged at the International Statistical Institute Session (ISI) in 1967 in Sydney in consultation with Professors Linnik, Lukacs, Moran and Rao. The plans for the overall distributions program emerged at the International Statistical Ecology Symposium in 1969 in New Haven and at the subsequent ISI Session in London in consultation with several friends and colleagues. As you can see, while the administrative plans started separately, it has been possible to realize the two related activities under the single theme of a Nato Advanced Study Institute. The common trend for the entire two week program is now research-review with emphasis on exposition, except for a small component of the second week's program. In accordance with the Nato proposal, the program was required to have only Patil, Kotz and Ord as three main lecturers. I am glad to say that it has been possible to broaden the stage and share it with you all. The more the merrier!! Let us enjoy it and let us make it immensely fruitful and rewarding.

In conclusion, we may pause and ponder on the following: What and how can we contribute to scientific progress in "useful manner"? Is our field still full of vigor? Has it fulfilled so far its potential? Are we using the right tools or are we sometimes carried away by mathematical considerations at the expense of real world contact? Or perhaps conversely, are we oversimplifying our scientific assumptions because of limitations of our analytical abilities? The atmosphere and the arrangements are being provided by the organizers, but only you, and we all together, can bring success to the program by active participation and interaction. I wish you my best. Please let me know if I can be of assistance. Thank you very much.

REFERENCES

[1] Aitchison, J. and Brown, J. A. C. (1957). The Lognormal Distribution. Cambridge University Press.

[2] Cresswell, W. L. and Frogatt, P. (1963). The Causation of Bus Driver Accidents. Oxford University Press.

[3] Cohen, J. E. (1971). Casual Groups of Monkeys and Men. Harvard University Press.

[4] Douglas, J. B. Analysis With Standard Contagious Distributions. In preparation.

[5] Ehrenberg, A. S. C. (1972). Repeat-Buying: Theory and Applications. Harvard University Press.

[6] Elderton, W. P. and Johnson, N. L. (1969). Systems of Frequency Curves. Cambridge University Press.

[7] Hahn, G. J. and Shapiro, S. S. (1967). Statistical Models in Engineering. John Wiley and Sons.

[8] Haight, F. A. (1967). Handbook of the Poisson Distribution. John Wiley and Sons.

[9] Hald, A. (1965). Statistical Theory With Engineering Applications. John Wiley and Sons.

[10] Herdan, G. (1966). The Advanced Theory of Language As Choice and Chance. Springer-Verlag, New York.

[11] Johnson, N. L. and Kotz, S. (1969). Distributions in Statistics, Vol. 1: Discrete Distributions. John Wiley and Sons.

[12] Johnson, N. L. and Kotz, S. (1970). Distributions in Statistics, Vol. 2: Continuous Distributions. John Wiley and Sons.

[13] Johnson, N. L. and Kotz, S. (1970). Distributions in Statistics, Vol. 3: Continuous Distributions. John Wiley and Sons.

[14] Johnson, N. L. and Kotz, S. (1972). Distributions in Statistics, Vol. 4: Multivariate Distributions. John Wiley and Sons.

[15] Kagan, A. M., Linnik, Y. V. and Rao, C. R. (1972). Characterization Problems in Mathematical Statistics. John Wiley and Sons.

[16] Kemp, A. W. and Kemp, C. D. Hypergeometric Families of Distributions and Applications. In preparation.

[17] Kendall, M. G. (1960). Advanced Theory of Statistics, Vol. 1: Distribution Theory. Griffin, London.

[18] Lancaster, H. O. (1969). The Chi-Squared Distribution. John Wiley and Sons.

[19] Lord, F. and Novick, M. R. (1968). Statistical Theories of Mental Test Scores. Addison Wesley.

[20] Lukacs, E. and Laha, R. G. (1964). Applications of Characteristic Functions. Griffin, London.

[21] Lukacs, E. (1970). Characteristic Functions. Hafner Publishing Company, New York.

[22] Mardia, K. V. (1970). Families of Bivariate Distributions. Griffin, London.

[23] Mardia, K. V. (1972). Statistics of Directional Data. Academic Press.
[24] Molenaar, W. (1970). Approximations to the Poisson, Binomial and Hypergeometric Distribution Functions. Mathematisch Centrum, Amsterdam.
[25] Moran, P. A. P. (1968). An Introduction to Probability Theory. Clarendon Press, Oxford.
[26] Mosteller, F. and Wallace, D. L. (1964). Inference and Disputed Authorship: The Federalist. Addison Wesley.
[27] Ord, J. K. (1972). Families of Frequency Distributions. Griffin, London.
[28] Patil, G. P. (ed.) (1965). Classical and Contagious Discrete Distributions. Statistical Publishing Society, Indian Statistical Institute, Calcutta.
[29] Patil, G. P. and Joshi, S. W. (1968). A Dictionary and Bibliography of Discrete Distributions. Oliver & Boyd, Edinburgh and Hafner, New York.
[30] Patil, G. P. (ed.) (1971). Random Counts in Scientific Work, Vol. 1: Random Counts in Models and Structures. The Pennsylvania State University Press.
[31] Patil, G. P. (ed.) (1971). Random Counts in Scientific Work, Vol. 2: Random Counts in Biomedical and Social Sciences. The Pennsylvania State University Press.
[32] Patil, G. P. (ed.) (1971). Random Counts in Scientific Work, Vol. 3: Random Counts in Physical Sciences, Geosciences and Business. The Pennsylvania State University Press.
[33] Patil, G. P., Pielou, E. C., and Waters, W. E. (eds.) (1971). Spatial Patterns and Statistical Distributions, Vol. 1: Statistical Ecology. The Pennsylvania State University Press.
[34] Patil, G. P., Boswell, M. T., Ratnaparkhi, M. V., and Kotz, S. A Dictionary and Bibliography of Characterizations of Statistical Distributions and Their Applications. In preparation.
[35] Patil, G. P., Boswell, M. T., and Ratnaparkhi, M. V. A Classified Bibliography of Continuous Distributions. In preparation. To be published in the International Statistical Institute Series on Statistical Bibliographies.
[36] Patil, G. P. and Ratnaparkhi, M. V. A Dictionary of Continuous Distributions. (planned)
[37] Pielou, E. C. (1969). An Introduction to Mathematical Ecology. John Wiley and Sons.
[38] Rao, C. R. (1970). Linear Statistical Inference and Its Applications. John Wiley and Sons.
[39] Steindl, J. (1965). Random Processes and the Growth of Firms (A Study of the Pareto Law). Griffin, London.
[40] Wilks, S. S. (1960). Mathematical Statistics. John Wiley and Sons, New York.
[41] Williams, C. B. (1964). Patterns of Balance in Nature. Academic Press.

SOME RECENT ADVANCES WITH POWER SERIES DISTRIBUTIONS

S. W. Joshi

Department of Mathematics, University of Texas at
Austin, Texas, U.S.A.

SUMMARY. Some recent results in power series distributions (psd's on the following topics are discussed: (i) minimum variance unbiased estimation, (ii) elementary integral expressions for the distribution function, and (iii) sum-symmetric powers series distributions which is a multivariate extension of univariate psd's.

KEY WORDS. Power series distributions, minimum variance unbiased estimation, integral expressions, distribution function, tail probabilities, truncated power series distributions, Hermite distribution, binomial distribution, Poisson distribution, negative binomial distribution, logarithmic distribution, sum-symmetric power series distributions.

1. INTRODUCTION. Let T be a countable set of real numbers with no finite limit point, a(.) a positive real valued function on T and $f(\theta) = \sum_{x \in T} a(x)\, \theta^x$, $f(\theta)$ being convergent over $\{\theta : 0 \leq \theta \leq \rho\}$. Then for $\theta \in \Omega = \{\theta : 0 < \theta < \rho\}$, the parameter space, the probability function (pf)

$$p(x) \begin{cases} a(x)\ \theta^x/f(\theta) & x \in T \\ 0 & \text{otherwise} \end{cases} \tag{1}$$

defines a generalized power series distribution [Patil (1963)] with the series function (sf) $f(\theta)$. T is its range. If $T \subset I = \{0, 1, 2, \ldots\}$ then we have the power series distribution (psd)

G. P. Patil et al. (eds.), Statistical Distributions in Scientific Work, Vol. 1, 9-17.

considered by Kosambi (1949) and Noack (1950) and studied extensively, among others, by Patil (1959), (1961), (1962), (1963), etc. Four well-known members of psd's are: (i) binomial, (ii) Poisson, (iii) negative binomial and (iv) logarithmic distributions with sf's, respectively, $(1+\theta)^n$, e^θ, $(1-\theta)^{-k}$ and $-\log(1-\theta)$. Here we wish to review some recent developments connected mainly with the author's work on the following topics: (i) minimum variance unbiased (MVU) estimation, (ii) elementary integral expressions for the distribution function of psd's, and (iii) a multivariate extension, namely, the sum-symmetric power series distributions (sspsd's) introduced by Patil (1968). Throughout, except in the last section, we shall assume $T \subseteq I$ and extend the definition of a(.) to I with $a(x) = 0$ for $x \notin T$.

2. PSD'S AND MVU ESTIMATION. As is easily checked a psd is a regular family of exponential distributions suitable reparametrized. Hence, by Lehman-Scheffe-Rao-Blackwell results the unique MVU estimator of a parameter function can be constructed if there exists one. However, the form (1) of the pf rather than the standard Pitman-Koopman form facilitates further investigation and we can obtain quite interesting results so that MVU estimation can be considered as an important feature of psd's. Specifically, we can obtain necessary and sufficient conditions for existence of the MVU estimator and an explicit expression for the same whenever it exists.

We shall use the following terminology.

(i) If $A^{(i)} \subseteq I$, $i = 1, 2, \ldots, n$, then the sum of $A^{(1)}$, $A^{(2)}, \ldots, A^{(n)}$ is $\Sigma A^{(i)} = \{\sum_{i=1}^{n} a^{(i)} : a^{(i)} \in A^{(i)}, i = 1, 2, \ldots, n\}$; $n[A] = \sum_{i=1}^{n} A^{(i)}$ if $A^{(i)} = A \subseteq I$ for $i = 1, 2, \ldots, n$.

(ii) For $a \in I$ the tailset of a is $S_a = \{x \in I : x \geq a\}$.

(iii) If $f(\theta) = \sum_{x=0}^{\infty} a(x)\,\theta^x$ then the index set of $f(\theta)$ is $W[f(\theta)] = \{x \in I : a(x) \neq 0\}$. Thus the range of a psd with sf $f(\theta)$ is $W[f(\theta)]$.

(iv) A real valued parameter function $g(\theta)$ (or $g(\alpha, \theta)$ whichever is appropriate) is said to be MVU estimable if it has a MVU estimator based on a random sample $X_1, X_2, \ldots, X_n$ of size n from an appropriate distribution.

(v) $Z = X_1 + X_2 + \ldots + X_n$ and $Y_1 = \min(X_1, X_2, \ldots, X_n)$.

Theorem 1 [Patil (1963)]. The necessary and sufficient condition for $g(\theta)$ to be MVU estimable on the basis of a random sample of size n from the psd given by (1) is that $g(\theta)\, f_n(\theta)$ (where $f_n(\theta) = (f(\theta))^n$) admit a power series expansion, say, $\Sigma\, c(z, n)\, \theta^z$ and $W[g(\theta)\, f_n(\theta)] \subseteq W[f_n(\theta)]$. Whenever it exists the MVU estimator of $g(\theta)$ is given by $h(z, n) = c(z, n)/b(z, n)$ for $z \in W[g(\theta)\, f_n(\theta)]$ and, $= 0$ otherwise, where $b(z, n)$ is the coefficient of θ^z in the expansion of $f_n(\theta)$ in powers of θ.

The theorem can be proved if one considers that $g(\theta)$ has an unbiased estimator $h(Z, n)$ based on a complete sufficient statistic Z if and only if

$$\frac{\sum_{z \in n[T]} h(z,n)\, b(z,n)\theta^z}{f_n(\theta)} = g(\theta)$$

since Z has the psd with sf $f_n(\theta)$.

The following conclusions can be drawn immediately.

(i) θ is MVU estimable if and only if for some n $n[T] = S_{\min(n[T])}$ and whenever it exists, the MVU estimator of θ is $h(z, n) = b(z-1, n)/b(z, n)$ if $Z \in n[T] + \{1\}$ and, $= 0$ otherwise.

(ii) θ is not MVU estimable if T is finite.

The above results bring out the curious fact that the MVU estimability of $g(\theta)$ depends only on the structure of the range of the psd and has nothing to do with the specific form of the psd as determined by $a(x)$. Patil (1963) has given more applications and also has given an interesting sufficient condition for MVU estimability of θ in terms of a number theoretic concept of Schnirelmann density.

Now consider the psd given by (1) truncated to the left of a point α where $\alpha \in T$. This left-truncated distribution has the pf

$$p(x; \alpha, \theta) = \begin{cases} a(x)\, \theta^x / f'(\alpha, \theta) & x \in S_\alpha \cap T \\ 0 & \text{otherwise} \end{cases} \qquad (2)$$

where $f'(\alpha, \theta) = f'(\theta) = \Sigma\, a(x)\, \theta^x$, Σ extending over $x \in S_\alpha \cap T$.

(2) is again a psd with sf $f'(\theta)$ and its MVU estimation is covered by the discussion so far if α is assumed to be a known constant.

Suppose now that both α and θ are unknown parameters and that we wish to obtain a MVU estimator of a real valued parameter function $g(\alpha, \theta)$. Clearly, (Y_1, Z) is a complete sufficient statistic for (α, θ). If $g(\alpha, \theta)$ has an MVU estimator, it must be a function of (Y_1, Z) alone. Thus we need the pf of (Y_1, Z) which can be obtained as

$$p(y_1, z) = P[Y_1 \geq y_1, Z = z] - P[Y_1 \geq y_1^*, Z = z]$$

where y_1^* denotes the smallest member of T greater than y_1, $y_1 \in T$. But

$$P[Y_1 \geq y_1, Z = z] = \frac{d(y_1,z,n)\theta^z}{f_n'(\alpha,\theta)}$$

where $d(y_1, z, n)$ is given by

$$(f'(\alpha,\theta))^n \equiv f_n'(\alpha,\theta) = \Sigma\, d(\alpha,z,n)\theta^z$$

or $$d(\alpha,z,n) = \Sigma \prod_{i=1}^{n} a(x_i)$$

the latter summation extending over $\sum_{i=1}^{n} x_i = z$ subject to $x_i \geq \alpha$, $i = 1, 2, \ldots, n$. Thus

$$p(y_1,z) = \frac{b(y_1,z,n)\theta^z}{f_n'(\alpha,\theta)} \qquad \alpha \leq y_1 \leq [z/n]$$
$$ny_1 \leq z$$

where $b(y_1, z, n) = d(y_1, z, n) - d(y_1^*, z, n)$. Now the MVU estimator of $g(\alpha, \theta)$, say $h(Y_1, Z)$ exists if and only if it satisfies

$$\sum_{Z=n\alpha}^{\infty} \sum_{y_1=\alpha}^{[z/n]} h(y_1, z)\, b(y_1, z, n)\theta^z = g(\alpha, \theta)\, f_n'(\alpha, \theta)$$

from which we get the result:

Theorem 2 [Joshi and Park (1974)]. For the psd given by (2), a function $g(\alpha, \theta)$ is MVU estimable on the basis of a random sample of size n if and only if for every $\alpha \in T$, $g(\alpha, \theta)\ f'_n(\alpha, \theta)$ admits a power series expansion in θ and $W[g(\alpha, \theta)\ f'_n(\alpha, \theta)] \subseteq W[f'_n(\alpha, \theta)]$. Whenever $g(\alpha, \theta)$ is MVU estimable the MVU estimator is given by

$$h(y_1, z) = \begin{cases} \dfrac{c(y_1, z, n) - c(y_1^*, z, n)}{b(y_1, z, n)} & \text{if } z \in W[g(y_1, \theta) f'_n(y_1, \theta)] \\ 0 & \text{otherwise} \end{cases}$$

where $c(y_1, z, n)$ = coefficient of θ^z in the expansion of $g(y_1, \theta)\ f'_n(y_1, \theta)$ in powers of θ.

We can deduce the following:

(i) α is always MVU estimable.

(ii) θ is MVU estimable if and only if $T = S_a$ for some $a \in I$.

(iii) The pf $p(\alpha+k;\ \alpha, \theta)$ and the distribution function $P(\alpha+k;\ \alpha, \theta) = \sum_{r=0}^{k} p(\alpha+r;\ \alpha, \theta)$ are always MVU estimable.

Expressions for the MVU estimators for the above functions and the variance of the MVU estimator of α are given by Joshi and Park (1974). It is quite easy to construct the explicit MVU estimator for α and θ of the truncated geometric distribution. We give below recurrence relations between coefficients $b(y_1, z, n)$ for the following four left-truncated psd's which can be used to evaluate MVU estimators when they exist.

(i) Poisson. Recurrence relation can be obtained using results of Ahuja and Enneking (1972). The following relations have been obtained using their technique.

(ii) Binomial. sf $(1+\theta)^m$, sample size n.

$$(z+1)b(\alpha,z+1,n) = (nm-z)b(\alpha,z,n) + nm\binom{m-1}{\alpha}b(\alpha,z-\alpha,n-1)$$

(iii) Negative binomial. sf $(1-\theta)^{-k}$

$$(z+1)b(\alpha,z+1,n) = (kn+z)b(\alpha,z,n) + n(k+\alpha)\binom{k+\alpha-1}{\alpha}b(\alpha,z-\alpha,n-1)$$

(iv) Logarithmic. sf $-\log(1-\theta)$.

$$(z+1)b(\alpha,z+1,n) = z\, b(\alpha,z,n) + n\, b(\alpha,z-\alpha,n-1).$$

Joshi and Park have also considered the right-truncated psd's, the main conclusion being that the right truncation point β is not MVU estimable.

3. ELEMENTARY INTEGRAL EXPRESSIONS. In this section we shall consider the psd's with the range $T = I_n \equiv \{0, 1, 2, \ldots, n\}$ or $T = I$. We define $T' = I_{n-1}$ if $T = I_n$, and $T' = I$ if $T = I$. We can easily establish the following proposition.

Proposition 3 [Joshi (to appear)]. Given the psd defined by (1), there exists a family of densities of absolutely continuous distributions

$$g(z; x) = \begin{cases} c(x; z)\ a(x)\ z^x/f(z) & 0 < z < \rho \\ 0 & \text{otherwise} \end{cases}$$

where $x \in T'$ such that

$$P(x; \theta) = \sum_{r=0}^{x} p(r) = \int_\theta^\rho g(z; x)\, dz$$

if and only if $f(\rho) = \infty$. $c(x; z)$ is given by

$$c(x; \theta)\ a(x)\theta^x = \frac{1}{\theta} \sum_{r=0}^{x} [\mu(\theta)-r]\ a(r)\theta^r$$

where $\mu = \mu(\theta)$ is the mean of the psd.

The three well known results relating Poisson to gamma, binomial to beta, and negative binomial to beta can be immediately derived from the proposition. As a new example, we can also see:

$$e^{-\alpha\theta-\theta^2/2} \sum_{r=0}^{x} \frac{H_r^*(\alpha)\theta^r}{r!}$$

$$= \int_\theta^\infty [H_{x+1}^*(\alpha) + z\, H_x^*(\alpha)]\, e^{-\alpha z-z^2/2}\, \frac{z^x}{x!}\, dz.$$

On the left hand side above is the distribution function of the

Hermite distribution [Kemp and Kemp (1965)]. $H_n^*(\alpha)$ is the modified Hermite polynomial of degree n defined by $H_n^*(\alpha) = i^{-n} H_n(i\alpha)$, $H_n(\alpha)$ being the usual Hermite polynomial of degree n.

As another application of the above proposition we have the following.

Suppose X_1, X_2 are independent random variables. X_i has a psd with sf $f_i(\theta_i)$ and the range $T_i (i = 1, 2)$. Suppose $f_1(\rho_1) = \infty$. Let $g_1(z; r)$ be the probability density function (pdf) corresponding to the psd with sf $f_1(\theta_1)$ as given by the proposition, with $r \in T_1'$ as the integer valued parameter. Consider another pdf g(z; s) as the mixture

$$g(z; s) = \sum_{r \in T_2} a_2(r)\theta_2^r \, g_1(z; s+r)/f_2(\theta_2)$$

where s is a non-negative integer with $s+r \in T_1'$ for every $r \in T_2$. If Z has the pdf g(z; s) then we can see that $P[Z \geq \theta_1] = p[X_1 - X_2 \leq s]$. This is a generalization of a connection between the non-central chi-square distribution and the distribution of the difference between two independent Poisson variates as observed by Johnson (1959).

4. SUM-SYMMETRIC POWER SERIES DISTRIBUTIONS. Analogous to the univariate psd given by (1) consider a subset T of $I^s = \{\underset{\sim}{x} = (x_1, x_2, \ldots, x_s) : x_i \in I, i = 1, 2, \ldots, s\}$, $s \geq 1$, and $a(\underset{\sim}{x})$, a non-negative real valued function on I^s which is positive on and only on T. If $f(\underset{\sim}{\theta}) = \Sigma\, a(\underset{\sim}{x})\, \theta_1^{x_1} \theta_2^{x_2} \ldots \theta_s^{x_s}$ is convergent and positive for $\underset{\sim}{\theta} \in \Omega$, Ω being the parameter space of s-dimensional vectors with positive components, an s-dimensional psd is defined by the pf

$$p(\underset{\sim}{x}) = \frac{a(\underset{\sim}{x})\, \theta_1^{x_1} \theta_2^{x_2} \ldots \theta_s^{x_s}}{f(\underset{\sim}{\theta})} \qquad \underset{\sim}{x} \in T.$$

If $f(\underset{\sim}{\theta}) = g(\theta_1 + \theta_2 + \ldots + \theta_s)$ for some g(.), Patil (1968) calls the psd a sum-symmetric psd (sspsd). Suppose $g(\theta) = \Sigma\, a(z)\, \theta^z$. Then, clearly, the pf of the s-dimensional sspsd is

$$p(\underset{\sim}{x}) = \frac{\dfrac{(x_1+x_2+x_3+\ldots+x_s)!}{x_1!x_2!\ldots x_s!}\, a(x_1+x_2+\ldots+x_s)\, \theta_1^{x_1}\, \theta_2^{x_2} \ldots \theta_s^{x_s}}{g(\theta_1+\theta_2+\ldots+\theta_s)} \tag{3}$$

Properties of the sspsd are studied in detail in Patil (1968), Joshi and Patil (1971) and (1972). Multinomial, multiple Poisson and negative multinomial distributions are sspsd's with $g(\theta_1+\theta_2+\ldots+\theta_s)$ being, respectively, $(1+\theta_1+\theta_2+\ldots+\theta_s)^n$, ex $\{\theta_1+\theta_2+\ldots+\theta_s\}$, and $(1-\theta_1-\theta_2-\ldots-\theta_s)^{-k}$.

The following can be considered a basic property of the sspsd's.

Property 4. The class of sspsd's consists of the power series mixtures on the parameter n of the singular multinomial distribution.

Thus, if given n, $\underset{\sim}{X} = (X_1, X_2, \ldots, X_s)$ has the probability generating function (pgf)

$$G(z_1,z_2,\ldots,z_s) = (p_1z_1+p_2z_2+\ldots+p_sz_s)^n$$

and n is a random variable with a psd with sf $g(\theta)$, then the unconditional distribution of $\underset{\sim}{X}$ is the sspsd given by (3) with $\theta_i = \theta p_i$, $i = 1, 2, \ldots, s$.

The pgf of the sspsd is given by

$$G(z_1,z_2,\ldots,z_s) = \frac{g(\theta_1z_1+\theta_2z_2+\ldots+\theta_sz_s)}{g(\theta_1+\theta_2+\ldots+\theta_s)}$$

from which the moments can be obtained. From these, the following two properties can be obtained with notation that $E[X_i] = \mu_i$, $Var[X_i] = \sigma_{ii}$, $Covar[X_i, X_j] = \sigma_{ij}$, $E[\Sigma X_i] =$, $Var[\Sigma X_i] = \sigma^2$.

Property 5. For an s-dimensional non-singular sspsd $\sigma_{ii} \lesseqgtr \mu_i$ if and only if $\sigma^2 \lesseqgtr \mu$ and also, $\sigma_{ii} \lesseqgtr \mu_i$ if and only if $\sigma_{jj} \lesseqgtr \mu_j$, $i, j = 1, 2, \ldots, s$. Further, correlations between pairs of components are either all positive, all negative, or all zero.

Property 6. For the s-variate sspsd, the multiple correlation coefficient between X_1 and $(X_2, X_3, \ldots, X_s)$ equals the correlation coefficient between X_1 and $X_2 + X_3 + \ldots + X_s$ in numerical magnitude.

Again, perhaps, like the univariate psd's the MVU estimation and maximum likelihood estimation are the most interesting structural features of the sspsd. Thus, for example, Patil (1968) has shown that for the sspsd, $\frac{\tilde{\theta}_i}{\tilde{\theta}} = \frac{\bar{x}_i}{\bar{x}}$ and $\frac{\hat{\theta}_i}{\hat{\theta}} = \frac{\bar{x}_i}{\bar{x}}$ where ~ and ^ denote, respectively, the maximum likelihood and MVU estimates, $\theta = \theta_1 + \theta_2 + \ldots + \theta_s$, $\bar{x}_i$ = sample mean for the ith component of the sspsd and $\bar{x} = \Sigma\, \bar{x}_i$ = the sample mean for the component-sum. Joshi and Patil (1972) have proved that if each θ_i is estimable then the proportionality $\frac{\tilde{\theta}_i}{\tilde{\theta}} = \frac{\bar{x}_i}{\bar{x}}$ characterizes the sspsd's among the multivariate psd's. Further, among the bivariate psd's the sspsd's have the characteristic property that $(\widehat{\theta_i/\theta}) = \bar{x}_i/\bar{x}$.

REFERENCES

[1] Ahuja, J. C. and Enneking, E. A. (1972). J. Amer. Stat. Assoc. 67, 232.

[2] Johnson, N. L. (1959). Biometrika 46, 352-363.

[3] Joshi, S. W. Integral expressions for the tail probabilities of the power series distributions. (To appear in Sankhyā.)

[4] Joshi, S. W. and Park, C. J. (1974). Minimum variance unbiased estimation for truncated power series distributions. Sankhyā Ser A, in press.

[5] Joshi, S. W. and Patil, G. P. (1971). Sankhyā Ser A, 33, 175-184.

[6] Joshi, S. W. and Patil, G. P. (1972). Sankhyā Ser A, 34, 377-386.

[7] Kemp, C. D. and Kemp, A. W. (1965). Biometrika 52, 381-394.

[8] Kosambi, D. D. (1949). Proc. National Inst. Sci. India 15, 109-113.

[9] Noack, A. (1950). Ann. Math. Statist. 21, 127-132.

[10] Patil, G. P. (1959). Contributions to estimation in a class of discrete distributions. Ph.D. Thesis, University of Michigan, Ann Arbor, Michigan.

[11] Patil, G. P. (1961). Sankhya 23, 269-280.

[12] Patil, G. P. (1962). Ann. Inst. Statist. Math. 14, 179-182.

[13] Patil, G. P. (1963). Ann. Math. Statist. 34, 1050-1056.

[14] Patil, G. P. (1968). Sankhyā Ser B 30, 355-366.

MULTIPARAMETER STIRLING AND C-TYPE DISTRIBUTIONS

T. Cacoullos

Statistical Unit, University of Athens, Athens, Greece

SUMMARY. The distribution of the sufficient statistic for a one-parameter power series distribution (PSD) truncated on the left at several known or unknown points is derived. The special cases of the logarithmic series, the Poisson, and the binomial and negative binomial distributions lead to multiparameter first- and second-type Stirling distributions and C-type distributions, respectively. Certain new kinds of numbers are defined in terms of partial finite differences of corresponding multiparameter Stirling and C-numbers. Minimum variance unbiased estimators are provided for the parametric function θ^{α} (α positive integer, θ the parameter of the PSD) also in the case of unknown truncation points, and for the probability function only when the truncation points are known.

KEY WORDS. Power series distributions, multiple truncation, minimum variance unbiased estimation, multiparameter Stirling and C-numbers.

1. INTRODUCTION. The distributions considered in this paper are closely connected with the problem of best estimation, i.e., minimum variance unbiased estimation (mvue), of certain functions of the parameter θ of a power series distribution (PSD) truncated on the left at several known or unknown truncation points. More specifically, given k independent random samples X_{ij}, j = $1,\ldots, n_i$, $i=1,\ldots,k$, with probability functions

G. P. Patil et al. (eds.), Statistical Distributions in Scientific Work, Vol. 1, 19-30.

$$p_i(t) = P[X_{ij} = t] = \frac{a_i(t)\,\theta^t}{f_i(\theta, r_i)}, \quad t = r_i, r_i+1, \ldots \tag{1.1}$$

where

$$f_i(\theta, r_i) = \sum_{t=r_i}^{N} a_i(t)\,\theta^t, \quad \text{(N finite or infinite)} \tag{1.2}$$

it is desired to estimate some function of θ including $p_i(x)$ itself.

Estimability conditions for certain functions of the parameter θ of a one-parameter PSD with probability function

$$P[X = t] = \frac{a(t)\,\theta^t}{f(\theta)}, \quad t \in T \;(\theta > 0) \tag{1.3}$$

where $T \subset I_o$, I_o the set of non-negative integers, were given by Patil (1963). Thus it was shown that, on the basis of a random sample of size $n \geq 1$ from (1.3), θ is estimable, i.e., it has an unbiased estimator if, and only if,

$$n[T] = S_{\min(n[T])}, \tag{1.4}$$

where

$$n[T] = T_1 + T_2 + \ldots + T_n \text{ with } T_i = T, \; (i = 1, \ldots, n)$$

and S_a denotes the tail set of a, that is,

$$S_a = \{x : x \in I_o \text{ and } x \geq a\}$$

In particular, (1.4) implies that θ is not estimable whenever T is a finite set, e.g., $\theta = p/q$ in the binomial case. On the other hand, θ is estimable if

$$T = S_r, \quad r \in I_o$$

where r denotes the truncation point, assumed known. This situation includes the Poisson, the logarithmic series and the negative binomial distributions truncated on the left at a known point. In fact, Roy and Mitra (1957) provide the mvue of θ^α for any integer $\alpha > 0$ in the form

$$\hat{\theta}_\alpha = \frac{A(x-\alpha, n, r)}{A(x, n, r)} \tag{1.5}$$

where

$$A(x,n,r) = \Sigma\, a(x_1) \ldots a(x_n), \tag{1.6}$$

the summation extending over all n-tuples $(x_1,\ldots,x_n)$ of integers $x_i \geq r$ with $x_1 + \ldots + x_n = x$.

The corresponding case when the truncation point r is unknown was studied by Joshi (1972), who showed that on the basis of a random sample, $x_1,\ldots,x_n$ from (1.3), the mvue of r is

$$\hat{r}(y,x) = y - \frac{A(x,n,y+1)}{A(x,n,y) - A(x,n,y+1)} \tag{1.7}$$

where

$$x = x_1 + \ldots + x_n \text{ and } y = \min(x_1,\ldots,x_n)$$

Interestingly enough, though θ is not estimable in the case of a finite-range distribution, it was shown that the probability function itself (hence also the corresponding distribution function) of a PSD truncated on the left is estimable, both when the truncation point r is known, Patil (1963), or unknown, Joshi (1972 and (1974).

The main difficulty, reflected even in the treatment of some special cases of truncation away from zero (r = 1), Tate and Goen (1958), Cacoullos (1961), Patil and Bildikar (1966), Cacoullos and Charalambides (1972), lies in deriving an explicit or computationally attractive expression for the distribution of the sufficient statistic, the sum of the observations; equivalently, to get around the evaluation of expressions such as A(x,n,r) in (1.6).

The above difficulty was resolved by employing exponential generating function (egf) techniques, which emerge quite naturally in the treatment of the aforementioned distribution problem, the n-fold convolution of a PSD truncated on the left at r ($r \geq 1$). Indeed, Charalambides (1972), (1973a), (1973b) and Cacoullos and Charalambides (1972) show how the egf technique provides a unified approach for the solution to the problem of mvue of the estimable functions of θ in (1.3), both when r is known or unknown. At the same time, the treatment of individual families of PSD's leads to corresponding extensions of certain kinds of numbers; the Poisson family is associated with generalized Stirling numbers of the second kind, whereas the logarithmic series distribution with Stirling numbers of the first kind (the well known Stirling numbers correspond to the simple case of truncation away from zero, i.e., r = 1); finally, the binomial and negative binomial cases led to the introduction, Cacoullos and Charalambides (1972), and extension, Charalambides (1972), (1973a), of a new kind of numbers called C-numbers.

The more general situation of truncation at several known truncation points, that is, (1.1) with $a_i(x) = a(x)$ $i = 1,\ldots,k$, motivated the introduction of certain multiparameter Stirling and C-numbers, Cacoullos (1973a) and (1973b). The mvue of θ^α and the probability function itself were given in terms of these multiparameter numbers, Cacoullos (1973c).

Here we consider the relevant distribution and estimation problems under the assumption of (1.1), especially, when the truncation points $r_1,\ldots,r_k$ are unknown, $a_i(x) = a(x)$, $i = 1,\ldots,k$ and $N = \infty$. When $\underset{\sim}{\rho} = (r_1,\ldots,r_k)$ is known, the mvue of $p_i(t)$ is also considered. In Section 3, the special cases of the logarithmic series, the Poisson and the binomial and negative binomial distributions are investigated. Some of the results of these sections are based on properties of the multiparameter Stirling and C-numbers, studied in Cacoullos (1973b).

2. GENERAL RESULTS.

For our purposes we require the following:

Definition 2.1. Let $G(y)$ be a real-valued function defined for all k-dimensional vectors $\underset{\sim}{y} = (y_1,\ldots,y_k)$ with integer components. Let $\Delta_{\underset{\sim}{y}} G(\underset{\sim}{y})$ denote the kth order partial difference of $G(\underset{\sim}{y})$ defined by

$$\Delta_{\underset{\sim}{y}} G(\underset{\sim}{y}) = \Delta y_k \, \Delta y_{k-1} \cdots \Delta y_1 \, G(\underset{\sim}{y}) \tag{2.1}$$

where Δy_i denotes the usual forward-difference operator operating on y_i, that is, $\Delta y_i \, G(\underset{\sim}{y}) = G(\underset{\sim}{y}+\underset{\sim}{e}_i) - G(\underset{\sim}{y})$, $\underset{\sim}{e}_i$ denoting the k-component vector $(0,\ldots,0,1,\ldots,0)$ with zero elements except 1 at the ith position.

We require the following lemma, of interest in itself.

Lemma 2.1. The probability function of the integer-valued vector random variable $\underset{\sim}{Y} = (Y_1,\ldots,Y_k)$ is given by

$$p(\underset{\sim}{y}) = P[\underset{\sim}{Y} = \underset{\sim}{y}] = (-1)^k \, \Delta_{\underset{\sim}{y}} \, Q(\underset{\sim}{y}) \tag{2.2}$$

where $Q(\underset{\sim}{y})$ denotes the tail probability function of $\underset{\sim}{Y}$, i.e.,

$$Q(\underset{\sim}{y}) = P[Y_1 \geq y_1, Y_2 \geq y_2, \ldots, Y_k \geq y_k] \equiv P[\underset{\sim}{Y} \geq \underset{\sim}{y}]$$

Proof. For $k = 2$, we have

$$P[Y_1 = y_1, Y_2 = y_2] = Q(y_1,y_2)-Q(u_1+1,u_2)-Q(u_1,u_2+1)+Q(u_1+1,u_2+1),$$

which in view of (2.1) gives (2.2). For $k > 2$ the proof is analogous (cf. the kth order partial difference of a continuous distribution function $F(\underset{\sim}{y})$ involved in the definition of the density at $\underset{\sim}{y}$).

Now as regards the distribution theory in the case of a known vector $\underset{\sim}{\rho} = (r_1,r_2,\ldots,r_k)$ of truncation points, we can state the following theorem, which is an immediate extension of Theorem 2.1 of Cacoullos (1973a).

Theorem 2.1. The sum

$$X = \sum_{i=1}^{k} \sum_{j=1}^{n_i} X_{ij}$$

of the observations of the k samples X_{ij}, $i=1,\ldots,k$, $j=1,\ldots,n_i$ from $p_i(t)$ is a complete sufficient statistic for θ and has the probability function

$$P[X = x] = \frac{c(x;\underset{\sim}{\nu},\underset{\sim}{\rho})}{g_{\underset{\sim}{\nu}}(\theta;\underset{\sim}{\rho})} \frac{\theta^x}{x!}, \quad x = m,m+1,\ldots \tag{2.3}$$

where we set $m = n_1r_1 + \ldots + n_kr_k = (\underset{\sim}{\nu},\underset{\sim}{\rho})$ and $g_{\nu}(u;\underset{\sim}{\rho})$ is the egf of the numbers $c(x;\underset{\sim}{\nu},\underset{\sim}{\rho})$ with the representations

$$g_{\nu}(u;\underset{\sim}{\rho}) \equiv \sum_{x=m}^{\infty} c(x;\underset{\sim}{\nu},\underset{\sim}{\rho}) \frac{u^x}{x!} = \prod_{i=1}^{k} \frac{1}{n_i!} [f_i(u,r_i)]^{n_i} \tag{2.4}$$

$$c(x;\underset{\sim}{\nu},\underset{\sim}{\rho}) = \frac{x!}{n_1! \ldots n_k!} \sum_{x} \prod_{i=1}^{k} \prod_{j=1}^{n_i} a_i(x_{ij}) \tag{2.5}$$

the summation extending over all n-tuples ($n = n_1 + \ldots + n_k$) of integers x_{ij} such that

$$x_{ij} \geq r_i \quad \text{and} \quad \sum_{i=1}^{k} \sum_{j=1}^{n_i} x_{ij} = x.$$

If, in addition to θ, $\underset{\sim}{\rho}$ is also unknown, then we have

Proposition 2.1. A complete sufficient statistic for the parameter set $(\theta,\underset{\sim}{\rho})$ of (1.1) is the set of statistics

$$(X,\underset{\sim}{Y}) = (X,Y_1,\ldots,Y_k)$$

where

$$Y_i = \min(X_{i1}, \ldots, X_{in_i}) \quad i = 1,\ldots,k$$

$$X = \sum_{i=1}^{k} \sum_{j=1}^{n_i} X_{ij}.$$

<u>Theorem 2.2.</u> The distribution of the sufficient statistic $(X,\underset{\sim}{Y})$ is a PSD with probability function

$$p(x,\underset{\sim}{y};\theta,\underset{\sim}{\rho}) = \frac{(-1)^k \Delta\underset{\sim}{y}\, c(x;\underset{\sim}{y},\underset{\sim}{\nu})}{g_{\nu}(\theta;\rho)} \frac{\theta^x}{x!}, \; x \geq (\underset{\sim}{\nu},\underset{\sim}{y}),\; \underset{\sim}{y} \geq \underset{\sim}{\rho} \tag{2.6}$$

where $g_{\underset{\sim}{\nu}}(\theta,\underset{\sim}{\rho})$ is given by (2.4).

<u>Proof.</u> Setting

$$Q(\underset{\sim}{y};x) = P[\underset{\sim}{Y} \geq \underset{\sim}{y}\;,\; X = x]\;,$$

we have by Lemma 2.1

$$P[X = x\;,\; \underset{\sim}{Y} = \underset{\sim}{y}] = (-1)^k \Delta\underset{\sim}{y}\; Q(\underset{\sim}{y};x). \tag{2.7}$$

However, we have

$$Q(\underset{\sim}{y};x) = P[\underset{\sim}{Y} \geq \underset{\sim}{y},\; X = x] = P[X_{ij} \geq y_i,\; X = x,\; i=1,\ldots,k,$$
$$j=1,\ldots,n_i]$$

$$= \frac{\theta^x}{\prod_{i=1}^{k} [f_i(\theta,r_i)]^{n_i}} \sum_{\underset{\sim}{y}} \prod_{i=1}^{k} \prod_{j=1}^{n_i} a_i(x_{ij})$$

where the summation extends over all n-tuples of integers x_{ij} such that

$$x_{ij} \geq y_i \quad \text{and} \quad \sum_{i=1}^{k} \sum_{j=1}^{n_i} x_{ij} = x.$$

Hence, by (2.4) and (2.5), Q(y;x) can be written as

$$Q(y;x) = \frac{c(x,\underset{\sim}{y},\underset{\sim}{\nu})}{g_{\underset{\sim}{\nu}}(\theta;\underset{\sim}{\rho})}\;,\; \frac{\theta^x}{x!} \tag{2.8}$$

However, since $\Delta_{\underset{\sim}{y}}$ operates only on $\underset{\sim}{y}$, we have

$$\Delta_{\underset{\sim}{y}} Q(\underset{\sim}{y};x) = \frac{1}{g_{\underset{\sim}{\nu}}(\theta,\underset{\sim}{\rho})} \frac{\theta^x}{x!} \Delta_{\underset{\sim}{y}} c(x;\underset{\sim}{y},\underset{\sim}{\nu});$$

thus (2.7) yields (2.6).

Let us now consider the estimation of θ. From the results of Joshi (1972), extended by Charalambides (1973b), θ^α is estimable for any integer $\alpha > 0$, and we have the following [cf. (1.5)].

Theorem 2.3. On the basis of k independent random samples X_{ij}, $j=1,\ldots,n_i$, $i=1,\ldots,k$ where X_{ij} follows (1.1) with $N = \infty$, there exists a unique mvue of θ^α which is a function of (X,Y). This is given by

$$\hat{\theta}_\alpha(x,\underset{\sim}{y}) = \frac{\Delta_{\underset{\sim}{y}} c(x-\alpha;\underset{\sim}{y},\underset{\sim}{\nu})}{\Delta_{\underset{\sim}{y}} c(x;\underset{\sim}{y},\underset{\sim}{\nu})} (x)_\alpha \tag{2.9}$$

for $\underset{\sim}{y} \geq \underset{\sim}{\rho}$, $x \geq (\underset{\sim}{\nu},\underset{\sim}{y})$

Proof. The condition of unbiasedness

$$E[\hat{\theta}_\alpha(x,\underset{\sim}{y})] = \theta^\alpha$$

yields, by (2.6),

$$\sum_{x=m}^{\infty} \sum_{y} \hat{\theta}(x,\underset{\sim}{y}) \Delta_{\underset{\sim}{y}} c(x;\underset{\sim}{y},\underset{\sim}{\nu}) \frac{\theta^x}{x!} = \sum_{x=m}^{\infty} \sum_{\underset{\sim}{y}} \Delta_{\underset{\sim}{y}} c(x;\underset{\sim}{y},\underset{\sim}{\nu}) \frac{\theta^{x+\alpha}}{x!}$$

where the second summation extends over all y such that $m \leq (\underset{\sim}{\nu},\underset{\sim}{y}) \leq x$. Equating the coefficients of θ^x on each side gives (2.9).

This, for $\alpha = 1$, gives the mvue $\hat{\theta}(x,\underset{\sim}{y})$ of θ,

$$\hat{\theta}(x,\underset{\sim}{y}) = \frac{\Delta_{\underset{\sim}{y}} c(x-1;\underset{\sim}{y},\underset{\sim}{\nu})}{\Delta_{\underset{\sim}{y}} c(x;\underset{\sim}{y},\underset{\sim}{\nu})} x \tag{2.10}$$

The variance v of $\hat{\theta}$ is also estimable with mvue

$$\hat{v}(x,\underset{\sim}{y}) = [\hat{\theta}(x,\underset{\sim}{y})]^2 - \hat{\theta}_2(x,\underset{\sim}{y}) \tag{2.11}$$

However, since by (2.9)

$$\hat{\theta}_2(x,\underset{\sim}{y}) = \hat{\theta}(x,\underset{\sim}{y})\ \hat{\theta}(x-1,\underset{\sim}{y})$$

we can write (2.11) as

$$\hat{v}(x,\underset{\sim}{y}) = \hat{\theta}(x,\underset{\sim}{y})\ [\hat{\theta}(x,\underset{\sim}{y}) - \hat{\theta}(x-1,\underset{\sim}{y})] \tag{2.12}$$

In general, it can easily be verified that

$$\hat{\theta}_\alpha(x,\underset{\sim}{y}) = \prod_{j=0}^{\alpha-1} \hat{\theta}(x-j,\underset{\sim}{y}). \tag{2.13}$$

This points out a difference between the mvue $\hat{\theta}_\alpha$ of θ^α and the mle θ^*_α say, of θ^α, which would satisfy, instead of (2.13),

$$\theta^*_\alpha(x,\underset{\sim}{y}) = [\theta^*_1(x,\underset{\sim}{y})]^\alpha$$

The mvue of θ^α when the truncation vector $\underset{\sim}{\rho}$ is assumed known is easily found to be [cf. Cacoullos (1973a)]

$$\hat{\theta}_\alpha(x) = \frac{c(x-\alpha,\underset{\sim}{\rho},\underset{\sim}{\nu})}{c(x,\underset{\sim}{\rho},\underset{\sim}{\nu})}\ (x)_\alpha \qquad \text{for } x \geq m+\alpha \tag{2.14}$$

Indeed, a comparison between Theorem 2.1 and Theorem 2.2, shows that the role of $c(x;\underset{\sim}{\rho},\underset{\sim}{\nu})$, when $\underset{\sim}{\rho}$ is known is played by $(-1)^k\ \Delta_y\ c(x;\underset{\sim}{y},\underset{\sim}{\nu})$ when $\underset{\sim}{\rho}$ is estimated, the probability function $P[X = x] = Q(\underset{\sim}{\rho};x)$ being replaced by $P[X = x,\ \underset{\sim}{Y} = \underset{\sim}{y}] = (-1)^k\ \Delta_{\underset{\sim}{y}} Q(\underset{\sim}{y};x)$.

Finally, we give the mvue of the probability function $p_i(t)$ itself, which, as already remarked in Section 1, is estimable even when the distribution range is finite [N finite in (1.2)]. When $\underset{\sim}{\rho}$ is known the mvue of $p_i(t)$ is [cf. Cacoullos (1973)],

$$p^*_i(t;x) = a_i(t)\ \frac{c(x-t;\underset{\sim}{\rho},\underset{\sim}{\nu}-\underset{\sim}{e}_i)}{c(x;\underset{\sim}{\rho},\underset{\sim}{\nu})}\ \frac{(x)_t}{n_i}\ ,\quad x \geq m-r_i+t,\ (t \geq r_i) \quad i=1,\ldots,k \tag{2.15}$$

<u>Note.</u> In Cacoullos (1973c) we considered $p_i(t)$ when the $a_i(t)$ in (1.1) are the same for all i, but the extension for (1.1) is immediate. The mvue of $p_i(t)$ when $\underset{\sim}{\rho}$ is unknown and of $\underset{\sim}{\rho}$ itself will be considered in a subsequent paper.

3. MULTIPARAMETER STIRLING AND C-TYPE DISTRIBUTIONS. The special cases of logarithmic series, Poisson, binomial and negative binomial distributions give rise to natural extensions of the Stirling numbers of the first and second kind and the so-called C-numbers, as illustrated in the following table: Multiparameter Stirling and C-Numbers. The simple C-numbers, i.e., for $k=1$, $r_1 = r=1$, were introduced by Cacoullos and Charalambides (1972) in relation to mvue problems for the binomial and negative binomial distributions truncated away from zero.

The representations of each kind of these numbers are readily obtained from the general expression (2.5) by substituting the corresponding appropriate $a(x_{ij})$ in each case. Moreover, by analogy to Patil and Wani (1965), we introduce the following.

Definition 3.1. The distributions (2.4) and (2.6) are called

(a) multiparameter first-type Stirling distributions when

$$c(x;\underset{\sim}{y},\underset{\sim}{\nu}) = |s(x;\underset{\sim}{y},\underset{\sim}{\nu})|$$

(b) multiparameter second-type Stirling distributions when

$$c(x;\underset{\sim}{y},\underset{\sim}{\nu}) = S(x;\underset{\sim}{y},\underset{\sim}{\nu})$$

(c) multiparameter C-type distributions when

$$c(x;\underset{\sim}{y},\underset{\sim}{\nu}) = C(x;\underset{\sim}{y},\underset{\sim}{\nu},\underset{\sim}{\sigma}) \quad \text{or} \quad |C(x;\underset{\sim}{y},\underset{\sim}{\nu},-\underset{\sim}{\sigma})|$$

The mvue of the Poisson parameter λ and the parameter θ of the negative binomial distribution can be written in a more suggestive way from the point of view of comparisons with the corresponding estimators in the usual case of no truncation. This is possible because of the recurrence relations

$$S(x;\underset{\sim}{y},\underset{\sim}{\nu}) = n\,S(x-1;\underset{\sim}{y},\underset{\sim}{\nu}) + \sum_{i=1}^{k}\binom{x-1}{r_i-1}S(x-r_i;\underset{\sim}{y},\underset{\sim}{\nu}-\underset{\sim}{e}_i), \tag{3.1}$$

$$|C(x;\underset{\sim}{y},\underset{\sim}{\nu},-\underset{\sim}{\sigma})| = [(\underset{\sim}{\sigma},\underset{\sim}{\nu}) + x]\,|C(x-1;\underset{\sim}{y},\underset{\sim}{\nu}-\underset{\sim}{e}_i,-\underset{\sim}{\sigma})$$

$$+ \sum_{i=1}^{k}\binom{x-1}{r_i-1}(s_i+r_i-1)_{r_i}\,|C(x-r_i;\underset{\sim}{y},\underset{\sim}{\nu}-\underset{\sim}{e}_i,-\underset{\sim}{\sigma})|, \tag{3.2}$$

established in Cacoullos (1973b). Thus, using (3.1), the mvue of λ as obtained from (2.9) with $c(x;y,\nu) = S(x;y,\nu)$ take the form

MULTIPARAMETER STIRLING AND C-NUMBERS

Sampled distribution	Logarithmic series distribution	Poisson	Binomial	Negative binomial
$f_i(\theta, r_i)$ $i=1,\ldots,k$	$-\log(1-\theta) - \sum_{j=1}^{r_i-1} \frac{\theta^j}{j}$	$e^\theta - \sum_{j=0}^{r_i-1} \frac{\theta^j}{j!}$	$(1+\theta)^{s_i} - \sum_{j=0}^{r_i-1} \binom{s_i}{j} \theta^j$	$(1-\theta)^{-s} - \sum_{j=0}^{r_i-1} \binom{-s_i}{j} (-\theta)^j$
$a_i(t)$, $(t \geq r_i)$	$\frac{1}{t}$	$\frac{1}{t!}$	$\binom{s_i}{t}$	$\binom{-s_i}{t} (-1)^t$
Parameter θ	$0 < \theta < 1$	$\lambda > 0$	p / q	p or q
Multiparameter numbers: $c(x; \underset{\sim}{y}, \underset{\sim}{\nu})$	Signless Stirling number of the first kind: $s(x; \underset{\sim}{y}, \underset{\sim}{\nu})$	Stirling number of the second kind: $S(x; \underset{\sim}{y}, \underset{\sim}{\nu})$	C-number $C(x; \underset{\sim}{y}, \underset{\sim}{\nu}, \underset{\sim}{\sigma})$ $\underset{\sim}{\sigma} = (s_1, \ldots, s_k)$	C-number $C(x; \underset{\sim}{y}, \underset{\sim}{\nu}, -\underset{\sim}{\sigma})$ $\underset{\sim}{\sigma} = (s_1, \ldots, s_k)$

$$\hat{\lambda}(x,\underset{\sim}{y}) = \frac{x}{n}\left[1 - \sum_{i=1}^{k}\binom{x-1}{r_i-1}\frac{\Delta_{\underset{\sim}{y}} S(x-r_i;\underset{\sim}{y},\underset{\sim}{\nu}-\underset{\sim}{e}_i)}{\Delta_{\underset{\sim}{y}} S(x;\underset{\sim}{y},\underset{\sim}{\nu})}\right] = c_1\hat{\lambda}_o ,$$

where $\hat{\lambda}_o = x/n = \bar{x}$ is the usual mvue of λ under no truncation and c_1, $0 \le c_1 \le 1$, may be regarded as the correction factor by which $\hat{\lambda}_o$ must be multiplied to yield the corresponding mvue of λ under truncation on the left at unknown truncation points. Similarly, for the negative binomial we have by (3.2) and (2.9)

$$\hat{\theta}(x,\underset{\sim}{y}) = \frac{x}{(\underset{\sim}{\sigma},\underset{\sim}{\nu}) + x-1}\left[1 - \sum_{i=1}^{k}\binom{x-1}{r_i-1}\frac{\Delta_{\underset{\sim}{y}}\,|C(x-r_i;\underset{\sim}{y},\underset{\sim}{\nu}-\underset{\sim}{e}_i,-\underset{\sim}{\sigma})|}{(s_i+r_i-1)_{r_i}\,\Delta_{\underset{\sim}{y}}\,|C(x;\underset{\sim}{y},\underset{\sim}{\nu},-\underset{\sim}{\sigma})|}\right],$$

that is,

$$\hat{\theta}(x,\underset{\sim}{y}) = c_2\,\hat{\theta}_o = c_2\cdot\frac{x}{(\underset{\sim}{\sigma},\underset{\sim}{\nu})+x-1}$$

where $\hat{\theta}_o$ denotes the usual mvue of θ and c_2, $0 \le c_2 \le 1$, is a correction factor.

For combinatorial interpretations of the multiparameter Stirling and C-numbers in terms of urn and cell-occupancy models, we refer to Cacoullos (1973b). There, some asymptotic relations between the signless Stirling numbers $|s(x;\underset{\sim}{y},\underset{\sim}{\nu})|$ and the signless C-numbers $|C(x;\underset{\sim}{y},\underset{\sim}{\nu},-\underset{\sim}{\sigma})|$, as well as between the Stirling numbers $S(x;\underset{\sim}{y},\underset{\sim}{\nu})$ and the C-numbers $C(x;\underset{\sim}{y},\underset{\sim}{\nu},\underset{\sim}{\sigma})$ are also given. They reflect the limiting relations between the negative binomial and the logarithmic series distributions in the former case, and between the binomial and the Poisson in the latter case.

REFERENCES

[1] Cacoullos, T. (1961). Ann. Math. Statist. 32, 904-905.
[2] Cacoullos, T. (1973a). Bull. Intern. Statist. Inst. 45, 231-237.
[3] Cacoullos, T. (1973b). Multiparameter Stirling and C-numbers. Submitted to SIAM J. Appl. Math.

[4] Cacoullos, T. (1973c). Minimum variance unbiased estimation for multiply truncated power series distributions. Submitted to SIAM J. Appl. Math.

[5] Cacoullos, T. and Charalambides, Ch. (1972). On minimum variance unbiased estimation for truncated binomial and negative binomial distributions. To appear in Ann. Inst. Statist. Math., Tokyo.

[6] Cacoullos, T. and Charalambides, Ch. (1972). Minimum variance unbiased estimation for truncated discrete distributions. Colloquia Mathematica Societatis Janos Bolyai, 9 European Meeting of Statisticians, Budapest.

[7] Charalambides, Ch. (1972). Minimum variance unbaised estimation for a class of left-truncated discrete distributions. Doctoral Thesis, Athens University, Athens, Greece. (In Greek)

[8] Charalambides, Ch. (1973a). The generalized Stirling and C-numbers. Sankhyā Ser A (to appear).

[9] Charalambides, Ch. (1973b). Minimum variance unbaised estimation for a class of left-truncated discrete distributions. Sankhyā Ser A (to appear).

[10] Joshi, S. W. (1972). Minimum variance unbiased estimation for truncated distributions with unknown truncation points. Tech. Rep., University of Austin, Texas.

[11] Joshi, S. W. (1975). In Statistical Distributions in Scientific Work, Volume I, Models and Structures, Patil, Kotz, and Ord (eds.). D. Reidel, Dordrecht and Boston, pp. 9-17.

[12] Patil, G. P. (1963). Ann. Math. Statist. 34, 1050-1056.

[13] Patil, G. P. and Bildikar, S. (1966). Sankhyā Ser A 28, 239-250.

[14] Patil, G. P. and Wani, J. K. (1965). Sankhyā Ser A 27, 271-280.

[15] Riordan, J. (1958). An Introduction to Combinatorial Analysis. Wiley, New York.

[16] Roy, J. and Mitra, S. K. (1957). Sankhyā Ser A 18, 371-278.

[17] Tate, R. F. and Goen, L. R. (1958). Ann. Math. Statist. 29, 755-765.

MODELS FOR GAUSSIAN HYPERGEOMETRIC DISTRIBUTIONS

Adrienne W. Kemp and C. D. Kemp

University of Bradford, Bradford, England

SUMMARY. The four distributions discussed here belong to two partially overlapping classes. The class(es) to which each belongs is found to determine the kinds of models giving rise to it. Urn, contagion, stochastic and STER process models are considered together with conditionality, weighting and mixing models.

KEY WORDS. Hypergeometric distributions, urn models, contagion, stochastic processes, STER distributions, conditionality, weighted distributions, mixed distributions.

1. INTRODUCTION. Four mathematically distinct distributions will be discussed. They are Kemp and Kemp's (1956) Types IA, II/IIIA and IV distributions, and a particular case of Gurland's (1958) negative-binomial ^ beta distribution which, for the purposes of this paper, we shall call Type D.

The p.g.f.'s of these four distributions are:-

Type IA $$\frac{{}_2F_1[-n,-a;\ b-n+1;\ s]}{{}_2F_1[-n,-a;\ b-n+1;\ 1]} = {}_2F_1[-n,-a;-a-b;\ 1-s],$$

Type II/IIIA $$\frac{{}_2F_1[-n,a;\ -b-n+1;\ s]}{{}_2F_1[-n,a;\ -b-n+1;\ 1]} = {}_2F_1[-n,a;\ a+b;\ 1-s],$$

G. P. Patil et al. (eds.), Statistical Distributions in Scientific Work, Vol. 1, 31-40.

$$\text{Type IV} \quad \int_0^1 \left(\frac{1-q}{1-qs}\right)^k \frac{q^{a-1}(1-q)^{b-1}}{B(a,b)}\, dq = \frac{{}_2F_1[k,a;\ a+b+k;\ s]}{{}_2F_1[k,a;\ a+b+k;\ 1]},$$

$$\text{Type D} \quad \int_0^1 (1+p-ps)^{-k} \frac{p^{a-1}(1-p)^{b-1}}{B(a,b)}\, dp = {}_2F_1[k,a;\ a+b;\ s-1]\ .$$

Types IA, II/IIIA and IV have p.g.f.'s of the form ${}_2F_1[\alpha,\beta;\ \gamma;\ s]/{}_2F_1[\alpha,\beta;\ \gamma;\ 1]$, and are particular cases of the Gaussian hypergeometric probability (G.h.p.) distributions; for conditions on the parameters see Kemp and Kemp (1956). Types IA, II/IIIA and D have p.g.f.'s of the form ${}_2F_1[\alpha,\beta;\ \gamma;\ \lambda(s-1)]$ where $\lambda = \pm 1$. We shall call this second class of distributions Gaussian hypergeometric factorial-moment (G.h.f.) distributions since their factorial-moment generating functions can be represented by Gaussian hypergeometric series.

Although these two classes of distributions overlap, nevertheless the distinction between them is valuable because the kinds of models available for these four distributions depend on the class/classes to which they belong.

2. SOME PARTICULAR INSTANCES OF THE DISTRIBUTIONS. Consideration of alternative names (e.g. hypergeometric, binomial$\wedge$beta) for the four distributions will be left until later in the paper.

This section mentions a number of common discrete distributions which are obtainable either as limiting forms or as particular instances of G.h.p. and G.h.f. distributions. The models which will be developed in subsequent sections can be adapted so as to give rise to these limiting forms and particular instances.

Suitable limiting processes, see Kemp (1968), yield the binomial, negative-binomial, Poisson, hyper-Poisson, Poisson$\wedge$beta and Laplace-Haag distributions.

The discrete rectangular, and also Chung and Feller's [Feller (1957), p. 77] distribution of leads in coin-tossing are particular instances of Type II/IIIA; their p.g.f.'s are ${}_2F_1[1,1-n;\ 1-n;\ s]/{}_2F_1[1,1-n;\ 1-n;\ 1] = {}_2F_1[1,1-n;\ 2;\ 1-s]$ and ${}_2F_1[-n,\ \frac{1}{2};\ -n+\frac{1}{2};\ s]/{}_2F_1[-n,\frac{1}{2};\ -n+\frac{1}{2};\ 1] = {}_2F_1[-n,\frac{1}{2};\ 1;\ 1-s]$ respectively, where n is a positive integer (as throughout the paper).

The Waring distribution and Marlow's factorial distribution are particular instances of Type IV; their p.g.f.'s are respectively $_2F_1[a,1; x+1; s]/_2F_1[a,1; x+1; 1]$, $x > a > 0$, and $_2F_1[m-n+1,1; m+1; s]/_2F_1[m-n+1,1; m+n; 1]$, $m + 1 > n > 1$.

3. URN MODELS AND CONTAGION MODELS. For integer values of the parameters Types IA and IIA are particular Polya distributions (Type IA is hypergeometric and Type IIA negative hypergeometric), whilst Types IIIA and IV are instances of the inverse Polya distribution (Type IIA is inverse hypergeometric and Type IV inverse hypergeometric with additional replacements). Thus Polya urn models lead to the integer cases of Types IA, II/IIIA and IV, as detailed in Kemp and Kemp (1956). The Type IA urn model underlies acceptance (inspection) sampling for attributes, capture-recapture sampling, and the exact test for a 2×2 contingency table. Irwin (1954) derived the Type IIA urn model in a study on infectious diseases, and the Type IIIA urn model also arises in inspection sampling. See also Sarkadi (1957) and Bosch (1963).

The dual nature of Types IA and II/IIIA as both G.h.p. and G.h.f. distributions can be better understood by considering contagion models with interchangeable events (i.e. with interchangeability of order of events). We present the topic in terms of attacks of an infectious disease, although formulation in terms of accidents is equally possible; there are many physical applications.

Suppose that there are a finite number n of opportunities for an attack; then the probability that at least one more attack will occur given that k attacks have already occurred is called the contagion function, and is equal to (n-k) times the probability of attack at the (k+1)'th opportunity, given that attacks have already occurred at the first k opportunities. Fréchet (1939, 1943) has shown by the principle of inclusion and exclusion that the contagion function is equal to the ratio of the (k+1)'th factorial moment to the k'th factorial moment.

If the contagion function is $(n-k)(Np-k)/(N-k)$, then the factorial-moment generating function (f.m.g.f.) becomes $_2F_1[-n, -Np; -N; -t]$ and we have the hypergeometric (Type IA) distribution.

For a contagion function equal to $(n-k)(Np-k)/(N+k)$ the f.m.g.f. is $_2F_1[-n,Np; N; -t]$ and the distribution is negative hypergeometric (Type II/IIIA).

Suppose now that the total number of opportunities becomes infinite, whilst the probabilities of attack at any opportunity tend to zero in such a way that the contagion function remains finite. Then the form of the contagion function still determines the kind of distribution, see Irwin (1953). If the contagion function is equal to $(a_1+k)(a_2+k)/(b+k)$ where $a_1, a_2, b > 0$, then the f.m.g.f. becomes ${}_2F_1[a_1, a_2; b; t]$ and the distribution is Type D.

It may be interesting to remark that, while studying the problem of measuring aggregation in ecological context, Patil and Stiteler [(1972), (1974)] have introduced the ratio of the (k+1)'th factorial moment to the k'th factorial moment to correspond to what they call k'th level crowding, and have discussed further related structural problems of several important families of distributions.

4. EQUILIBRIUM STOCHASTIC PROCESSES. Equilibrium time-homogeneous processes can easily be constructed so as to yield desired distributions [e.g., see Boswell and Patil (1970, 1972)]. To obtain G.h.p. distributions, consider the Kolmogorov equations

$$\frac{dp_0(t)}{dt} = \mu_1 p_1(t) - \lambda_0 p_0(t),$$

$$\frac{dp_i(t)}{dt} = \mu_{i+1} p_{i+1}(t) - (\mu_i+\lambda_i) p_i(t) + \lambda_{i-1} p_{i-1}(t), \quad i \geq 1,$$

where λ_i and μ_i are such that $\lambda_{i-1}/\mu_i = (a_1+i-1)/(b+i-1)$. Then the equilibrium solution is given by $p_i/p_{i-1} = \lambda_{i-1}/\mu_i$, and the equilibrium distribution p.g.f. is ${}_2F_1[a_1,1; b; s]/{}_2F_1[a_1,1; b; 1]$. This is Type II/IIIA if $a_1 = -n$, $b < a$, and is Type IV if $0 < a_1 < b - 1$.

Suppose now that λ_i and μ_i satisfy the relationship $\lambda_{i-1}/\mu_i = (a_1+i-1)(a_2+i-1)/(b+i-1)i$; the resultant equilibrium distribution now has the p.g.f. ${}_2F_1[a_1,a_2; b; s]/{}_2F_1[a_1,a_2; b; 1]$ and can be Type IA, II/IIIA or IV by suitable choice of parameters.

5. STER DISTRIBUTIONS. The use of STER models arises in connection with an inventory decision problem. If demand is a discrete random variable with p.g.f. $G(s) = \sum_{i \geq 0} p_i s^i$, then the

corresponding STER distribution has probabilities $f(j)$ which are Sums successively Truncated from the Expectation of the Reciprocal of the demand variable, i.e. $f(j) = \sum_{i>j} p_i/i)/(1-p_0)$, $j = 0,1,\ldots$, Bissinger (1965). The relationship between the two p.g.f.'s is

$$H(s) = \sum_{j\geq 0} f(j)\, s^j = \frac{1}{(1-s)(1-p_0)} \int_s^1 \frac{G(s)-p_0}{s}\, ds,$$

Patil and Joshi (1968). Note that zero-modification of the demand variable does not affect the STER distribution. Kemp and Kemp (1969) showed that if $G(s) = s\ {}_pF_q[(a);\ (b);\ \lambda(s-1)]$ then $H(s) = {}_{p+1}F_{q+1}[(a),1;\ (b),2;\ \lambda(s-1)]$.

Adapting this result to the distributions of interest we find that if demand has the p.g.f. $s\ {}_2F_1[a_1,2;\ b;\ \lambda(s-1)]$ then the corresponding STER variable has the p.g.f. ${}_2F_1[a_1,1;\ b;\ \lambda(s-1)]$; so for $a_1 < 0 < b$, $\lambda = -1$, an initial shifted Type II/IIIA distribution has a corresponding Type II/IIIA STER distribution, and for $0 < a_1, b$, $\lambda = -1$, an initial shifted Type D distribution yields a Type D STER distribution.

Now suppose that demand has the p.g.f. $s\ {}_2F_1[a_1,a_2;\ \lambda(s-1)]$; the corresponding STER variable now has the p.g.f. ${}_2F_1[a_1,a_2;\ 2;\ \lambda(s-1)]$. Hence for $a_1 < 0 < a_2$, $\lambda = -1$, (or similarly $a_2 < 0 < a_1$, $\lambda = -1$), a shifted Type II/IIIA demand distribution gives rise to a Type II/IIIA distribution, and for $a_1, a_2 < 0$, $\lambda = +1$, a shifted Type D distribution yields a Type D STER distribution.

6. CONDITIONALITY MODELS.

In this section we extend certain well-known conditionality results which give rise to G.h.p. distributions as conditional distributions. Simple conditionality models for G.h.f. distributions do not seem to exist.

Let X and Y be two independent discrete random variables with p.g.f.'s $G_X(s) = \sum_i p_i s^i$ and $G_Y(s) = \sum_i q_i s^i$. Then the p.g.f. for X given X + Y is $G_{X|X+Y}(s) = \sum_i p_i q_{k-i} s^i / \sum_i p_i q_{k-i}$, where $X + Y = k$, and the p.g.f. for X given Y-X is $G_{X|Y-X}(s) = \sum_i p_i q_{k+i} s^i / \sum_i p_i q_{k+i}$, where $X - Y = k$; we note that conditional distributions for Y given Y-X do not need separate consideration since $G_{Y|Y-x}(s) = \sum_i p_{i-k} q_i s^i / \sum_i p_{i-k} q_i = s^k G_{x|Y-x}(s)$.

When X and Y have binomial distributions with parameters (n,p) and (m,p) respectively, then the conditional distribution for X given X + Y is of course hypergeometric (Type IA). And when X and Y are both negative-binomial with parameters (u,q) and (v,q) the conditional distribution for X given X + Y is negative hypergeometric (Type II/IIIA).

Suppose now that $G_x(s) = {}_1F_1[-n; u; -\lambda s]/{}_1F_1[-n; u; -\lambda]$ whilst $G_Y(s) = e^{\lambda(s-1)}$; we find that $G_{x|x+Y} = {}_2F_1[-n, -k; u; s]/{}_2F_1[-n,-k; u; 1]$, i.e. that the conditional distribution for X given X + Y is Type IA. The conditional distribution for Y given X + Y will also be Type IA.

Similarly, when $G_x(s) = {}_0F_1[\ ; u; \lambda s]/{}_0F_1[\ ; u; 1]$ and $G_Y(s) = {}_0F_1[\ ; v; \lambda s]/{}_0F_1[\ ; v; \lambda]$, then the conditional distribution for X given X + Y has the p.g.f. $G_{x|x+Y}(s) = {}_2F_1[-k,1-v-k; u; s]/{}_2F_1[-k,1-v-k; u; 1]$ and is again Type IA.

We note that when X is binomial and Y is negative-binomial, or when X is negative-binomial and Y is binomial, the conditional distribution for X given X + Y does not belong to the set of distributions under discussion.

Neither the Type IV nor the Type D distribution can arise from a model with X conditional on X + Y fixed, since both are non-terminating distributions.

Consider now distributions for X conditional on Y - X fixed. Of the four distributions of interest only Type IA arises in this manner; it does so in three ways.

Suppose that X and Y have binomial distributions with parameters (n,p) and (m,1-p) respectively; then the conditional distribution for X given Y - X has the p.g.f. $G_{x|Y-x}(s) = {}_2F_1[-n,k-m; k+1; s]/{}_2F_1[-n,k-m; k+1; 1]$, and is hypergeometric since $k < m$.

If X has the p.g.f. ${}_2F_0[-n,-u;\ ; \lambda s]/{}_2F_0[-n,-u;\ ; \lambda]$ where $0 < n < u$, and Y is Poisson with parameter λ^{-1}, then the p.g.f. for X given Y - X is

$G_{x|Y-x}(s) = {}_2F_1[k-n,k-u;\ k+1;\ s]/{}_2F_1[k-n,k-u;\ k+1;\ 1]$ which is again Type IA since $k < n$.

7. WEIGHTING MODELS. These give rise to distributions which have been modified by the method of ascertainment. Sometimes observations are only partially observable, or are partially destroyed, i.e. are sampled with unequal chances of observation, Rao (1965), Boswell and Patil (1971), Kemp (1973). If the model for the sampling chances is known, then the resultant weighted distribution can be deduced from the form of the original distribution.

Consider G.h.p. distributions. If the sampling chances (weights) w_r are proportional to r, then the distribution with p.g.f. ${}_2F_1[a_1,a_2;\ b;\ s]/{}_2F_1[a_1,a_2;\ b;\ 1]$ is ascertained as the distribution with p.g.f. $s\ {}_2F_1[a_1+1,a_2+1;\ b+1;\ s]/{}_2F_1[a_1+1,a_2+1;\ b+1;\ 1]$.

More generally, if $w_r \propto r!/(r-k)!$, then the same initial distribution will be ascertained as that with p.g.f. $s^k\ {}_2F_1[a_1+k,a_2+k;\ b+k;\ s]/{}_2F_1[a_1+k,a_2+k;\ b+k;\ 1]$, i.e. a G.h.p. distribution will be ascertained as a shifted G.h.p. distribution.

Note that if the original distribution is unknown, then, unless the weighting function is known explicitly, the extent of the distortion produced by the method of ascertainment is unrecognizable.

8. MIXING PROCESSES. Amongst the models which have already been considered, we have found that urn models, conditionality models and ascertainment weighting situations lead to G.h.p. distributions, whereas contagion processes and STER models relate to G.h.f. distributions. In this final section we examine mixing processes--these are found to yield both G.h.p. and G.h.f. distributions.

We begin by recalling the opening section of Kemp and Kemp's (1971) paper on the lost-games distribution; consideration was given there to mixing models for the Type IV distribution. We showed that the distribution can arise

(i) in two parameter-wise distinct ways by mising the inverse sampling form of the negative-binomial, using a Beta distribution,

(ii) by applying a simple transformation to the above mixing process, and so obtaining Irwin's (1968) compounded proneness and liability accident model, which mixes the Bliss and Fisher (1953) form of the negative binomial, using an F-distribution,

(iii) from a mixed Poisson model, by using a mixing distribution with a density function based on a Whittaker function, Dacey (1969),

(iv) in two parameter-wise distinct ways as a mixed confluent hypergeometric distribution.

The remaining distributions of interest are all G.h.f.

The usual binomial-beta derivation of Type II/IIIA may be rewritten as

$$\int_0^1 {}_1F_0[-n; \ ; p(1-s)] \frac{p^{c-1}(1-p)^{s-c-1}}{B(c,d-c)} dp = {}_2F_1[-n,c; d; 1-s],$$

whilst the Gurland derivation of Type D is

$$\int_0^1 {}_1F_0[k; \ ; y(s-1)] \frac{y^{c-1}{}_1{-}y^{d-c-1}}{B(c,d-c)} dy = {}_2F_1[k,c; d; s-1] .$$

Type IA cannot arise in this manner.

The two above results may be generalized as follows:-

$$\int_0^1 {}_2F_1[a_1,a_2; c; y(s-1)] \frac{y^{c-1}(1-y)^{d-c-1}}{B(c,d-c)} dy = {}_2F_1[a_1,a_2; d; (s-1)]$$

and $\int_0^1 {}_2F_1[a_1,d; b; y(s-1)] \frac{y^{c-1}(1-y)^{d-c-1}}{B(c,d-c)} dy$

$$= {}_2F_1[a_1,c; b; \lambda(s-1)].$$

From $d > c > 0$ and $\lambda = \pm 1$ it can be deduced that Types II/IIIA and D can both arise by both processes, Type D in two mathematically distinct ways.

It can be shown that gamma-mixing a generalized hypergeometric factorial-moment distribution yields another such distribution, and that as a particular case

$$\int_0^\infty {}_1F_1[a_1; b; \lambda(s-1)] \frac{e^{-y}y^{c-1}}{\Gamma(c)} dy = {}_2F_1[a_1,c; b; \lambda(s-1)]$$

where $b > a_1 > 0$, $\lambda > 0$. Taking $\lambda = +1$ gives a gamma-mixed Poisson beta model for the Type D distribution (interchanging the rôles of a a_1 and c gives another, similar model). Suppose, however, that $a_1 = -m$, $b = -n$ where m, n are integers, and $\lambda > 0$; the distribution to be mixed is now Laplace-Haag, see Fréchet (1943), pp. 148-156. Gamma-mixing this Laplace-Haag distribution, with the additional restriction $\lambda = +1$, gives the negative hypergeometric (Type II/IIIA).

The derivation of Type IV as a mixed negative-binomial indicated that it could also be obtained as a mixed Poisson. The same argument applies to Type D. It is possible to show that

$$\int_0^\infty e^{x(s-1)} \frac{\Gamma(b)x^{\frac{a_1+a_2-3}{2}} e^{-x/2}}{\Gamma(a_1)\,\Gamma(a_2)} W_{\frac{a_1+a_2+1}{2} - b,\ \frac{a_1-a_2}{2}}(s)\,dx$$

$$= {}_2F_1[a_1,a_2;\ b;\ s-1],$$

where a_1, a_2, $b > 0$.

REFERENCES

[1] Bissinger, B. H. (1965). In Classical and Contagious Discrete Distributions, G. P. Patil (ed.). Statistical Publishing Society, Calcutta and Pergamon, New York, 15-17.

[2] Bliss, C. I. and Fisher, R. A. (1953). Biometrics 9, 176-200.

[3] Bosch, A. J. (1963). Statistica Neerlandica 17, 201-213.

[4] Boswell, M. T. and Patil, B. P. (1970). In Random Counts in Scientific Work, Vol. 1, Random Counts in Models and Structures, G. P. Patil (ed.). Penn State University Press, University Park, Pa., 3-22.

[5] Boswell, M. T. and Patil, G. P. (1971). In Statistical Ecology, Vol. 1, Spatial Patterns and Statistical Distributions, Patil, Pielou and Waters (eds.). Penn State University Press, University Park, Pa., 99-130.

[6] Boswell, M. T. and Patil, G. P. (1972). In Stochastic Point Processes, P. A. W. Lewis (ed.). Wiley, New York, 285-298.

[7] Dacey, M. F. (1969). Geog. Anal. 1, 283-317.

[8] Feller, W. (1957). An Introduction to Probability Theory and Its Applications, Vol. 1, 2nd ed. Wiley, New York.

[9] Fréchet, M. (1939). Les probabilités associées à un système d'événements compatibles et dépendants. 1. Evénements en nombre fini fixe. Actualités Sci. Indust. No. 859. Hermann, Paris.
[10] Fréchet, M. (1943). Les probabilités associées à un système d'événements compatibles et dépendants. 2. Cas particuliers et applications. Actualités Sci. Indust. No. 942. Hermann, Paris.
[11] Gurland, J. (1958). Biometrics 14, 229-249.
[12] Irwin, J. O. (1953). J. Roy. Statist. Soc. B 15, 87-89.
[13] Irwin, J. O. (1954). Biometrika 41, 266-268.
[14] Irwin, J. O. (1968). J. Roy. Statist. Soc. A 131, 205-225.
[15] Kemp, A. W. (1968). Sankhyā A 30, 401-410.
[16] Kemp, C. D. (1973). Accid. Anal. Prev. 5, 371-373.
[17] Kemp, A. W. and Kemp, C. D. (1971). Zastos. Mat. 12, 167-173.
[18] Kemp, C. D. and Kemp, A. W. (1956). J. Roy. Statist. Soc. B 18, 202-211.
[19] Kemp, C. D. and Kemp, A. W. (1969). Bull. I.S.I. 43, 336-338.
[20] Patil, G. P. and Joshi, S. W. (1968). A Dictionary and Bibliography of Discrete Distributions. Oliver and Boyd, Edinburgh and Hafner, New York.
[21] Patil, G. P. and Stiteler, W. M. (1972). Bull. I.S.I. 44, 55-81.
[22] Patil, G. P. and Stiteler, W. M. (1974). Res. Popul. Ecol. 15, 238-254.
[23] Rao, C. R. (1965). In Classical and Contagious Discrete Distributions, G. P. Patil (ed.). Statistical Publishing Society, Calcutta and Pergamon, New York, 320-332.
[24] Sarkadi, K. (1957). Magyar Tud. Akad. Mat. Kutató Int. Közl. 2, 59-69.

ON THE PROBABILISTIC STRUCTURE AND PROPERTIES OF DISCRETE LAGRANGIAN DISTRIBUTIONS*

P. C. Consul and L. R. Shenton

University of Calgary, Alberta, Canada
University of Georgia, Athens, Georgia, U.S.A.

SUMMARY. Lagrange distributions have their origin in the Lagrange expansion for the root of an equation. The basic formula is discussed, and its relation to convolution probability functions. General formulae are given for cumulants and non-central moments. Lastly, a Lagrange-type distribution is shown to hold for a certain queueing process.

KEY WORDS. Lagrange distribution, Lagrange expansion, moments, cumulants, queueing.

1. INTRODUCTION. A new class of discrete probability distributions under the title "Lagrangian Distributions" has been recently introduced into the literature by Consul and Shenton (1971, 1973). The particular title was chosen by them on account of the generation of these probability distributions by the well known Lagrange expansion of a function f(x) as a power series in y when y = x/g(x). It has been well illustrated by the authors (1971) that this class consists of many families and that the double binomial family of these discrete Lagrangian distributions has, as particular cases, many interesting members such as the Borel-Tanner distribution [Tanner (1961)], Haight distribution

---

*The authors acknowledge with thanks the financial support of the National Research Council of Canada, the North Atlantic Treaty Organization, Brussels, and the University of Georgia, Athens, Georgia, U.S.A.

G. P. Patil et al. (eds.), Statistical Distributions in Scientific Work, Vol. 1, 41-57.

[Haight (1961)], generalized negative binomial distribution [Jain and Consul (1971)], and generalized Poisson distribution [Consul and Jain (1972)]. All these distributions are found to be of relevance in queueing theory and possess some interesting properties [Consul and Shenton (1973)].

In this expository note we shall consider the probabilistic structure of Lagrangian distributions and discuss some of their important properties and applications.

2. THE BASIC LAGRANGIAN DISTRIBUTIONS. If g(s) is a probability generating function (pgf) defined on some or all non-negative integers such that $g(0) \neq 0$, then the transformation

$$s = u\, g(s) \tag{1}$$

defines, for the numerically smallest non-zero root s, a new pgf $s = \phi(u)$ whose expansion in powers of u is given by Lagrange's formula as

$$s = \phi(u) = \sum_{x=1}^{\infty} \frac{u^x}{x!} \left[\left(\frac{\partial}{\partial s}\right)^{x-1} [g(s)]^x \right]_{s=0} \tag{2}$$

Hereinafter, we shall refer to the above pgf $\phi(u) = E(u^X)$ as the <u>basic Lagrangian pgf</u> and the discrete distribution represented by it, viz.

$$P(X=x) = \frac{1}{x!} \left(\frac{\partial}{\partial s}\right)^{x-1} [g(s)]^x \Bigg|_{s=0} \qquad x \in N \tag{3}$$

as the <u>basic Lagrangian probability distribution</u> (basic LPD) defined on N, the set of positive integers. By using (2) and (3) one can write some interesting examples of such basic LPD; e.g.

(i) the geometric distribution is the basic LPD generated on taking g(s) equal to the linear binomial element $p + q\,s$, $p + q = 1$, $0 < p < 1$. In fact, the transformation $s = u(p + qs)$ gives $s = u\,p(1 - qu)^{-1} = \Sigma\, u^x\, pq^{x-1}$;

(ii) the Borel distribution is the basic LPD generated by $g(s) = e^{m(s-1)}$, $(m > 0)$ the pgf of the Poisson distribution;

(iii) the Haight distribution is the basic LPD generated by $g(s) = p(1-qs)^{-1}$, $(p + q = 1)$ the pgf of the geometric distribution where $p = \alpha(1+\alpha)^{-1}$;

(iv) the probability distribution of the first visit to +1 in a simple unsymmetric random walk on the set N of integers is the basic LPD generated by taking for $g(s)$ the binomial element $g(s) = p + qs^2$, $p + q = 1$. Evidently, the transformation formula $s = u(p + qs^2)$ provides the pgf

$$s = \frac{1 - \sqrt{(1 - 4u^2pq)}}{2qu}$$

which is well known;

(v) the Consul distribution

$$p(X=x) = \frac{1}{x}\binom{mx}{x-1}\left(\frac{\theta}{1-\theta}\right)^{x-1}(1-\theta)^{mx}, \quad x \in N \tag{4}$$

where m is a positive integer when $0 < \theta < 1$ and $m < 0$ when $\theta < 0$, is the basic LPD generated on taking $g(s) = (1-\theta+\theta s)^m$, the pgf of the binomial distribution for a positive integer m and the negative binomial distribution for $m < 0$.

Theorem 2.1. The probability $P(X=n)$ in the basic LPD, defined by (3), is equal to $n^{-1}\,P(X^{(n)}=n-1)$ where $X^{(n)}$ denotes the variate whose pgf is the n-fold convolution of the probability distribution (pd) given by $g(s)$. The proof is straightforward. The theorem is useful in understanding the role of the basic Lagrangian transformation and in writing down the probability elements of the convolutions of basic LPD's.

3. LAGRANGIAN DISTRIBUTIONS IN GENERAL. If $g(s)$ and $f(s)$ are two given pgf's in s defined on some or all non-negative integers and $g(0) \neq 0$, then the transformation $s = u\,g(s)$ gives, for the smallest non-zero root s, a power series $\psi(u)$ for $f(s)$ and thus defines the general Lagrangian probability distribution (LPD) by

$$\begin{cases} P(Z=x) = \dfrac{1}{x!}\dfrac{\partial^{x-1}}{\partial s^{x-1}}\left\{[g(s)]^x\,\dfrac{\partial f(s)}{\partial s}\right\}_{s=0}, & x=1, 2, \ldots \\ P(Z=0) = f(0). \end{cases} \tag{5}$$

From this, two simple theorems follow:

Theorem 3.1. The Lagrangian distributions obtained from (5) by taking $f(s) = s^n$ are the n-fold convolutions of the basic LPD (3).

Theorem 3.2. The general LPD with pgf $g(s)$ as a power series in u is the basic LPD (3) displaced one step to the left.

A result which is not so obvious is contained in

Theorem 3.3. The general LPD with pgf $f(s) = \psi(u)$ as a power series in $u = s/g(s)$ is obtained by randomizing the index parameter n in the probability distribution (pd) with pgf s^n (as a power series in u) according to the pd with pgf $f(s)$ in s.

If $\{f_r\}$, $r = 0, 1, 2, \ldots$ represent the successive probabilities in the pd of the pgf $f(s)$ in s, then

$$f(s) = \sum_{n=0}^{\infty} f_n s^n = \sum_{n=0}^{\infty} f_n \cdot [\phi(u)]^n$$

$$= f_0 + \sum_{n=1}^{\infty} nf_n \left[\sum_{x=n}^{\infty} \frac{u^x}{(x-n)!x} \left\{ \partial^{x-n} [g(s)]^x \right\}_{s=0} \right] \quad (\partial \equiv \partial/\partial s).$$

Rearranging the summatory term it follows that

$$f(s) = f_0 + \sum_{x=1}^{\infty} \frac{u^x}{x!} \partial^{x-1} [g(s)]^x \left\{ \sum_{n=1}^{x} nf_n s^{n-1} \right\} \Bigg|_{s=0}$$

$$= f_0 + \sum_{x=1}^{\infty} \frac{u^x}{x!} \partial^{x-1} \left\{ [g(s)]^x \, \partial f(s) \right\} \Bigg|_{s=0}$$

from which the result follows.

Corollary: The negative binomial distribution is a special case of the general LPD.

It can be considered as a randomization of parameter n in $f(s) = (q + ps)^n$ where $s = u\,g(s) = u(q + ps) = u\,q(1 - up)^{-1}$ is the pgf of a basic LPD, namely the geometric distribution. In fact $\psi(u) = f(s) = q^n(1-up)^{-n}$.

Theorem 3.4. For a given transformation $s = u\,g(s)$ all Lagrangian probability distributions are closed under convolution.

A particular case of this theorem was proved by Consul and Shenton (1971).

Let X_1 and X_2 be two independent Lagrangian random variables, not necessarily identically distributed, and let their pgf's in u be $f_1(s)$ and $f_2(s)$ where $s = u\, g(s)$; then

$$\psi_1(u) = f_1(s) = \sum_{h=0}^{\infty} \frac{u^h}{h!} \partial^{h-1} \left\{ (g(s))^h \, \partial f_1(s) \right\} \Big|_{s=0}$$ and

$$\psi_2(u) = f_2(s) \sum_{k=0}^{\infty} \frac{u^k}{k!} \partial^{k-1} \left\{ (g(s))^k \, \partial f_2(s) \right\} \Big|_{s=0} ,$$ in

which $\partial^{-1} \cdot \partial$ is taken to be unity.

Now $X_1 + X_2$ has the pgf $\psi_1(u)\, \psi_2(u)$. From $s = u\, g(s)$ the product $f_1(s)\, f_2(s)$ of the above two expansions can be easily written in the Lagrangian form

$$\sum_{r=0}^{\infty} \frac{u^r}{r!} \partial^{r-1} \left\{ (g(s))^r \, \partial \left\{ f_1(s)\, f_2(s) \right\} \right\} \Big|_{s=0}$$ which must

be equal to $\psi_1(u)\, \psi_2(u)$. The result is easily generalized.

Corollary. The above simple theorem provides us with the following interesting differentiation formula which seems to be new:

$$\partial^{n-1}[(g(s))^n \, \partial\{f_1(s)\, f_2(s)\}] = \sum_{h=0}^{n} \binom{n}{h} \partial^{h-1} \{(g(s))^h \, \partial f_1(s)\} \, \partial^{n-h-1}[(g(s))^{n-h} \, \partial f_2(s)] \tag{6}$$

4. CUMULANTS OF THE GENERAL LPD. The basic LPD and the general LPD are such that the determination of the central moments is somewhat complicated even for special cases [see for example, Consul and Jain (1972)]. In this section we obtain specific expressions for the cumulants of any order for all general LPD's.

The following notations will be used henceforth. The symbols after the semicolon are for the basic LPD.

Distribution	LPD (General;Basic)	Transformer	Transformed
pgf	$\psi(u)$; $\phi(u)$	$g(s)$	$f(s)$; s
Cumulants	$\{L_r\}$; $\{D_r\}$	$\{G_r\}$	$\{F_r\}$; $F_1 = 1$, $F_{r>1} = 0$

Non-central moments $\{l_r\}$; $\{d_r\}$ $\{g_r\}$ $\{f_r\}$; $f_1 = 1$

$f_{r>1} = 0$

The relations for the cumulants of the LPD are derived by the mapping $s = \exp S$, $u = \exp \beta$ so that the transformations $s = u\,g(s)$ and $f(s) = E(u^x) = \psi(u)$ become, after taking logarithms,

$$S = \beta + \ln g(e^S) \tag{7a}$$

$$\text{and} \quad \ln f(e^S) = \sum_{r=1}^{\infty} L_r \beta^r/r! \;, \tag{7b}$$

which may be written as

$$S = \beta + \sum_{r=1}^{\infty} G_r S^r/r! \tag{8a}$$

$$\text{and} \quad \sum_{r=1}^{\infty} F_r S^r/r! = \sum_{r=1}^{\infty} L_r \beta^r/r! \;. \tag{8b}$$

The process of solving these two equations for the cumulants $L_1, L_2, L_3, \ldots$ is rather complicated, but the task becomes some what simpler by introducing the cumulants $\{D_r\}$, $r = 1, 2, 3, \ldots$ defined by (8a) and

$$S = \sum_{r=1}^{\infty} D_r \beta^r/r! \tag{9}$$

Thus, the relation (8b) can be written as

$$\sum_{r=1}^{\infty} L_r \cdot \beta^r/r! = \sum_{m=1}^{\infty} F_m \left[\sum_{j=1}^{\infty} D_j \beta^j/j!\right]^m/m! \tag{10}$$

so that

$$\begin{aligned} L_r &= \frac{\partial^r}{\partial \beta^r}\left[\sum_{m=1}^{\infty} F_m \left(\sum_{j=1}^{\infty} D_j \beta^j/j!\right)^m/m!\right]_{\beta=0} \\ &= \sum_{m=1}^{r}\left[F_m \frac{\partial^r}{\partial \beta^r}\left(\sum_{j=1}^{r} D_j \beta^j/j!\right)^m/m!\right]_{\beta=0} \end{aligned} \tag{11}$$

The coefficient of F_m, $m = 1, 2, 3, \ldots$ can be obtained by expanding the multinomial $\left(\sum_{j=1}^{r} D_j \beta^j/j!\right)^m$ and evaluating the coefficient of β^r. This can be achieved in a systematic manner by considering all possible partitions of m.

Thus the cumulants L_r, $r = 1, 2, 3, \ldots$ of the general LPD become

$$L_r = \sum_{m=1}^{r} F_m \left[\sum \frac{r!}{\pi_1! \pi_2! \ldots \pi_r!} \prod_{i=1}^{r} \left(\frac{D_i}{i!}\right)^{\pi_i} \right] \tag{12}$$

where the second summation is taken over all partitions π_1, π_2, $\ldots$, π_r of m such that $\pi_1 + 2\pi_2 + \ldots + r\pi_r = r$.

The first few cumulants of the L.D. can now be written down as particular cases of (12) and are of the form

$$\begin{cases} L_1 = F_1 D_1 \\ L_2 = F_1 D_2 + F_2 D_1^2 \\ L_3 = F_1 D_3 + 3F_2 D_1 D_2 + F_3 D_1^3 \\ L_4 = F_1 D_4 + 3F_2 D_2^2 + 4F_2 D_1 D_3 + 6F_3 D_1^2 D_2 + F_4 D_1^4 . \\ \ldots\ldots\ldots\ldots\ldots\ldots \end{cases} \tag{13}$$

Though the cumulants $\{D_r\}$ are not yet determined explicitly, note that the relations (12) and (13) are similar to the expressions for non-central moments in terms of cumulants given on pages 68-69 of Kendall and Stuart (1963). Also note that (12) resembles Faà de Bruno's (1855) expression for the nth derivative of a composite function.

Since the cumulants, given by (12) and (13), of the general LPD depend upon the cumulants D_i, $i = 1, 2, 3, \ldots$ of the basic LPD, we shall now derive simple expressions for D_i.

<u>4.1 Cumulants of the Basic LPD.</u> In this case, the equations (7a) and (8a) remain unaltered while (7b) changes to (9).

On eliminating S from (8a) by using the value of S from (9), we get

$$\sum_{j=1}^{\infty} D_j \beta^j/j! = \beta + \sum_{r=1}^{\infty} G_r \left(\sum_{j=1}^{\infty} D_j \beta^j/j!\right)^r / r! \quad . \tag{14}$$

If we define $D_1^* = D_1 - 1$, $D_j^* = D_j$, $j \geq 2$, then (14) becomes

$$\sum_{j=1}^{\infty} D_j^* \beta^j/j! = \sum_{r=1}^{\infty} G_r \left(\sum_{j=1}^{\infty} D_j \beta^j/j! \right)^r / r! \quad ,$$

which is precisely the relation between $\{L_r\}$ and $\{F_r\}$ given in (10). Hence the cumulants of the basic LPD are given by (12) on replacing L_r and F_m by D_r^* and G_m respectively.

As illustrations we have

$$\left\{\begin{array}{l} D_1^* = G_1 D_1 \\ D_2 = G_1 D_2 + G_2 D_1^2 \\ D_3 = G_1 D_3 + 3G_2 D_1 D_2 + G_3 D_1^3 \\ D_4 = G_1 D_4 + 3G_2 D_2^2 + 4G_2 D_1 D_3 + 6G_3 D_1^2 D_2 + G_4 D_1^4 \\ \cdots\cdots\cdots\cdots\cdots\cdots\cdots\cdots , \end{array}\right. \tag{15}$$

which can be solved successively to yield

$$\left\{\begin{array}{l} D_1 = 1/(1-G_1) \\ D_2 = G_2/(1-G_1)^3 \\ D_3 = G_3/(1-G_1)^4 + 3G_2^2/(1-G_1)^5 \\ D_4 = G_4/(1-G_1)^5 + 10G_3G_2/(1-G_1)^6 + 15G_2^3/(1-G_1)^7 \\ \cdots\cdots\cdots\cdots\cdots\cdots\cdots\cdots . \end{array}\right. \tag{16}$$

Since the cumulants $\{D_r\}$ are now completely determined, the cumulants $\{L_r\}$ of all general LPD's are determined by (12) and (13) in terms of the cumulants $\{G_r\}$ and $\{F_r\}$, which are supposed to be known. These formulae can be used to write down the cumulants of any order for the special LPD's.

4.2 <u>Non-Central Moments of the General LPD</u>. Though one can evaluate the non-central moments of the general LPD from its cumulants, simple expressions for these can be obtained independently as well. Rewrite (7b) as

$$\begin{cases} f(e^S) = \sum_{r=0}^{\infty} l_r \, \beta^r/r! \\ \text{or} \\ \sum_{r=0}^{\infty} l_r \, \beta^r/r! = \sum_{m=0}^{\infty} f_m \, S^m/m! \quad . \end{cases} \tag{17}$$

Replacing S by the expression (9) in powers of β and the basic LPD cumulants $\{D_r\}$, relation (17) becomes

$$\sum_{r=0}^{\infty} l_r \, \beta^r/r! = \sum_{m=0}^{\infty} f_m \left(\sum_{k=1}^{\infty} D_k \, \beta^k/k! \right)^m m! \quad ,$$

where $l_0 = f_0 = 1$ by definition. Hence

$$\sum_{r=1}^{\infty} l_r \, \beta^r/r! = \sum_{m=1}^{\infty} f_m \left(\sum_{k=1}^{\infty} D_k \, \beta^k/k! \right)^m /m! \quad ,$$

which corresponds exactly to (10) if the non-central moments l_r and f_m are replaced by the cumulants L_r and F_m respectively. Thus, the formula for the non-central moments of the general LPD is (12) with l_r for L_r, f_m for F_m.

5. DOUBLE BINOMIAL FAMILY OF GENERAL LPD AND RECURSIVE RELATIONS FOR CUMULANTS AND MOMENTS. The definition of the general LPD does not appear to be very exciting; nonetheless, it generates many interesting and useful discrete probability distributions. The double binomial family is obtained by taking $g(s) = (q+ps)^m$ and $f(s) = (q'+p's)^n$ with the restrictions $0<p, p'<1$, $mp<1$, $p+q=1$, $p'+q'=1$, so that the pd is given by

$$\begin{cases} P(X=0) = (q')^n \\ P(X=x) = \frac{n}{x} \, q'^n \, (pq^{m-1})^x \sum_{r=0}^{k} \binom{n-1}{r} \binom{mx}{x-r-1} \left(\frac{qp'}{q'p}\right)^{r+1} \quad x=1,2,3,\ldots \end{cases} \tag{18}$$

where $k = \min(x-1, n-1)$.

The double binomial family of LPD give a large number of discrete distributions as special cases [Consul and Shenton (1971)], but two of these have been studied by Consul and Jain (1971, 1972, 1973) and seem to be worthy of more detailed study.

For this special family of general LPD it is possible to derive alternative expressions for the cumulants and non-central moments which are not so complicated as (12) and (15) and are of a recursive nature.

Case 1. When p' is a continuous differentiable function of p.

Under the transformations $s = \exp S$, $u = \exp \beta$ our basic transformation $s = u\, g(s)$ becomes

$$S = \beta + m \ln(q+pe^S) \quad , \tag{19}$$

which defines S as a function of β, m, p. Taking partial derivatives of S with respect to β and p we get on simplification

$$\frac{1-m}{mq}\frac{\partial S}{\partial \beta} = \frac{1}{mq} - \frac{1}{q+p(1-m)e^S} \qquad \text{and}$$

$$\frac{p(1-m)}{m(1-mp)}\frac{\partial S}{\partial p} = \frac{1}{1-mp} - \frac{1}{q+p(1-m)e^S} \quad ,$$

so that there is the following partial differential relation

$$(1-mp)\frac{\partial S}{\partial \beta} - pq\frac{\partial S}{\partial p} = 1 \quad . \tag{20}$$

In terms of the non-central moments of the double binomial LPD, the pgf f(s) becomes

$$(q'+p'e^S)^n = \sum_{k=0}^{\infty} 1_k\, \beta^k/k! = H(\beta) \quad , \qquad \text{(say)} \tag{21}$$

where S is a function of β and p while H is a function of β, p and p'. Defining $Q = p'+p'e^S$ and differentiating (12) partially w.r.t. β and p, we have

$$np'\, Q^{n-1}\, e^S \frac{\partial S}{\partial \beta} = \frac{\partial H(\beta)}{\partial \beta} \quad , \tag{22a}$$

$$n\, Q^{n-1}\left[(e^S-1)\frac{\partial p'}{\partial p} + p'\, e^S\frac{\partial S}{\partial p}\right] = \frac{\partial H(\beta)}{\partial p} \quad . \tag{22b}$$

On eliminating $\frac{\partial S}{\partial \beta}$ and $\frac{\partial S}{\partial p}$ between (22a), (22b) and (20) and on simplification

$$n\,Q^{n-1}\left(pq\,\frac{\partial p'}{\partial p} - p'q'\right) + n\left(p' - pq\,\frac{\partial p'}{\partial p}\right)H(\beta) =$$

$$p'(1-mp)\,\frac{\partial H(\beta)}{\partial \beta} - pqp'\,\frac{\partial H(\beta)}{\partial p} \quad . \tag{23}$$

Now, if we suppose $pq(\partial p'/\partial p) = p'q'$, i.e., $p' = ap/(q+ap)$ where a is any real positive number, the expression (23) yields

$$np'H(\beta) = (1-mp)\,\frac{\partial H(\beta)}{\partial \beta} - pq\,\frac{\partial H(\beta)}{\partial p} \quad . \tag{24}$$

On equating the coefficients of $\beta^k/k!$ in (24), the expression gives a recursive relation for the non-central moments of the probability distribution (18) in the form

$$(1-mp)l_{k+1} = np'l_k + pq\,\frac{\partial l_k}{\partial p} \ , \quad k = 0, 1, 2, \dots \quad . \tag{25}$$

Similarly, starting with the expression (7b) for the cumulants of a double binomial general LPD we obtain the recursive relation

$$\begin{cases} (1-mp)L_{k+1} = pq\,\dfrac{\partial L_k}{\partial p} \ , & k = 1, 2, 3, \dots \\ (1-mp)L_1 = np' \ . & \end{cases} \tag{26}$$

Case 2. When p and p' are independent.

As $\partial p'/\partial p = 0$, the relation (23) takes the form

$$nH(\beta) - nq'\,Q^{n-1} = (1-mp)\,\frac{\partial H(\beta)}{\partial \beta} - pq\,\frac{\partial H(\beta)}{\partial p} \ , \tag{27}$$

wherein the term $nq'\,Q^{n-1}$ can be eliminated with the help of the partial derivative of (21) w.r.t. p', namely $nQ^{n-1} = nH(\beta) - p'[\partial H(\beta)/\partial p']$. Thus the recursive formula for non-central moments, on simplification and equating the coefficients of $\beta^k/k!$, becomes

$$(1-mp)l_{k+1} = np'l_k + pq\,\frac{\partial l_k}{\partial p} + p'q'\,\frac{\partial l_k}{\partial p'} \tag{28}$$

for $k = 0, 1, \dots$.

Similarly, the corresponding recursive formula for the cumulants becomes

$$\begin{cases} (1-mp)L_{k+1} = pq\dfrac{\partial L_k}{\partial p} + p'q'\dfrac{\partial L_k}{\partial p'} , & k = 1, 2, 3, \ldots \\ (1-mp)L_1 = np' . & \end{cases} \tag{29}$$

For the special case of the generalized Poisson distribution, which is given by $g(s) = e^{\theta(s-1)}$, $f(s) = e^{M(s-1)}$ one can easily see that the relations (28) and (29) will become modified. Thus, the recursive relations for the cumulants and non-central moments of the generalized Poisson distribution become

$$\begin{cases} (1-\theta)L_{k+1} = \theta\dfrac{\partial L_k}{\partial \theta} + M\dfrac{\partial L_k}{\partial M} , & k = 1, 2, 3, \ldots \\ (1-\theta)l_{k+1} = M\, l_k + \theta\dfrac{\partial l_k}{\partial \theta} + M\dfrac{\partial l_k}{\partial M} , & k = 0, 1, 2, \ldots \end{cases} \tag{30}$$

with $L_1 = M/(1-\theta)$.

6. THE FIRST BUSY PERIOD FOR A SINGLE SERVER QUEUE WITH BERNOULLIAN INPUT. Let us consider there are k customers waiting for service in a queue at a counter, when the service is initially started. More customers are coming in during the service period of these k customers and are joining the queue under the discipline of "first come, first served". The server will be idle if and only if there is no customer in the queue waiting for service, and thus the number of customers served in the First Busy Period (FBP) will be (k+r), provided r additional customers had arrived and been served in the FBP before the server became idle. To obtain the expected length and variance of the random variable X, the number of customers served in the FBP, the probability distribution of X is needed; but this is dependent upon the system of arrivals and the service intervals of each customer. The problem has been considered for Poisson arrivals and constant service times by Borel and Tanner (1961) and Prabhu (1965). We shall consider a more general problem where the input of the customers is Bernoullian or negative binomial, and the initial number of customers waiting for service in a queue is also a discrete random variable. Also we shall provide a very simple method to determine the probability distribution of the number of customers served in the FBP.

Let us suppose that a number of electronic machines, which frequently need exactly the same type of service (constant service time t) are sold by a company in two adjacent cities A and B, which are serviced by one electronic engineer who lives in city A.

All the machines (say k) needing service in city B are brought every day to city A and wait for service, the service beginning when the engineer starts his day's work. All the machines needing service in city A are to be brought to the engineer after each service interval t and are to be serviced in the order of their arrival, but after the k machines of city B have been serviced. The FBP of this engineer would fit exactly the problem under consideration.

Let p be the probability of any machine in city A needing service during the service interval t of one machine and let a fixed number, say m, of machines be in use in the factory. If any machines break down in any service interval t, they are replaced immediately before the next service interval begins, so that the work does not suffer. As the service times are constant and the machines breaking down in each interval are independent of the other intervals, the probability distribution of different numbers of breakdowns during the service period of k machines of city B will be given by the successive terms of the expansion $(q+p)^{mk}$, where $q = 1-p$. Since p is usually small, the number of machines breaking down in each interval t will never be excessive. Though this method of generating a Bernoullian input seems to be somewhat artificial, yet it is more logical, and leads to a natural generation of Poisson input (when p is small and m is large such that mp is constant) and negative binomial input (when the machines come from several factories of city A where some sort of contagion plays its role on account of varying degree of uses of the machines).

In a similar manner the number k can be considered to be a random variable having a binomial, Poisson or negative binomial distribution. We shall consider the problem in two parts: (i) when k is a fixed number, (ii) when k is a random variable.

Theorem 6.1. The probability distribution of the customers served in the FBP initiated by k customers is the kth convolution of the basic LPD having $g(s) = (q+ps)^m$ and is given by

$$P(N=x) = f_x^{(k)} = \frac{k}{x}\binom{mx}{x-k} p^{x-k} q^{(m-1)x+k}, \quad x=k, k+1, \ldots$$

where N is the r.v. denoting the total number of customers in the FBP.

Proof. Evidently $f_x^{(k)} = 0$ for $x<k$ and for $x=k$ the FBP terminates with the service of the kth customer. Thus,

$$f_k^{(k)} = \text{Prob. of no new arrival} = q^{mk} \quad . \tag{31}$$

If $r(\geq 1)$ new customers arrive during the service period of k customers, then the FBP will continue beyond time kt and (N-k) new customers will be served before the end of the FBP. This gives the relation

$$f_x^{(k)} = \sum_{r=1}^{x-k} \binom{mk}{r} p^r q^{mk-r} f_{x-k}^{(r)} , \quad x \geq k+1 . \tag{32}$$

For x = k+1, the relations (32) and (31) give

$$f_{k+1}^{(k)} = m\,k\,p\,q^{m(k+1)-1} , \tag{33}$$

and the use of (32), (31) and (33) for x = k+2 provides

$$\begin{aligned} f_{k+2}^{(k)} &= \binom{mk}{2} p^2 q^{mk-2} f_2^{(2)} + \binom{mk}{1} p\,q^{mk-1} f_2^{(1)} \\ &= \frac{k}{k+2} \binom{mk+2m}{3} p^2 q^{m(k+2)-2} . \end{aligned} \tag{34}$$

By using the above four relations one can similarly show that

$$f_{k+3}^{(k)} = \frac{k}{k+3} \binom{mk+3m}{3} p^3 q^{m(k+3)-3} . \tag{35}$$

The results (34) and (35) suggest that the general expression for the probability distribution $f_x^{(k)}$, given by (32), can be reduced to the form

$$f_x^{(k)} = \frac{k}{x} \binom{mx}{x-k} p^{x-k} q^{(m-1)x+k}, \quad x=k,\ k+1,\ \ldots \tag{36}$$

Assuming that (36) holds for some $x(\geq k)$, we obtain $f_{x+1}^{(k)}$ from the relations (32) and (36) in the form

$$f_{x+1}^{(k)} = \sum_{r=1}^{x-k+1} \binom{mk}{r} p^r q^{mk-r} f_{x-k+1}^{(x)}$$

$$= \sum_{r=1}^{x-k+1} \binom{mk}{r} p^r q^{mk-r} \frac{r}{x-k+1} \binom{m(x-k+1)}{x-r-k+1} p^{x-r-k+1} q^{(m-1)(x-k+1)+r}$$

$$= p^{x-k+1} q^{(m-1)(x+1)+k} \frac{mk}{x+1-k} \sum_{i=0}^{x-k} \binom{mk-1}{i} \binom{m(x+1-k)}{x-k-i}$$

which, by using the combinatorial formula $\sum_{j=0}^{n} \binom{a}{j}\binom{c}{n-j} = \binom{a+c}{n}$,

becomes $f_{x+1}^{(k)} = \frac{k}{x+1}\binom{m(x+1)}{x+1-k} p^{x+1-k} q^{(m-1)(x+1)+k}$. This establishes our result (36) and the theorem on the queueing model.

We assume that the input is Bernoullian as described above so that the result of Theorem 6.1 can be used in the derivation.

Let us suppose that k is a binomial variate, given by $P(X=k) = \binom{n}{k} P^k Q^{n-k}$, $P+Q=1$, where P is the probability of a customer needing service during any fixed interval in the city B which itself has a total of n potential customers.

Evidently, the probability $P(Y=y)$ that the FBP consists of y customers only is given by

$$P(Y=y) = \sum_{k=1}^{\min(n,y)} \binom{n}{k} P^k Q^{n-k} f_y^{(k)}, \quad y = 1, 2, \ldots .$$

By using (36) for the value of $f_y^{(k)}$ and on simplification the probability of the customers served in the FBP becomes

$$P(Y=y) = \frac{n}{y} Q^n (pq^{m-1})^y \sum_{k=1}^{z} \binom{n-1}{k-1}\binom{my}{y-k}\left(\frac{Pq}{pQ}\right)^k, \quad y = 1, 2, 3, \ldots \tag{38}$$

where $z = \min(n,y)$.

The FBP is non-existent when k=0. Accordingly, the probability of not having the FBP initially is Q^n.

The pd (38) with $z = \min(n,y)$ can also be expressed in the hypergeometric form

$$P(Y=y) = \frac{n}{my+1}\binom{my+1}{y}\left(\frac{Pq}{pQ}\right)(pq^{m-1})^y Q^n \; {}_2F_1\left(1-y, 1-n; my-y+2; \frac{Pq}{pQ}\right) \tag{39}$$

for $y = 1, 2, 2, \ldots$ and $P(Y=0) = Q^n$.

The pd given in (38) or (39) is the double binomial family of the general LPD as indicated earlier.

All the moments of this pd exist when $mp < 1$. The expected number and the variance of the customers served in the FBP in this general case are $nP(1-mp)^{-1}$ and $nP(Q-mQp+mpq)(1-mp)^{-3}$.

Corollary 1. When P=p, the pd, (38) or (39), of the customers served in the FBP reduces to the general LPD model

$$p(Y=y) = \frac{n}{n+my}\binom{n+my}{y} p^y q^{n+my-y}, \quad y = 0, 1, 2, 3, \ldots \tag{40}$$

which was independently defined and studied by Jain and Consul (1971) as a generalization of the negative binomial distribution.

Corollary 2. If k is a Poisson r.v. with mean M per unit service interval, then the pd of the number of customers served in the FBP can be derived from (38) by taking its limit for $P \to 0$, $n \to \infty$ such that $nP=m$, and is given by another general LPD model

$$P(Y=y) = e^{-M} \frac{(Mq^m)^y}{y!} \, {}_2F_0(1-y, -my, \frac{p}{Mq}), \quad y = 1, 2, 3, \ldots . \tag{41}$$

Corollary 3. If the initial number k of customers is a Poisson r.v. with mean M per unit service interval and the subsequent arrivals are also Poissonian with mean m per unit service interval, then the pd of the customers served in the FBP can be derived from (38) and is the general LPD model

$$P(Y=y) = \frac{M(M+ym)^{y-1}}{y!} e^{-M-ym}, \quad y = 1, 2, 3, \ldots \tag{42}$$

and $P(Y=0) = e^{-M}$ is the probability of no FBP.

This probability model was also studied in detail earlier by Consul and Jain (1972, 1973). The basic LPD models discussed should be useful in any situation which involves random counts [see G. P. Patil (1970)]. In general it is conceivable that discrete data occurring in ecology, epidemiology, and meteorology could be statistically modeled on one of the family of Lagrange distributions.

REFERENCES

[1] Faà de Bruno. (1855). Quart. J. Math. 1, 359-360.
[2] Consul, P. C. and Jain, G. C. (1972). Technometrics 15, 659-669.

[3] Consul, P. C. and Jain, G. C. (1973). On some interesting properties of the generalized Poisson distribution. Biometrische Zeitschrift.
[4] Consul, P. C. and Shenton, L. R. (1971). SIAM J. Appl. Math. 23, 239-248.
[5] Consul, P. C. and Shenton, L. R. (1973). Comm. in Statist. 2, 263-272.
[6] Haight, F. A. (1961). Biometrika 48, 167-173.
[7] Kendall, M. G. and Stuart, A. (1963). The Advanced Theory of Statistics, Vol. 1. Charles Griffin and Company, Ltd., London.
[8] Jain, G. C. and Consul, P. C. (1971). SIAM J. Appl. Math. 21, 501-513.
[9] Patil, G. P. (ed.) (1970). Random Counts in Scientific Work, Vol. 1. (Random Counts in Models and Structures.) The Pennsylvania State University Press, University Park and London.
[10] Prabhu, N. U. (1965). Queues and Inventories. John Wiley, New York.
[11] Tanner, J. C. (1961). Biometrika 48, 222-224.

ESTIMATION OF PARAMETERS ON SOME EXTENSIONS OF THE KATZ FAMILY OF DISCRETE DISTRIBUTIONS INVOLVING HYPERGEOMETRIC FUNCTIONS*

John Gurland and Ram Tripathi

Department of Statistics, University of Wisconsin, Madison, Wisconsin, U.S.A.

SUMMARY. A two-parameter family of discrete distributions developed by Katz (1963) is extended to three- and four-parameter families whose probability generating functions involve hypergeometric functions. This extension contains other distributions appearing in the literature as particular cases. Various methods of estimating the parameters are investigated and their asymptotic efficiency relative to maximum likelihood estimators compared.

KEY WORDS. Discrete distributions, hypergeometric functions, estimation, efficiency.

1. INTRODUCTION. The probability functions of the two-parameter Katz (1963) family of discrete distribuitons satisfy the first order recurrence relation

$$\frac{P_{j+1}}{P_j} = \frac{\alpha+\beta j}{j+1} \quad \begin{array}{l} j = 0, 1, 2, \ldots \\ \alpha > 0;\ \beta < 1. \end{array} \tag{1.1}$$

---

*This research was supported in part by the Wisconsin Alumni Research Foundation and by the Air Force Office of Scientific Research, AFOSR 72-2363B.

G. P. Patil et al. (eds.), Statistical Distributions in Scientific Work, Vol. 1, 59-82.

Henceforth, we shall refer to it as K family. The probability generating function (p.g.f.) obtained from (1.1) is

$$G(z) = \frac{{}_2F_1(\alpha/\beta,1;1,\beta z)}{{}_2F_1(\alpha/\beta,1;1,\beta)} = \frac{{}_1F_0(\alpha/\beta,z)}{{}_1F_0(\alpha/\beta,\beta)} \tag{1.2}$$

where ${}_2F_1$, ${}_1F_0$ are particular cases of the general hypergeometric function in (1.3).

The p.g.f. (1.2) of the K family is a special case of Kemp's (1968) more general family with the p.g.f.

$$G(z) = \frac{{}_pF_q(a_1,a_2,\ldots,a_p;b_1,b_2,\ldots,b_q,\lambda a)}{{}_pF_q(a_1,a_2,\ldots,a_p;b_1,b_2,\ldots,b_q,\lambda)} \tag{1.3}$$

where

$${}_pF_q(a_1,a_2,\ldots,a_p;b_1,b_2,\ldots,b_q,\lambda z) = \sum_{j=0}^{\infty} \frac{(a_1)_j(a_2)_j\ldots(a_p)_j}{(b_1)_j(b_2)_j\ldots(b_q)_j}\frac{\lambda^j z^j}{j!} \tag{1.4}$$

and $$(a_i)_j = a_i(a_i+1)(a_i+2)\ldots(a_i+j-1). \tag{1.5}$$

Dacey (1972) has investigated this general family of Kemp in considerable detail and has constructed a method for identifying various members of this family.

In the present paper, we propose some three-parameter extensions of the K family. It will also be seen that one of these extensions reduces to the Crow-Bardwell (1963) family of discrete distributions as a particular case. It is also possible to extend these families further to ones involving more than three parameters, but for the purpose of practical applications, we shall confine our attention to families involving no more than three.

In Section 4, we shall present some methods of estimating the parameters in the CB (Crow-Bardwell) family and some extensions of the CB family which may be regarded as limiting cases of the extended K family. The proposed methods are based on generalized minimum chi square as developed by Barankin and Gurland (1951). These methods yield estimators which have rather high asymptotic relative efficiency (a.r.e.) and which are comparatively simple in form. An additional advantage of this method is that it provides a statistic for testing the goodness of fit as a by-product. This test statistic is free of some shortcomings of the traditional Pearson's χ^2 test statistic [cf. Hinz and Gurland (1970)]. An example illustrating these techniques will be applied to some data.

The a.r.e. of the proposed estimators will be obtained for the CB family in particular, and some extended forms of it. To achieve this we develop methods of computing derivatives of the confluent hypergeometric function with respect to any one of its parameters. In a separate paper, corresponding results for some other extensions of the K family, not considered here, will be presented.

2. EXTENSIONS OF THE K FAMILY. As evident in (1.1), the ratio $\frac{P_{j+1}}{P_j}$ in the K family is expressible as a ratio of linear terms in j with the denominator factor j+1. One possible extension of the K family is to express the ratio as

$$\frac{P_{j+1}}{P_j} = \frac{\beta j+\alpha}{j+\lambda}, \quad \begin{array}{l} j = 0,1,\ldots \\ \alpha > 0,\ \beta < 1,\ \lambda > 0. \end{array} \tag{2.1}$$

The p.g.f. corresponding to (2.1) is

$$G(z) = \frac{{}_2F_1(\alpha/\beta,1;\lambda,\beta z)}{{}_2F_1(\alpha/\beta,1;\lambda,\beta)} \ . \tag{2.2}$$

It is interesting to note that the CB family with two parameters is a particular case of (2.2) when $\beta = 0$ in the limit, i.e.,

$$\lim_{\beta=0} \frac{{}_2F_1(\alpha/\beta,1;\lambda,\beta z)}{{}_2F_1(\alpha/\beta,1;\lambda,\beta z)} = \frac{{}_1F_1(1;\lambda,\alpha z)}{{}_1F_1(1;\lambda,\alpha)} \ . \tag{2.3}$$

If P_j be the coefficient of z^j in the limiting p.g.f. (2.3), then

$$\frac{P_{j+1}}{P_j} = \frac{\alpha}{j+\lambda} \tag{2.4}$$

which is the ratio for the CB family.

Other particular cases of (2.2) or of a shifted version with p.g.f.

$$G(z) = \frac{z^s {}_2F_1(\alpha/\beta,1,\lambda+s,\beta z)}{{}_2F_1(\alpha/\beta,1,\lambda+s,\beta)}$$

yield distributions developed by various authors, including Irwin (1963), Good (1953). Other extensions of families involving hypergeometric functions are discussed in Ord (1967).

3. EXTENSIONS INVOLVING THE CONFLUENT HYPERGEOMETRIC FUNCTION.

The ratio $\frac{P_{j+1}}{P_j}$ for the CB family is expressible as

$$\frac{P_{j+1}}{P_j} = \frac{\alpha}{j+\lambda}\ , \quad j = 0,1,\ldots \quad \alpha > 0,\ \lambda > 0. \tag{3.1}$$

The p.g.f. obtained from (3.1) is in terms of the confluent hypergeometric series and is given by

$$G(z) = \frac{{}_1F_1(1;\lambda,\alpha z)}{{}_1F_1(1;\lambda,\alpha)}\ , \quad \lambda > 0,\ \alpha > 0. \tag{3.2}$$

As an extension of the CB family with recurrence relation (3.1), let us consider the recurrence relation

$$\frac{P_{j+1}}{P_j} = \frac{\alpha(j+\gamma)}{(j+\lambda)(j+1)} = \frac{\text{linear in } j}{\text{quadratic in } j} \tag{3.3}$$

which reduces to (3.1) when $\gamma = 1$. From (3.3), we obtain $G(z) = \sum_{j=0}^{\infty} z^j P_j$ which can be written as

$$G(z) = \frac{F(\gamma,\lambda,\alpha z)}{F(\gamma,\lambda,\alpha)}\ . \tag{3.4}$$

This is a valid p.g.f. provided either

(a) $\alpha > 0,\ \gamma > 0,\ \lambda > 0$

or

(b) $\alpha > 0,\ \gamma < 0,\ \lambda < 0,\ [\gamma] = [\lambda]$

where the notation $[\gamma]$ denotes the integral part of γ. The p.g.f. (3.4) reduces to that of the CB family for $\gamma = 1$. For any other value of γ, (3.4) gives an extension of the CB family which, for convenience, we shall denote by E_1CB. For $\gamma = \lambda$, the E_1CB family reduces to a Poisson distribution with mean α and is equidispersed. For $\gamma > \lambda$, the E_1CB family is overdispersed and for $\gamma < \lambda$, it is underdispersed.

For another extension of the CB family, which has the first order recurrence relation (3.1), let us consider a family which has a second order recurrence relation of the form

$$\alpha_0(j+2)(j+1)P_{j+2} + \alpha_1(j+\lambda-\alpha_0)P_{j+1} - \alpha_1^2(j+\gamma)P_j = 0 \quad j = 0,1,\ldots \tag{3.6}$$

The corresponding p.g.f. is

$$G(z) = \frac{F(\gamma,\lambda,\alpha_0+\alpha_1 z)}{F(\gamma,\lambda,\alpha_0+\alpha_1)} \quad . \tag{3.7}$$

This is a valid p.g.f. under any one of the following sets of conditions

(i) $\gamma, \lambda, \alpha_0, \alpha_1, > 0$

(ii) $\lambda > \gamma > 0, \alpha_1 > 0, \alpha_0$ arbitrary

(iii) $\gamma < 0, \lambda < 0, [\gamma] = [\lambda], \alpha_1 > 0, \alpha_1 + \alpha_0 \geq 0.$

This four-parameter extension of the CB family reduces to three parameters if we take $\gamma = 1$ in (3.7). For convenience, we shall denote this three-parameter extension by E_2CB. It has p.g.f.

$$G(z) = \frac{F(1,\lambda,\alpha_0+\alpha_1 z)}{F(1,\lambda,\alpha_0+\alpha_1)} \quad . \tag{3.8}$$

For $\lambda = 1$, the E_2CB distribution is equidispersed and reduces to a Poisson distribution with mean α_1. For $\lambda > 1$, it is overdispersed and for $\lambda < 1$ it is underdispersed.

Estimation of the parameters of the CB, E_1CB and E_2CB families will be considered in Section 4. Estimation of the parameters in the other extensions mentioned above, will be considered in a separate paper.

4. ESTIMATION OF PARAMETERS. Crow and Bardwell have considered various estimators of the parameters, including maximum likelihood, in the CB family. It is our purpose here to develop estimators of the parameters in the CB, E_1CB, and E_2CB families based on minimum chi square techniques. The asymptotic relative efficiency of these estimators as well as some of those obtained by Crow and Bardwell (1963) for their family will be considered in Section 5.

The minimum chi-square procedure we shall adopt for obtaining the estimators of the parameters $\theta_1,\theta_2,\ldots,\theta_q$ of a distribution with probability function $p(x|\theta), \theta' = (\theta_1,\theta_2,\ldots,\theta_q)$ is briefly as follows:

Suppose we find s functions $\tau_1,\tau_2,\ldots,\tau_s$ of moments and/or frequencies such that we can write

$$\tau = W\theta \tag{4.1}$$

where $\tau' = (\tau_1,\tau_2,\ldots,\tau_s)$ and $W_{s\times q}$ is a matrix of known constants, rank $(W) = q < s$. Let $t' = (t_1,\ldots,t_s)$ be a sample counterpart of τ and Σ_t be the covariance matrix of t. Let $\hat{\Sigma}_t$ be a consistent estimate of Σ_t. Then a generalized minimum chi square estimate $\hat{\theta}$ of θ is obtained by minimizing

$$Q = (t-W\theta)'\hat{\Sigma}_t^{-1}(t-W\theta) \tag{4.2}$$

with respect to θ, and is given by

$$\hat{\theta} = (W'\hat{\Sigma}_t^{-1}W)^{-1}(W'\hat{\Sigma}_t^{-1}t). \tag{4.3}$$

This technique will be applied here to obtain estimators of the parameters in the CB, E_1CB, and E_2CB families.

The above procedure for obtaining minimum chi-square estimators also yields a statistic for testing fit by utilizing the minimum value, $\hat{Q}$, say, of the quadratic form (4.2) with θ replaced by $\hat{\theta}$. The asymptotic distribution of this $\hat{Q}$ statistic is that of χ^2 with s-q d.f. Such a test will be included in the illustrative example in Section 6.

4.1. Minimum Chi-Square Estimators of the Parameters of the CB Family. In estimating parameters of the CB family, we shall make use of the recurrence relations involving factorial moments $\mu_{(r)}$ and probabilities P_r:

$$\mu_{(r+2)} + (\lambda+r-\alpha)\mu_{(r+1)} - \alpha(r+1)\mu_{(r)} = 0 \tag{4.4}$$

$$(r+\lambda)P_{r+1} - P_r = 0 \qquad r = 0,1,\ldots \tag{4.5}$$

Another useful relation obtained by summing (4.5) over r is

$$\mu_{(1)} + (\lambda-1)(1-P_0) = \alpha. \tag{4.6}$$

Let $$\eta_1 = \mu_{(1)}, \ \eta_r = \frac{\mu_{(r+1)}}{\mu_{(r)}} \qquad r = 1,2,\ldots \tag{4.7}$$

$$\xi_r = \frac{P_{r+1}}{P_r}, \quad r = 1,2,\ldots \tag{4.8}$$

Substituting for η_0, η_r and ξ_r in (4.4), (4.5) and (4.6), we obtain

$$\lambda\eta_r - \alpha(r+1+\eta_r) = -\eta_r(r+\eta_{r+1}) \quad r = 0,1,\ldots \tag{4.9}$$

$$\lambda(1-P_0) - \alpha = \eta_0 + 1-P_0 \tag{4.10}$$

$$\lambda\xi_r - \alpha = -r\xi_r \quad r = 0,1,2,\ldots \tag{4.11}$$

We shall hereafter refer to (4.9) as moment relations, (4.10) as the relation involving $\mu_{(1)}$ and P_0 and (4.11) as probability relations. We propose various methods of obtaining minimum chi square estimators for the parameters of the CB family by making use of the above relations. To describe these methods, we first take a desired number of moment and probability relations, say s, which may or may not include (4.10), and obtain solutions for λ and α. The number of solutions of both λ and α that we shall take will be the same as the number of equations considered in order to achieve linear independence of the solutions. Taking the elements of τ as these solutions expressed in terms of η_r, ξ_r, $r = 0,1,\ldots$ and P_0, we can write $\tau = W\theta$ where W is a s×2 matrix of known constants and θ is a 2×1 vector with $\theta_1 = \lambda$, $\theta_2 = \alpha$. Then, we can obtain the minimum chi square estimators of θ by means of the technique mentioned earlier.

These methods differ only in the number of moment and probability relations considered, but the essential technique remains the same. As an example, we shall present the details of a method called the method of 2 moment relations and 2 probability relations for obtaining minimum chi square estimators of the parameters of the CB family.

Let us consider the first two moment relations and first two probability relations:

$$\lambda\eta_0 - \alpha(1+\eta_0) = \eta_0\eta_1 \tag{4.12}$$

$$\lambda\eta_1 - \alpha(2+\eta_1) = -\eta_1(1+\eta_2) \tag{4.13}$$

$$\lambda\xi_0 - \alpha = 0 \tag{4.14}$$

$$\lambda\xi_1 - \alpha = -\xi_1 \tag{4.15}$$

From (4.13) and (4.14), we obtain

$$\tau_1 = \frac{\eta_1(1+\eta_2)}{\xi_0(2+\eta_1)-\eta_1} = \lambda, \quad \tau_2 = \frac{\eta_1\xi_0(1+\eta_2)}{\xi_0(2+\eta_1)-\eta_1} = \alpha . \tag{4.16}$$

From (4.12) and (4.13), we obtain

$$\tau_3 = \frac{\eta_1[1-\eta_0+\eta_2+\eta_0\eta_2-\eta_0\eta_1]}{2\eta_0-\eta_1} = \lambda . \tag{4.17}$$

From (4.13) and (4.15), we obtain

$$\tau_4 = \frac{\xi_1\eta_1\eta_2}{\xi_1(2+\eta_1)-\eta_1} = \alpha . \tag{4.18}$$

Thus, we have $\tau' = (\tau_1,\tau_2,\tau_3,\tau_4)$ and if we take

$$W' = \begin{bmatrix} 1 & 0 & 1 & 0 \\ 0 & 1 & 0 & 1 \end{bmatrix} , \quad \theta' = (\lambda,\alpha) \tag{4.19}$$

we have the linear relation

$$\tau = W\theta . \tag{4.20}$$

Let $t' = (t_1,t_2,t_3,t_4)$ be a sample counterpart of τ obtained by substituting sample moments and relative frequencies. We shall now indicate how to obtain Σ_t. Let J_1,J_2,J_3 be the Jacobians of the following transformations:

$$J_1\colon (\mu_1' \; \mu_2' \; \mu_3' \; P_0 \; P_1 \; P_2) \to (\mu_{(1)}\mu_{(2)}\mu_{(3)}\xi_0\xi_1)$$

$$J_2\colon (\mu_{(1)}\mu_{(2)}\mu_{(3)}\xi_0\xi_1) \to (\eta_0 \; \eta_1 \; \eta_2, \; \xi_0 \; \xi_1)$$

$$J_3\colon (\eta_0, \; \eta_1 \; \eta_2 \; \xi_0 \; \xi_1) \to (\tau_1 \; \tau_2 \; \tau_3 \; \tau_4)$$

and Σ be the covariance matrix of $(m_1' \; m_2' \; m_3' \; p_0 \; p_1 \; p_2)$ where m_i'

is the ith raw sample moment and p_j is the jth sample relative frequency. Then the covariance matrix Σ_t of t is

$$\Sigma_t = J_3J_2J_1 \, (J_3J_2J_1)'. \tag{4.21}$$

The elements of the matrixes J_1, J_2, J_3 and Σ are given below:

$$\underset{5\times6}{J_1} = \frac{\partial(\mu_{(1)},\mu_{(2)},\mu_{(3)},\xi_0,\xi_1)}{\partial(\mu_1',\mu_2',\mu_3',P_0,P_1,P_2)} = \begin{bmatrix} 1 & 0 & 0 & 0 & 0 & 0 \\ -1 & 1 & 0 & 0 & 0 & 0 \\ 2 & -3 & 1 & 0 & 0 & 0 \\ 0 & 0 & 0 & -\frac{\xi_0}{P_0} & \frac{1}{P_0} & 0 \\ 0 & 0 & 0 & 0 & \frac{\xi_1}{P_1} & \frac{1}{P_1} \end{bmatrix} \tag{4.22}$$

$$\underset{5\times5}{J_2} = \frac{\partial(\eta_0,\eta_1,\eta_2,\xi_0,\xi_1)}{\partial(\mu_{(1)},\mu_{(2)},\mu_{(3)},\xi_0,\xi_1)} = \begin{bmatrix} 1 & 0 & 0 & 0 & 0 \\ -\frac{\eta_1}{\mu_{(1)}} & \frac{1}{\mu_{(1)}} & 0 & 0 & 0 \\ 0 & -\frac{\eta_2}{\mu_{(2)}} & \frac{1}{\mu_{(2)}} & 0 & 0 \\ 0 & 0 & 0 & 1 & 0 \\ 0 & 0 & 0 & 0 & 1 \end{bmatrix} \tag{4.23}$$

$$\underset{4\times 5}{J_3} = \frac{\partial(\tau_1,\tau_2,\tau_3,\tau_4)}{\partial(\eta_0,\eta_1,\eta_2,\xi_0,\xi_1)}$$

$$= \begin{bmatrix} 0 & \dfrac{2\xi_0\tau_1}{\eta_1\{\xi_0(2+\eta_1)-\eta_1\}} & \dfrac{\eta_1}{\xi_0(2+\eta_1)-\eta_1} & -\dfrac{\tau_1(2+\eta_1)}{\xi_0\{\xi_0(2+\eta_1)-\eta_1\}} & 0 \\ 0 & \dfrac{2\xi_0\tau_2}{\eta_1\{\xi_0(2+\eta_1)-\eta_1\}} & \dfrac{\eta_1\xi_0}{\xi_0(2+\eta_1)-\eta_1} & -\dfrac{\tau_2\eta_1}{\xi_0\{\xi_0(2+\eta_1)-\eta_1\}} & 0 \\ \dfrac{\eta_1(\eta_2-\eta_1-1)-2\tau_3}{2\eta_0-\eta_1} & \dfrac{\eta_0(2\tau_3-\eta_1^2)}{\eta_1(2\eta_0-\eta_1)} & \dfrac{\eta_1(1+\eta_0)}{2\eta_0-\eta_1} & 0 & 0 \\ 0 & \dfrac{2\tau_4\xi_1}{\eta_1\{\xi_1(2+\eta_1)-\eta_1\}} & \dfrac{\tau_4}{\eta_2} & 0 & -\dfrac{\eta_1\tau_4}{\xi_1\{\xi_1(2+\eta_1)-\eta_1\}} \end{bmatrix} . \quad (4.24)$$

The elements of Σ can be obtained as follows:

$$\left.\begin{aligned}
\mathrm{Cov}(m_i', m_j') &= \frac{1}{n}\,[\mu'_{i+j} - \mu_i'\mu_j'] \\
\mathrm{Cov}(p_i, p_j) &= -\frac{P_iP_j}{n} \qquad i \neq j \\
\mathrm{Var}(p_i, p_i) &= \frac{P_i(1-P_i)}{n} \\
\mathrm{Cov}(m_i', p_j) &= \frac{1}{n}\,(j^i - \mu_i')P_j
\end{aligned}\right\} \quad . \tag{4.25}$$

It is now possible to obtain the minimum chi-square estimators $\hat{\theta}$ of θ as

$$\hat{\theta} = (W'\,\hat{\Sigma}_t^{-1}\,W)^{-1} \quad (W'\,\hat{\Sigma}_t^{-1}\,t). \tag{4.26}$$

We can obtain minimum chi-square estimators $\hat{\theta}$ of θ by a similar procedure involving any number of moment and probability relations including (4.10). A similar procedure will be adopted for estimating the parameters in the E_1CB and the E_2CB families.

4.2. Minimum chi square estimators for the parameters of the E_1CB family. Let us consider the recurrence relations for the factorial moments and the probabilities of the E_1CB family:

$$\left.\begin{aligned}
\mu_{(r+2)} + (\lambda+r-\alpha)\mu_{(r+1)} - \alpha(\gamma+r)\mu_{(r)} &= 0 \\
(r+1)(\lambda+r)P_{r+1} - \alpha(\gamma+r)P_r &= 0
\end{aligned}\right\} \tag{4.27}$$

$$r = 0,\ 1,\ \ldots\ .$$

Putting η_r and ξ_r in (4.27), we obtain

$$\theta_1\eta_r - \theta_2(r+\eta_r) - \theta_3 = -\eta_r(r+\eta_{r+1}) \tag{4.28}$$

$$\theta_1(r+1)\xi_r - r\theta_2 - \theta_3 = -r(r+1)\xi_r, \quad r = 0,1,\ldots \tag{4.29}$$

$$\theta_1 = \lambda, \qquad \theta_2 = \alpha, \qquad \theta_3 = \gamma\alpha . \tag{4.30}$$

Again, we shall call (4.28) and (4.29) respectively moment and probability relations. Various procedures involving a desirable number, say s, of moment and probability relations are proposed for estimating the parameters in the E_1CB family. The essential techniques involved for obtaining the minimum chi square estimator $\hat{\theta}$ of θ, $\theta' = (\lambda,\alpha,\gamma\alpha)$, remain the same as those mentioned in the section 4.1. From $\hat{\theta}$, we obtain the minimum chi square estimators of the parameters γ,λ and α in the E_1CB family as

$$\begin{aligned} \hat{\gamma} &= \hat{\theta}_3/\theta_1 \\ \hat{\lambda} &= \hat{\theta}_1 \\ \hat{\alpha} &= \hat{\theta}_2 . \end{aligned} \tag{4.31}$$

4.3. Minimum chi square estimators for the parameters of E_2CB family. Let us consider the recurrence relations for the factorial moments and for the probabilities of the E_2CB family:

$$(\alpha_0+\alpha_1)\mu_{(r+2)} + \alpha_1(\lambda-\alpha_0+r-\alpha_1)\mu_{(r+1)} - \alpha_1^2(r+1)\mu_{(r)} = 0 \tag{4.32}$$

$$\alpha_0(r+2)P_{r+2} + \alpha_1(r+\lambda-\alpha_0)P_{r+2} - \alpha_1^2 P_r = 0 , \quad r = 0,1,\dots . \tag{4.33}$$

On substituting for η_r and ξ_r in (4.32) and (4.33), we obtain the following moment and probability relations:

$$\theta_1\eta_r\eta_{r+1} + \theta_2\eta_r - \theta_3(r+1+\eta_r) = -\eta_r(r+\eta_{r+1}) \tag{4.34}$$

$$\theta_1(r+2)\xi_r\xi_{r+1} + \theta_2\xi_r - \theta_3 = -r\xi_r , \quad r = 0,1,\dots \tag{4.35}$$

where

$$\theta_1 = \frac{\alpha_0}{\alpha_1} , \quad \theta_2 = \lambda-\alpha_0, \quad \theta_3 = \alpha_1. \tag{4.36}$$

Various methods for obtaining minimum chi square estimators of the parameters of E_2CB family are possible by considering any number of moment and frequency relations. For the particular method selected, the minimum chi square estimator $\hat{\theta}$ of θ, $\theta' = (\frac{\alpha_0}{\alpha_1}, \lambda-\alpha_0, \alpha_1)$ can be obtained by means of the techniques illustrated in the section 4.1. From $\hat{\theta}$, we obtain

$$\hat{\alpha}_1 = \hat{\theta}_3, \quad \hat{\alpha}_0 = \hat{\theta}_1\hat{\theta}_3, \quad \hat{\lambda} = \hat{\theta}_2 + \hat{\theta}_1\hat{\theta}_3. \tag{4.37}$$

In section 5, we shall investigate the asymptotic relative efficiency of the estimators of the CB, E_1CB and E_2CB families obtained by various methods of estimation. We shall also present recommendations for some of the methods proposed based on their asymptotic relative efficiency.

5. ASYMPTOTIC RELATIVE EFFICIENCY OF MINIMUM CHI-SQUARE ESTIMATORS.

Let $\Sigma_{\hat{\theta}}$ denote the asymptotic covariance matrix of the minimum chi-square estimator $\hat{\theta}$ of θ. Then

$$\Sigma_{\hat{\theta}} = (W'\Sigma_t^{-1}W)^{-1} . \tag{5.1}$$

Let L be the likelihood function and

$$I = \left\{E(\frac{\partial \log L}{\partial \theta_i})\ (\frac{\partial \log L}{\partial \theta_i'})\right\} = \left\{-E(\frac{\partial^2 \log L}{\partial \theta_i \partial \theta_i'})\right\} \tag{5.2}$$

be the information matrix. Then the asymptotic relative efficiency of $\hat{\theta}$ is given by

$$\text{A.R.E.} = \frac{1}{|I|\cdot|\Sigma_{\hat{\theta}}|} . \tag{5.3}$$

In computing the information matrix (5.2) we require derivatives of the confluent hypergeometric function with respect to each of its parameters. Three different methods of obtaining numerical derivatives of the confluent hypergeometric function are possible:

(i) termwise differentiation
(ii) differentiation based on interpolation formulae
(iii) differentiation based on the method of extrapolation to the limit.

Comparing the results obtained by these three methods enables us to verify the results and obtain reliable values of the derivatives.

5.1. Asymptotic relative efficiency of some estimators of the parameters in the CB family. We shall now investigate the a.r.e. of some minimum chi-square estimators proposed here and compare their behavior with some estimators proposed by Crow and Bardwell (1963). In Table 1 appears the a.r.e. of

(i) simple estimator based on 2 moment relations (Crow and Bardwell (1963))
(ii) simple estimator based on 1 moment relation and another relation involving $\mu_{(1)}$ and P_0 (Crow and Bardwell (1968))
(iii) minimum chi-square estimator based on 2 moment relations and 1 probability relation
(iv) minimum chi-square estimator based on 2 moment relations and another relation involving $\mu_{(1)}$ and P_0.

An examination of Table 1 reveals that the a.r.e. of all the estimators except (i) are high for large values of λ and small values of α. The a.r.e. of (i) is very low except for large values of λ and small values of α. The estimator in (ii), however, has reasonably high efficiency throughout the parameter grid considered here. The estimator (iii) has somewhat low a.r.e. for small values of λ and large values of α, whereas, the estimator in (iv) has very high a.r.e. throughout the parameter grid considered. In fact, the estimator (iv) has uniformly higher a.r.e. than the estimators (i), (ii) and (iii), therefore, we recommend it for practical applications.

On comparing the a.r.e. of other estimators, we have found that very little is gained by adding more probability relations to the estimator (iii). Inclusion of higher order moment relations is not advisable due to extensive variation of corresponding sample moments.

Table 1

Asymptotic relative efficiency of some estimators for the parameters in the CB family

λ \ α		.7	1.0	2.5	7.5	10.0
.7	(i)	.24	.19	.15	.36	.45
	(ii)	.81	.75	.64	.78	.85
	(iii)	.96	.92	.64	.34	.46
	(iv)	.97	.94	.85	.92	.96
1.0	(i)	.37	.29	.20	.35	.44
	(ii)	.91	.88	.77	.80	.85
	(iii)	.97	.93	.66	.38	.45
	(iv)	.99	.98	.94	.94	.96
2.5	(i)	.77	.70	.45	.33	.40
	(ii)	1.00	.99	.97	.89	.88
	(iii)	.99	.97	.81	.39	.41
	(iv)	1.00	1.00	.99	.99	.98
7.5	(i)	.97	.96	.90	.61	.50
	(ii)	.98	.97	.93	.84	.85
	(iii)	1.00	1.00	.98	.73	.57
	(iv)	1.00	1.00	.99	.94	.91
10.0	(i)	.99	.98	.95	.76	.64
	(ii)	.98	.97	.92	.81	.78
	(iii)	1.00	1.00	.99	.86	.73
	(iv)	1.00	1.00	1.00	.96	.92

5.2. A.R.E. of some estimators for the parameters in the E_1CB family. In Table 2 appears the a.r.e. of

(i) simple estimator based on 3 moment relations
(ii) simple estimator based on 2 moment and 1 probability relations
(iii) minimum chi-square estimator based on 3 moment and 3 probability relations
(iv) minimum chi-square estimator based on 2 moment and 4 probability relations

for the parameters in the E_1CB family. The values in Table 2 suggest a definite pattern for the a.r.e. of the estimators

Table 2

A.R.E. of some estimators for the parameters in the E_1CB family

α		$\gamma = .5$				$\gamma = 1.2$				$\gamma = 2.3$				$\gamma = 5.2$			
		.7	1.00	2.5	5.0	.7	1.0	2.5	5.0	.7	1.0	2.5	5.0	.7	1.0	2.5	5.0
.7	(i)	.19	.10	.01	.003	.17	.10	.02	.01	.16	.10	.05	.06	.15	.12	.14	.22
	(ii)	.82	.74	.45	.29	.83	.76	.51	.34	.83	.76	.52	.25	.81	.72	.33	.05
	(iii)	1.00	1.00	.99	.91	1.00	1.00	.99	.99	1.00	1.00	.99	.88	1.00	1.00	.96	.64
	(iv)	1.00	1.00	.99	.88	1.00	1.00	.98	.83	1.00	1.00	.97	.80	1.00	1.00	.93	.62
1.00	(i)	.30	.18	.02	.004	.27	.17	.04	.02	.24	.16	.07	.07	.22	.18	.16	.22
	(ii)	.87	.80	.51	.29	.87	.81	.55	.34	.87	.81	.54	.25	.84	.75	.34	.05
	(iii)	1.00	1.00	1.00	.93	1.00	1.00	.99	.92	1.00	1.00	.99	.89	1.00	1.00	.97	.67
	(iv)	1.00	1.00	.99	.89	1.00	1.00	.99	.85	1.00	1.00	.97	.82	1.00	1.00	.94	.65
2.5	(i)	.60	.49	.16	.003	.59	.47	.18	.06	.57	.46	.20	.11	.52	.43	.28	.26
	(ii)	.96	.94	.78	.42	.97	.95	.77	.44	.97	.94	.70	.33	.93	.86	.45	.09
	(iii)	1.00	1.00	1.00	.97	1.00	1.00	1.00	.97	1.00	1.00	1.00	.95	1.00	1.00	.99	.81
	(iv)	1.00	1.00	.99	.95	1.00	1.00	.99	.93	1.00	1.00	.99	.90	1.00	1.00	.97	.79
5.0	(i)	.80	.72	.43	.17	.80	.72	.44	.20	.79	.72	.45	.24	.77	.70	.48	.36
	(ii)	.99	.99	.94	.74	.99	.99	.92	.67	.99	.98	.86	.53	.96	.92	.62	.21
	(iii)	1.00	1.00	1.00	1.00	1.00	1.00	1.00	.99	1.00	1.00	1.00	.99	1.00	1.00	1.00	.93
	(iv)	1.00	1.00	1.00	.99	1.00	1.00	1.00	.98	1.00	1.00	1.00	.97	1.00	1.00	.89	.92

considered here. These estimators have high a.r.e. for higher values of λ and small values of α. For every α, the value of the a.r.e. increases (to 1.00) as λ increases and for every λ it decreases as α increases. For any pair (λ,α), the value of the a.r.e. decreases as γ increases.

It is evident from the table that the estimator (ii) has higher efficiency than the estimator (i) for small values of γ but for high values of γ the estimator (ii) has substantially lower a.r.e. than the estimator (i). Both the estimators (i) and (ii) have lower a.r.e. than the estimators (iii) and (iv). From Table 2, we also see that the a.r.e. of the estimators (iii) and (iv) are very close and are both substantially higher than the values for (i) and (ii). Since for (iii) and (iv) the a.r.e. values for the most part are rather high, these estimators would appear to be preferable in practical applications for the range of the parameters indicated in the table.

Since the E_1CB family is a valid distribution for $\gamma < 0$, $\lambda < 0$, with $[\gamma] = [\lambda]$, it is interesting to compare the behavior of various estimators of the parameters in this region. After examining many estimators it was found that utilization of the probability relations is important in this region of the parameter space and further, that the efficiency decreases rapidly with numerical increase of the parameter values.

Table 3 provides a slight extension of Table 2 into this region of the parameter space but with some other estimators involving probability relations. The a.r.e. is given for the following types of estimators using minimum chi-square:

(i) based on 3 moment relations and 1 probability relation
(ii) based on 2 moment relations and 2 probability relations
(iii) based on 1 moment relation and 3 probability relations
(iv) based on 3 moment relations and 3 probability relations
(v) based on 2 moment relations and 4 probability relations.

It is clear from the table that estimators (iii), (iv) and (v) using more probability relations have higher efficiency. Further, the efficiency decreases as the numerical values of γ and λ increases. To achieve high efficiency for larger values of the parameters than appear in the table would require estimators utilizing more relations.

5.3. A.R.E. of some estimators for the parameters in the E_2CB family. Various estimators for the parameters in the E_2CB family were considered, including

(i) simple estimator based on 3 moment relations
(ii) simple estimator based on 2 moment and 1 probability relation
(iii) minimum chi-square estimator based on 3 moment and 3 probability relations
(iv) minimum chi-square estimator based on 2 moment and 4 probability relations.

Asymptotic relative efficiencies were computed for the grid of parameter values resulting from λ = .5, 1.2, 2.3, 5.5; α_0 = .7, 1.0, 2.5, 5.2, and α_1 = .7, 1.0, 2.5, 5.2. It was found that for the estimators considered, the a.r.e. increases for increasing values of λ for any pair (α_0,α_1). For any pair (α_0,λ), the a.r.e. for these estimators increases for increasing values of α_1, but for any pair (α_1,λ), it decreases for increasing α_0. It was also observed that estimators (i) and (ii) are for the most part quite similar in regard to a.r.e. values. Estimators (iii) and (iv) also behave similarly and have a.r.e. values generally higher than those for (i) and (ii). All four estimators, however, behave poorly when α_0 = 5.2 and α_1 = .7, 1.0. For practical purposes, estimators (iii) or (iv) appear to be preferable when $\alpha_0 < 1$, $\lambda > .5$ and all values of α_1 considered.

6. NUMERICAL EXAMPLE. For illustrative purposes, we include a numerical example in which data is fitted by the CB, the E_1CB and the E_2CB families, where the parameters are estimated by the minimum chi-square technique developed above.

Student's haemacytometer yeast cell count. Let us consider the well known data [Student (1970)] obtained from haemacytometer yeast cell counts. Table 4 gives the observed data, and the fits obtained by minimum chi-square estimators to the CB, E_1CB, and E_2CB families. The various relations on which these estimators are based are indicated in the footnote to the table. For comparison the fits obtained by Neyman (1939) to his Type A distribution and by Crow and Bardwell (1963) to their hyper-Poisson distribution are also included. The value of the Pearson's chi-square test statistic and the corresponding probability of exceeding it is included at the bottom of the table for all fits considered. For the test of fit based on minimum chi-square, the value of the relevant test statistic is also included along with the corresponding probability value.

From Table 4, it is evident that the fit (5) to the hyper-Poisson distribution by minimum chi-square is virtually the same as that obtained by maximum likelihood, and also closely resembles the other fits (3) and (4) obtained by Crow and Bardwell. It is interesting that the fit (6) to the E_1CB family is an improvement over the other fits in the table. This is reflected by the comparatively small value of the Pearson chi-square statistic and the especially small value of the minimum chi-square test statistic. It is also noteworthy that the estimates of the parameters γ and λ are negative.

The fit to the E_2CB family is rather curious in that the estimates of the parameters λ and α_0 are rather large. The expected frequencies, however, are comparable to those obtained by fitting the data to the CB family. The value of the Pearson chi-square test statistic, 3.98, is accordingly similar, but the corresponding probability value, .046, is much smaller due to the reduction in the number of d.f. On the other hand, the value of the minimum chi-square statistic, 1.021, corresponds to a probability value .796 based on 3 d.f.

For Table 4, the column headings correspond to the following:

(1) Fit obtained by Neyman (1939) to his type A distribution.
(2) Fit obtained by Crow and Bardwell (1963) to hyper-Poisson distribution based on maximum likelihood.

Table 3

A.R.E. of some estimators for the parameters in the E_1CB family for negative values of parameters

$\lambda \backslash \alpha$		.7	1.0	2.5	5.0	.7	1.0	2.5	5.0	.7	1.0	2.5	5.0
- .8	(i)	.20	.11	.04	.02								
	(ii)	.98	.94	.76	.44								
	(iii)	.98	.96	.80	.39								
	(iv)	1.00	1.00	.99	.85								
	(v)	1.00	1.00	.99	.84								
-1.8	(i)					.03	.01	.00	.00				
	(ii)					.17	.10	.02	.02				
	(iii)					.90	.82	.52	.38				
	(iv)					1.00	.99	.86	.68				
	(v)					1.00	1.00	.96	.91				
-2.8	(i)									.00	.00	.00	.00
	(ii)									.01	.01	.00	.00
	(iii)									.10	.07	.01	.00
	(iv)									.55	.34	.06	.03
	(v)									.99	.98	.79	.57

(3) Fit obtained by Crow and Bardwell (1963) to hyper-Poisson distribution based on 2 moment relations.

(4) Fit obtained by Crow and Bardwell (1963) to hyper-Poisson distribution based on 1 moment relation and another relation involving $\mu_{(1)}$ and P_0.

(5) Fit to the CB family obtained by minimum chi-square based on 2 moment relations and another relation involving $\mu_{(1)}$ and P_0.

(6) Fit to the E_1CB family obtained by minimum chi-square based on 1 moment and 3 probability relations.

(7) Fit to the E_2CB family obtained by minimum chi-square based on 3 moment and 3 probability relations.

Table 4

Distribution of yeast cells observed by "Student" in 400 squares of Heamacytometer fitted by Neyman Type A, the CB family and the E_1CB family

		(1)	(2)	(3)	(4)	(5)	(6)	(7)
No. of cells per square	Observed no. of squares	Neyman Type A	CB Maximum Likelihood	CB	CB	CB	E_1CB	E_2CB
0	213	214.8	215.0	216.6	214.1	214.3	220.1	219.6
1	128	121.3	120.3	118.5	120.7	120.7	129.0	116.6
2	37	45.7	46.7	46.3	47.0	46.9	32.7	44.9
3	18	13.7	13.9	14.1	14.0	13.9	14.9	14.0
4	3	3.6	3.3	3.5	3.4	3.3	2.8	3.8
5	1	.8	.7	.8	.7	.7	.4	.9
6	0	.1	.1	.2	.1	.1	.1	.2
Total	400	400.0	400.0	400.0	400.0	400.0	400.0	400.0
χ_p^2		3.45	3.74	3.82	3.73	3.75	1.59	3.98
d.f.		2	2	2	2	2	0	1
$P\chi_p^2$		.183	.162	.157	.163	.154	*	.046
χ_{Min}^2						.127	.0003	1.021

Table 4 (Continued)

		(1)	(2)	(3)	(4)	(5)	(6)	(7)
No. of cells per square	Observed no. of squares	Neyman Type A	CB Maximum Likelihood	CB	CB	CB	E_1CB	E_2CB
d.f.						1	1	3
$P\chi^2_{Min}$						.721	.985	.796
$\hat{\alpha}$			1.269	1.371	1.263	1.225	.653	$\hat{\alpha}_1$ = 5.729
$\hat{\lambda}$			2.268	2.509	2.241	2.228	-1.831	$\hat{\lambda}$ = 102.7
							$\hat{\gamma}$=-1.643	$\hat{\alpha}_0$ = 106.3

* Probability for zero degree of freedom can not be computed.

REFERENCES.

[1] Barankin, E. W. and Gurland, J. (1951). Univ. Calif. Pub. Statist. 1, 89-129.
[2] Crow, E. L. and Bardwell, G. E. (1963). Estimation of the parameters of the hyper-Poisson distributions in Classical and Contagious Discrete Distributions, G. P. Patil (ed.), Pergamon Press, New York, 127-140.
[3] Dacey, M. F. (1972). Sankhyā Ser. B 34, 243-250.
[4] Good, I. J. (1953). Biometrika 40, 237-264.
[5] Hinz, P. and Gurland, J. (1970). J. Amer. Stat. Assoc. 65, 887-903.
[6] Irwin, J. O. (1963). Inverse factorial series as frequency distributions in Classical and Contagious Discrete Distributions, G. P. Patil (ed.), Pergamon Press, New York, 159-174.
[7] Katz, L. (1963). Unified treatment of a broad class of discrete probability distributions in Classical and Contagious Discrete Distributions, G. P. Patil (ed.), Pergamon Press, New York, 175-182.
[8] Kemp, A. W. (1968). Sankhyā Ser. A 30, 401-410.
[9] McGuire, J. V., Brindley, T. A. and Bancroft, T. A. (1957). Biometrics 13, 65-78.
[10] Neyman, J. (1939). Ann. Math. Stat. 10, 35-57.
[11] Ord, J. K. (1967). Biometrika, 54, 649-656.
[12] Simon (1954). Biometrika, 42, 425-440.
[13] "Student" (1970). Biometrika 5, 351-360.

A CHARACTERISTIC PROPERTY OF CERTAIN GENERALIZED POWER SERIES DISTRIBUTIONS

G. P. Patil and V. Seshadri

The Pennsylvania State University, University Park, Pa., U.S.A. and McGill University, Montreal, Quebec, Canada.

SUMMARY. There are some probability distributions which remain invariant in their form under suitable transformations. In this paper, we show that the logarithmic series distribution and the geometric distribution enjoy the property of invariance of form as a characterizing property under a special type of transformation which we introduce below as a modulo sequence. Further, we provide a necessary and sufficient condition for the generalized power series distribution to have this characteristic property.

KEY WORDS. Modulo sequence, logarithmic series distribution, geometric distribution.

1. NOTATION AND TERMINOLOGY.

Definition 1. Let I^+ be the set of positive integers. A sequence of random variables $\{X_k\}$ is said to be a modulo sequence of random variables generated by X_1 if, for every $k \in I^+$, the random variable (rv) X_k has, as its distribution, the conditional probability distribution of X_1/k given that $X_1/k \in I^+$, that is, given that X_1/k has assumed a positive integral value, that is, given that $X_1 \equiv 0 \bmod (k)$. Further, the rv X_k is said to be the modulo (k) rv generated by the rv X_1.

Definition 2. Let T be a subset of the set I of non-negative integers. Define $f(\theta) = \Sigma\, a(x)\, \theta^x$ where the summation extends over T and $a(x) > 0$, $\theta \geq 0$ with $\theta \in \Theta$, the parameter space, such that $f(\theta)$ is finite and differentiable. Then a rv X

G. P. Patil et al. (eds.), Statistical Distributions in Scientific Work, Vol. 1, 83-86.

with probability function (pf)

$$\text{Prob}\,\{X = x\} = p(x;\theta) = a(x)\theta^x/f(\theta) \quad x \in T$$

is said to have the generalized power series distribution (GPSD) with range T and the series function (sf) $f(\theta)$.

Definition 3. The GPSD with range $T = I^+$ and the sf $f(\theta) = -\log(1-\theta) = 1/\alpha(\theta)$ is said to be the logarithmic series distribution (LSD) with parameter θ, $0 < \theta < 1$.

Definition 4. The GPSD with range $T = I^+$ and the sf $f(\theta) = \theta/(1-\theta) = 1/\beta(\theta)$ is said to be the geometric distribution (GD) with parameter θ, $0 < \theta < 1$.

2. RESULTS.

Theorem 1. Let $\{X_k\}$ be a modulo sequence of rv's generated by X_1. If X_1 has the LSD with parameter θ, then its modulo (k) rv X_k has the LSD with parameter θ^k for every $k \in I^+$.

Proof. From definition 1, we have $\text{Prob}\,\{X_k = x\} = \text{Prob}\,\{\frac{X_1}{k} = x/\frac{X_1}{k} \in I^+\}$, which because of definition 3

$$= \frac{\alpha(\theta)\ \theta^{kx}}{kx} \Big/ \sum_{x=1}^{\infty} \frac{\alpha(\theta)\ \theta^{kx}}{kx} = \frac{\alpha(\theta^k)\ (\theta^k)^x}{x}.$$

Hence the Theorem.

Theorem 2. Let $\{X_k\}$ be a modulo sequence of rv's generated by X_1. If X_1 has the GD with parameter θ, then its modulo (k) rv X_k has the GD with parameter θ^k for every $k \in I^+$.

Proof. Follows as for Theorem 1.

Theorem 3. Let $\{X_k\}$ be a modulo sequence of rv's generated by X_1 such that X_1 has a GPSD with parameter θ and that the modulo (k) rv X_k has the LSD with parameter θ^k for $k > 1$. Then X_1 has the unity - truncated LSD with parameter θ.

Proof. Let the sf of the GPSD of X_1 be

$$g(\theta) = \Sigma\, a(y)\theta^y \tag{1}$$

where $y \in T$ with T as yet unspecified.

Now, by hypothesis, $\text{Prob}\,\{X_k = x\}$

$$= \frac{\alpha(\theta^k)\ (\theta^k)^x}{x} \quad \text{for } k > 1 \tag{2}$$

which by definition 1 of the modulo sequence

$$= \text{Prob}\ \{\frac{X_1}{k} = x / \frac{X_1}{k} \in I^+\}$$

$$= \frac{a(kx)\theta^{kx}}{\sum_{x=1}^{\infty} a(kx)\theta^{kx}} \quad \text{for } k > 1 \tag{3}$$

Comparing (2) and (3), we note that, for a fixed θ,

$$xa(kx) = \text{constant for every } k > 1 \tag{4}$$

Putting $k = 2$ and $x = 1$ and $x = y$ we get from (4),

$$ya(2y) = a(2) \quad \text{for } y \geq 1 \tag{5}$$

Further, putting $k = y$ and $x = 1$ in (4) gives $a(y) = xa(yx)$ which by putting $x = w$ in turn gives

$$a(y) = 2a(2y) \quad \text{for } y > 1 \tag{6}$$

Finally, we get from (5) and (6) that

$$a(y) = \frac{2a(2)}{y} \quad \text{for } y > 1 \tag{7}$$

From (1) and (7) it follows that X_1 has the unity-truncated LSD with parameter θ, since a constant multiple of the sf of a GPSD does not affect its probability function.

Theorem 4. Let $\{X_k\}$ be a modulo sequence of rv's generated by X_1 such that X_1 has a GPSD with sf $g(\theta) = \Sigma\ a(y)\theta^y$ with $a(1) = 2a(2)$; and further that X_k has the LSD with parameter θ^k for $k > 1$. Then X_1 has the LSD with parameter θ.

Proof. Following the proof of Theorem 3, we have from (7) $a(y) = \frac{2a(2)}{y}$ for $y > 1$. Also by hypothesis $a(1) = 2a(2)$, therefore $a(y) = \frac{2a(2)}{y}$ for $y \geq 1$. Hence the Theorem.

Theorem 5. Let $\{X_k\}$ be a modulo sequence of rv's generated by X_1 such that X_1 has a GPSD with parameter θ and that the modulo (k) rv X_k has the GD with parameter θ^k for $k > 1$. Then X_1 has the unity - truncated GD with parameter θ.

Proof. Similar to the proof of Theorem 3.

Theorem 6. Let $\{X_k\}$ be a modulo sequence of rv's generated by X_1 such that X_1 has a GPSD with sf $g(\theta) = \Sigma a(y)\theta^y$ with $a(1) = a(2)$; and further that X_k has the GD with parameter θ^k for $k > 1$. Then X_1 has the GD with parameter θ.

Proof. Similar to the proof of Theorem 4.

Lastly, let $f_k(\theta^k) = \sum_{x=1}^{\infty} a(x)\,(\theta^k)^x$ be a sf in powers of θ^k for each $k \in I^+$. Let $\{X_k\}$ be a modulo sequence of rv's generated by X_1. We have then the following.

Theorem 7. A necessary and sufficient condition for the modulo (k) rv X_k to be a GPSD with the sf $f_k(\cdot)$ in powers of θ^k is that the coefficient $a(kx)$ decomposes into two suitable factors which separate k and x.

Proof. To prove the necessity, we have for all $x \in I^+$ and

$$\frac{a(kx)\theta^{kx}}{\sum_{x=1}^{\infty} a(kx)\theta^{kx}} = \frac{a(x)\,(\theta^k)^x}{f_k(\theta^k)}, \text{ therefore} \tag{8}$$

$$a(kx) = b(k)\,a(x), \tag{9}$$

where $b(k) = \sum_{x=1}^{\infty} a(kx)\theta^{kx} / f_k(\theta^k)$ with $b(1) = 1$

To prove the sufficiency, (8) clearly follows from (9).

Corollary 1. The LSD is characterized by $b(k) = 1/k$ where $b(k)$ is defined by (9) for $k \in I^+$.

Corollary 2. The GD is characterized by $b(k) = 1$ where $b(k)$ is defined by (9) for $k \in I^+$.

REFERENCES

[1] Patil, G. P. (1962). Ann. Inst. Statist. Math. Tokyo. 14, 179-182.

STABLE DISTRIBUTIONS: PROBABILITY, INFERENCE, AND APPLICATIONS IN FINANCE--A SURVEY, AND A REVIEW OF RECENT RESULTS

S. J. Press*

University of Chicago, Chicago, Illinois, U.S.A. and University of British Columbia, Vancouver, British Columbia, Canada

SUMMARY. This paper provides an overview of the stable probability laws. The probability theory underlying this class is reviewed from both a univariate and multivariate standpoint. Known results on estimating parameters of the laws are summarized and some new Monte Carlo results involving estimation by using sample characteristic functions is reported. The growing controversy regarding the application of these laws in the field of Finance is examined in terms of the most recent evidence.

KEY WORDS. Stable distributions, multivariate stable laws, applications in finance.

1. PROBABILITY. Suppose the random vectors $X_1,\ldots,X_n,\ldots$ are independent and have the same cdf., $F^*(x)$. Suppose for suitably chosen constants (a_n,b_n), the cdfs. of the standardized sums

$$Y_n = \frac{1}{b_n}\sum_{1}^{n} X_n - a_n ,$$

converge to a distribution function $F(x)$. $F^*(x)$ is said to be "attracted" to $F(x)$, and the set of all cdfs. attracted to $F(x)$ is called its "domain of attraction." Stable distributions comprise that class which have non-empty domains of attraction.

---

*I am grateful to B. Mandelbrot, K. V. Mardia, J. K. Ord, G. P. Patil, and W. Ziemba for helpful comments and suggestions, and to the National Science foundation for their financial support through Grant GS-42866.

G. P. Patil et al. (eds.), Statistical Distributions in Scientific Work, Vol. 1, 87-102.

An alternative definition is the following. A distribution function $F(y)$ is said to be stable if to every $b_1 > 0$, $b_2 > 0$, and real vectors (c_1, c_2) there corresponds a positive number b and a real vector c such that for every $y \equiv (y_j)$, $-\infty < y_j < \infty$,

$$F\left(\frac{y-c_1}{b_1}\right) * F\left(\frac{y-c_2}{b_2}\right) = F\left(\frac{y-c}{b}\right), \tag{1.1}$$

where * denotes the convolution operator.

Considerable probability theory has been established for the stable laws. Detailed developments may be found in Feller, Vol. 2 (1966), and Gnedenko and Kolmogorov (1954), for the univariate case; and in Press (1972), for the multivariate case. The main results are summarized below.

1a. Characteristic Function of Univariate Stable Laws. The univariate family of stable laws was first defined by Lévy in 1924. Lévy's approach was to start with the characteristic function representation of a univariate stable law,

$$\phi(t) \equiv \exp\{iat - |t\delta|^{\alpha}[1 + i\beta \frac{t}{|t|} \omega(t, \alpha)]\}, \tag{1.2}$$

where $\omega(t, \alpha) = \tan(\pi\alpha/2)$, for $\alpha \neq 1$, and $\omega(t, \alpha) = (2/\pi)\log|t|$, for $\alpha = 1$; $(t/|t|) \equiv 0$, at $t = 0$. The parameter α is generally referred to as the "characteristic exponent" of the law, $0 < \alpha \leq 2$; δ is called the scale parameter, $\delta \geq 0$; a is a location parameter, $-\infty < a < \infty$, and β is a symmetry parameter, $-1 \leq \beta \leq 1$ ($\beta = 0$ implies a distribution symmetric about a).

The characteristic exponent measures the "thinness" of the tails of the distribution. Thus, if a stable random variable is observed, the larger the value of α characterizing the law, the less likely it is to observe values of the random variable which are far from its central location. A small value of α will mean considerable probability mass in the tails of the distribution. An $\alpha = 2$ corresponds to a normal distribution (for any β), while an $\alpha = 1$ corresponds to a Cauchy distribution. Distributions corresponding to other values of α are not well known.

The scale parameter δ affords an opportunity to study the distribution without regard to the units of the random variable. Thus, if the stable random variable is measured in feet, for example, δ will be measured in feet. For the normal distribution, δ is proportional to the standard deviation (δ^2 is the semi-variance). For the symmetric Cauchy distribution δ is the semi-interquartile range. The class of distributions may be studied without regard to units by taking $\delta = 1$.

The location parameter a is the median of the distribution when $\beta = 0$ (for any α); of course for $\alpha = 2$ and any β, a is still the median. Thus, the univariate stable distribution is specified by the four parameters $(\alpha, \beta, a, \delta)$.

1b. Properties of Univariate Stable Laws.

A. All non-degenerate ($\delta > 0$) stable distributions are absolutely continuous and have continuously differentiable densities (this follows from the absolute integrability of $\phi(t)$ in (1.2)).

B. For $0 < \alpha \leq 1$ stable laws have no first or higher order integer moments; for $1 < \alpha < 2$, the stable laws have a first moment, and all moments or order θ, $\theta < \alpha$; for $\alpha = 2$, all moments exist [for proof, see Gnedenko (1939)].

C. The stable random variable is positive if and only if $\beta = -1$, $\alpha < 1$, and $a \geq 0$; it is negative if and only if $\beta = +1$, $\alpha < 1$, and $a \leq 0$. If $\beta < 0$ the law is skewed to the right while if $\beta > 0$ it is skewed to the left [see Feller (1966)].

The graph of the density of a positive stable random variable looks, in shape like that of typical log-normal and gamma variates, although, of course, the right tail of the stable variate is fatter.

D. All stable distributions are unimodal [Ibragimov and Chernin, (1959)]. If $0 < \alpha < 1$, the mode has the opposite sign from β, while for $1 \leq \alpha < 2$, the mode has the same sign as β.

E. All stable distributions are infinitely divisible. This result follows readily from the basic definitions.

F. Infinite series expansions for the densities of stable laws were given in 1952 by Berström, and by Feller in 1966.

The series representations were given in terms of a standardized parameterization somewhat different from that given in (1.2).

First translate the distribution to its center (equivalently, take $a = 0$ in (1.2)), and let $\phi^*(t) = \exp(-iat)\,\phi(t)$. Then, for $\alpha \neq 1$, (1.2) may be rewritten as

$$\log \phi^*(t) = -\frac{|t\delta|^{\alpha}}{\cos(\pi\alpha/2)}\left[\cos\frac{\pi\alpha}{2} + i\beta\frac{t}{|t|}\sin\frac{\pi\alpha}{2}\right].$$

Change to polar coordinates taking

$$\cos\frac{\pi\alpha}{2} \equiv r\cos\theta,\quad -\beta\sin\frac{\pi\alpha}{2} \equiv r\sin\theta,$$

so that

$$\log \phi^*(t) = -\frac{r|t\delta|^{\alpha}}{\cos(\pi\alpha/2)}\left[\cos\theta - i\frac{t}{|t|}\sin\theta\right],$$

where $r^2 \equiv \cos^2\frac{\pi\alpha}{2} + \beta^2\sin^2\frac{\pi\alpha}{2}$, and the sign of r is taken to ensure that $\cos\theta > 0$. Define $R \equiv r\delta^{\alpha}/\cos(\pi\alpha/2)$, so that $\log\phi^*(t) = -R|t|^{\alpha}[\cos\theta - i\frac{t}{|t|}\sin\theta]$. Bergström showed that by taking the scaling $R = 1$, the corresponding density may be written as a convergent series, for $0 < \alpha < 1$, as

$$p(x) = \frac{1}{\pi}\sum_{k=1}^{\infty}\frac{(-1)^k}{k!}\frac{\Gamma(\alpha k+1)}{x|x|^{\alpha k}}\sin k\left(\frac{\alpha\pi}{2} + \theta - \alpha\cdot\arg x\right), \tag{1.3}$$

and for $1 < \alpha \leq 2$ as

$$p(x) = \frac{1}{\pi}\sum_{k=1}^{\infty}\frac{(-1)^k}{k!}\frac{\Gamma[(k+1)/\alpha]}{\alpha}x^k\cos\left[k\left(\frac{\pi}{2} + \frac{\theta}{\alpha}\right) + \frac{\theta}{\alpha}\right], \tag{1.4}$$

where $\arg x = 0$ for $x > 0$ and $\arg x = \pi$ for $x < 0$.

G. If X follows a stable law with characteristic exponent α, $\lim_{x\to\infty} x^{\alpha}P\{X > x\} = C(\alpha)$ [see Feller (1966, p. 547)], where $C(\alpha)$ is a constant depending on α. For large x, however, a plot of log x vs log $P(X > x)$, on double-log paper, will yield a straight line with slope α.

H. If X follows a stable law with $1 \leq \alpha < 2$, plots of observed values of X on normal probability paper will have a sigmoid shape and will be convex to the left of the central region and concave to the right.

I. If $X_1,\ldots,X_n$ are independent and follow a stable law with the same (α,β) $\sum_1^n X_j$ follows the same law, except for a possible location and scale change. The proof follows directly from the definition in terms of characteristic functions.

<u>1c. Characteristic Function of Multivariate Stable Laws.</u> The definition of stable laws was extended to the multivariate case by Lévy in 1937. Let X denote a p-vector with characteristic function

$$\phi(t) \equiv Ee^{it'X},$$

where t denotes a p-vector and the prime denotes transpose. Lévy showed it follows from (1.1) that the log-characteristic function of a stable random vector must be representable in the form

$$\log \phi(t) = ia't-(t't)^{\alpha/2}\{f[t/(t't)^{1/2}] + ig[\alpha,t/(t't)^{1/2}]\} \tag{1.5}$$

where $f(\cdot)$ denotes a positive function of $[t/(t't)^{1/2}]$, and $g(\cdot)$ denotes a function of $[t/(t't)^{1/2}]$ which has one value for $\alpha \neq 1$ and another for $\alpha = 1$. It was shown by Press (1972) that for a subclass[1] of (1.5) the representation reduces for $\alpha \neq 1$, to

$$\log \phi(t) = ia't - \frac{1}{2}\sum_1^m (t'\Omega_j t)^{\alpha/2}\{1 + i\beta(t)\tan\frac{\pi\alpha}{2}\}, \tag{1.6}$$

and for $\alpha = 1$ to

$$\log \phi(t) = ia't - \frac{1}{2}\sum_1^m (t'\Omega_j t)^{1/2}\{1 + \frac{2i}{\pi}\beta_1(t)\} \tag{1.7}$$

where for every j, Ω_j is a pxp positive definite symmetric matrix, the "order" m = 1,2,...,

---

[1]It was pointed out by Professor Mervyn Stone, in a personal communication to the author, that because of an error in the derivation of (1.6) and (1.7) in [Press (1972)], the result in (1.6) and (1.7) yields only a subclass of (1.5) and not the complete class.

$$\beta(t) \equiv \frac{\sum_1^m (t'\Omega_j t)^{\alpha/2}(-w_j' t)/|w_j' t|}{\sum_1^m (t'\Omega_j t)^{\alpha/2}}$$

$$\beta_1(t) \equiv \frac{\sum_1^m (t'\Omega_j t)^{1/2}[w_j' t/|w_j' t|]\log|w_j' t|}{\sum_1^m (t'\Omega_j t)^{1/2}},$$

and $w_j' w_j = 1$ for every j, so that $-1 \leq \beta(t) \leq 1$. For $\beta(t) = 0$ we obtain the class of symmetric stable laws with log-characteristic functions

$$\log \phi(t) = ia't - \frac{1}{2}\sum_1^m (t'\Omega_j t)^{\alpha/2} . \tag{1.8}$$

The order of the law may be interpreted as the number of independent factors into which a stable random variable may be decomposed. Special cases, including independence, are discussed in Press (1972). A broad subclass of symmetric stable laws which are included in (1.6) and (1.7) is the one of order one,

$$\log \phi(t) = ia't - \frac{1}{2}(t'\Omega t)^{\alpha/2}. \tag{1.9}$$

1d. Properties of Multivariate Stable Laws. The properties stated below either follow from or are demonstrated in Press (1972).

A. All non-degenerate multivariate stable distributions are absolutely continuous and have continuously differentiable densities.

B. The moment properties of the components of stable random vectors follow the univariate results for moments (see 1b.B).

C. The random vector X: pxl follows a multivariate stable law with characteristic exponent α if and only if every linear combination of the components of X follows a univariate stable law with characteristic exponent α.

D. The laws corresponding to the characteristic functions in (1.5) have non-empty domains of attraction; conversely, every law obtained as the limit law of standardized sums of independent and identically distributed random vectors is expressible as in (1.5); see also Rvačeva (1962).

E. All multivariate stable laws are infinitely divisible.

F. The marginal distributions corresponding to a multivariate stable law are all stable.

G. If a random vector follows a multivariate stable law which is symmetric (as defined in (1.8)), every linear combination of its components follows a univariate symmetric stable law.

H. For $\phi(t)$ defined by (1.8), $\log|\phi(t)|$ is concave if and only if $1 \leq \alpha \leq 2$, see Press (1972). An extension of this result is given in Ziemba (1974).

I. If X and Y are independent, p-variate stable vectors, and A: qxp, B: qxp, $q \leq p$, rank (A) = rank (B) = q, AX + BY + c follows a stable law of the same form as in (1.8), except with new parameters. Explicit formulas may be obtained by multiplying the characteristic functions of X and Y.

J. A multivariate symmetric stable law of order one centered at zero has elliptical probability density contours.

2. INFERENCE. The inference problems associated with stable laws are not straightforward. The reasons for the difficulties are that:

(a) although densities exist they are not generally available in closed form (see (1.3) and (1.4) for infinite series representations).

(b) there are no minimal sufficient statistics;

(c) because for many purposes the most interesting behavior of stable laws is found in the tails of the distributions, very large samples must be taken before a sufficient number of observed values are found in the tails to permit credible inferences.

The symmetric univariate stable laws were tabulated by Fama and Roll (1968) who integrated a truncation of the series in (1.4) term-by-term, for the symmetric case of $\beta = 0$ and $1 < \alpha < 2$, and computed numerical values of the resulting cdf.

Bartels in 1972 extended the earlier Fama/Roll work by tabulating the fractiles of the general asymmetric stable distributions for characteristic exponents in the range $1 \le \alpha \le 2$. Fractiles for the one-sided laws ($\alpha < 1, \beta = -1, a = 0, \delta = 1$) were provided by Paulson et al. (1974). Holt and Crow (1973) provided plots of the densities and tabulated the fractiles of the cdf. for $\alpha = .25, .5, \ldots, 1.75$, and $\beta = -1, -.75, \ldots, +.75, +1$, for $a = 0$ and $\delta = 1$.

Mandelbrot, in 1963 suggested a large sample method for estimating α which was based upon Property 1b.G, above. This method was applied graphically by Fama in 1965, to estimate α in symmetric stable distributions for $1 < \alpha < 2$.

In 1971, Fama and Roll proposed the use of sample fractiles to estimate α and δ, for $1 < \alpha < 2$. For δ, they proposed the estimator

$$\hat{\delta} = \frac{1}{2(.827)} [\hat{x}_{72} - \hat{x}_{28}],$$

where $\hat{x}_h$ is defined by $F_n(\hat{x}_h) = h/100$, and $F_n(x)$ denotes the sample cdf. Several estimators based on order statistics were suggested for estimating α, and monte carlo studies were used to compare them, and to develop their properties. The method they finally decided was best involved searching the table of standardized symmetric stable cdfs.

Maximum likelihood estimates (MLE's) of α and δ (assuming $\beta = a = 0$) were obtained by Du Mouchel in 1971. In one simulation study he drew 10 samples of size 1000 from stable distributions with $\alpha = 1.5$ and 1.9 and $\delta = 1$, and in another study he drew 10 samples of size 100 with $\alpha = 1.5$ and $\delta = 1$. Maximum likelihood estimators of (α, δ) were computed for each sample. The computation was cleverly carried out by classifying the observed data into class interval groups and developing the likelihood function for the resulting multinomial density function as a function of the probabilities for each class interval.[2]

---

[2]By using this approach if ten class intervals are used, only ten class interval probabilities must be evaluated, whereas by not grouping and using the direct approach, one probability is required for each observation. If probabilities are determined by infinite series representation, the direct approach requires that n infinite series by multiplied to evaluate the likelihood function for a sample of size n; the grouping approach would require the same number of probabilities regardless of sample size.

The class interval probabilities for extreme value observations were obtained by means of the Bergström series expansions [see (1.4)]. The class interval probabilities for the more central observations were obtained by integrating the characteristic function numerically using the Fast Fourier Transform algorithm to improve efficiency. MLE's were obtained using the Newton-Raphson method of iterating into a solution. This procedure, though proven feasible, is quite complicated, and limited in application. Du Mouchel (1973) shows moreover, that if both α and a are unknown, the likelihood function will not have a maximum value; moreover, the likelihood function will get ever larger as α approaches zero. If the estimator of α is bounded away from zero, however, the MLE can then be found, and the resulting restricted MLE of all four parameters of the stable law will then jointly follow an asymptotic multivariate normal distribution.

In 1972, Press proposed several estimators based upon the sample characteristic function. Let $x_1,\ldots,x_n$ denote independent observations from the distribution in (1.2). The sample characteristic function is given by

$$\hat{\phi}(t) = \frac{1}{n} \sum_{j=1}^{n} \exp(itx_j). \tag{2.1}$$

$\{\hat{\phi}(t), -\infty < t < \infty\}$ is a stochastic process (non-stationary) with the useful property that $0 < |\hat{\phi}(t)| \leq 1$. So all of the moments of $\hat{\phi}(t)$ are finite and all simple functions of $\hat{\phi}(t)$ have limiting normal distributions. Because of these properties and because $\hat{\phi}(t)$ is so easily computed, estimators based on $\phi(t)$ hold great promise at this time. Estimators of $\omega \equiv (\alpha, a, \delta, \beta)$ suggested in Press (1972) included

(a) the minimum distance estimator. Define

$$g(\omega) \equiv \sup_t |\phi(t; \omega) - \hat{\phi}(t)|.$$

Then define the estimator $\hat{\omega}$ by

$$g(\hat{\omega}) = \min_\omega [g(\omega)].$$

A variant which might be even simpler to evaluate is found from

$$g^*(\omega) \equiv \sup_t |\log \phi(t; \omega) - \log \hat{\phi}(t)|,$$

with the estimator $\tilde{\omega}$ defined by

$$g^*(\tilde{\omega}) = \min_{\omega}[g^*(\omega)].$$

Both $\hat{\omega}$ and $\tilde{\omega}$ are strongly consistent estimators.

(b) The minimum rth mean distance estimator. Define

$$h(\omega) \equiv \int_{-\infty}^{\infty} |\phi(t;\ \omega) - \hat{\phi}(t)|^r\ W(t)dt,$$

where $W(t)$ denotes a suitable convergence factor and r denotes a preassigned number. The estimator is defined by

$$h(\hat{\omega}_r) = \min_{\omega}[h(\omega)]\ .$$

A "log-variant," as in the case of the minimum distance estimator, could be tried as well.

(c) Moment estimators. In this approach, $\hat{\phi}(t)$ is evaluated at four preselected values of $t = (t_1,t_2,t_3,t_4)$. Setting $\phi(t_k) = \hat{\phi}(t_k)$, $k = 1,\ldots,4$, and solving for the four parameters yields "moment estimators." The method extends, to the multivariate case in a straightforward way [see Press (1972) for details].

The author first used the sample characteristic function in estimation in 1964 where the approach was used to estimate parameters of a compound Poisson process. The moment estimation approach was first suggested by Kleinman in 1965 for the univariate case in which $\beta = 0$.

The problem with the moment estimation method is the lack of direction currently available for how to pre-select the values $(t_1,\ldots,t_4)$. In an effort to shed some light on this question the author ran a simulation for known $(a = 0, \beta = 0)$, and letting $\delta = 1$, and $\alpha = 1.0, 1.1, 1.3, 1.5, 1.7, 1.9, 2.0$, for sample sizes of 20, 40, 60, 80, 100, 200, 250. The values of (t_1,t_2) were preselected, a sample was drawn, the moment estimator of α was evaluated, and the error $\varepsilon(t_1,t_2) = \hat{\alpha} - \alpha$ was found. The process was repeated for 100 samples and the averages $\overline{\varepsilon(t_1,t_2)}$, $\overline{|\varepsilon(t_1,t_2)|}$ were computed. Then (t_1,t_2) were changed until $\overline{|\varepsilon(t_1,t_2)|}$ was minimized; the minimizing values of

(t_1^*, t_2^*) were determined, along with the average estimators, $\hat{\alpha}$, at the minimizing values, $\overline{|\varepsilon(t_1,t_2)|^2}$, and values of $\overline{|\varepsilon(t_1,t_2)|}$ at values of (t_1,t_2) near the minimizing value. For lack of space, tables for n = 20 and 250 only are given here following the references. A study of the results of the simulation suggested the following conclusions:

(1) For moderate or large sample sizes the mean squared error of estimation of α is quite negligible at optimal (t_1,t_2), while for sample sizes as small as 20, the mean squared error increases to about 12%.

(2) Average absolute error of estimation for α is about 8% for a sample size of 20, while it drops to less than 1% for n = 250.

(3) By selecting operating values $(t_1,t_2) \equiv (0.2, 1.5)$, for all α not near $\alpha = 2$, a rule of thumb with small estimation error will result (even in small samples).

(4) Estimates of α tend generally to be upward biased.

(5) The moment estimator of α at values of (t_1,t_2) distant from the optimum point is quite stable (in that the mean absolute error remains fairly constant), as long as α is not near two; estimator becomes quite sensitive as α approaches two (where other methods of estimation are more appropriate).

The rth mean distance estimator proposed in Press (1972) has now been studied by Paulson, Holcomb, and Leitch (1974). For their study they chose a quadratic mean distance estimator with a normal cdf. as a convergence factor, and considered samples of sizes at least 50 using 50 replications. A gradient projection method was used for optimization. The numerical results were extremely gratifying showing that the quadratic mean estimation method is a very effective means of estimating all four parameters of a univariate stable distribution. The authors claim they can show analytically that the rth mean distance estimator is a convex function of the parameters, and that the asymptotic distribution of the resulting estimators is normal.

3. APPLICATIONS IN FINANCE.

Stable distributions in Finance were first suggested by Bachelier in 1900 (for $\alpha = 2$), and by Mandelbrot in 1963 (for general α). Mandelbrot proposed that this class of distributions could be used to approximate the behavior of changes of speculative security prices, while Bachelier proposed the narrower notion that price changes followed a normal distribution. Fama in 1965 tested many securities empirically and found symmetric, non-normal behavior with log fat tails. While non-stable distribution models have been proposed [see e.g. Press (1967, 1968, 1970), Blattberg and Gonedes (1974), Praetz (1972)] to explain the behavior of changes of security prices, since there are many long tail distributions which are non-stable, the question of which of all the models is most reasonable, by some criterion, has not yet been resolved. Officer (1972) and more recently Hsu, Miller, and Wichern (1974), have both found the stable distribution hypothesis to be inconsistent, in several respects, with the empirically observed behavior of security price changes. Moreover, Fieletz and Smith (1972) have found, as had Mandelbrot, that security price change distributions tend to be skewed and not symmetric, as claimed by Fama in 1965. Teichmoeller in 1971, suggested that stable laws were in fact better representations of security price change behavior than mixtures of normals, while Press in 1972 raised some doubts above these conclusions. Selection of a portfolio of variable price assets which jointly follow a stable distribution was studied by Fama (1965), Press (1972), Chapter 12; Samuelson (1967); and Ziemba (1974).

The real question of substance in these applications is of course the one of determining the most appropriate "natural" mechanism underlying security price change behavior. The model of Bachelier (1900) was mechanistic, as were the stable law model of Mandelbrot (1963), and the Compound Events model of Press (1967). Other models have been (and will be) found to be good fits, for given sets of securities, even though these models may not have a very credible mechanistic basis. The fact remains, however, that regardless of how good a fit stable distributions turn out to be for security price data, their possible application in Finance has spurred an interest in these laws which does not promise to subside in the immediate future.

REFERENCES

[1] Bachelier, L. (1900). Theorie de la Speculation. Gauthier-Villars, Paris.

[2] Bartels, Robert H. (1972). Stable Distributions in Economics The University of Sydney, Sydney, Australia.

[3] Bergström, H. (1952). Arkiv. for Matematik II, 375-378.

[4] Blattberg, R. C. and N. J. Gonedes (1974). J. Bus. 47, 244-280.

[5] Du Mouchel, William H. (1971). Stable Distributions in Statistical Inference. Dissertation, Yale University, New Haven, Conn.

[6] Du Mouchel, William H. (1973). Ann. Statist. 1, 948-957.
[7] Fama, E. (1965). J. Bus. 38, 34-105.
[8] Fama, E. (1965). Manag. Sci. 2, 404-419.
[9] Fama, E. and Roll, R. (1968). J. Amer. Statist. Assoc. 63, 817-836.
[10] Fama, E. and Roll, R. (1971). J. Amer. Statist. Assoc. 66, 331-338.
[11] Feller, W. (1966). An Introduction to Probability Theory and Its Applications, Vol. 2. John Wiley and Sons, Inc., New York.
[12] Fieletz, B. D. and E. W. Smith (1972). J. Amer. Statist. Assoc. 67, 813-814
[13] Gnedenko, B. V. (1939). Bull. Soc. Math. France 52, 49-85.
[14] Gnedenko, B. V. and Kolmogorov, A. N. (1954). Limit Distributions for Sums of Independent Random Variables. Addison-Wesley Publishing Co., Cambridge, Mass.
[15] Holt, D. R. and E. L. Crow. (1973). J. Res., Nat. Bur. of Stand. 77B, 143-198.
[16] Hsu, D. A., Miller, R. B., and Wichern, D. W. (1974). J. Amer. Statist. Assoc. 69, 108-113.
[17] Ibragimov, I. A. and Chernin, K. E. (1959). SIAM Soc. J. 4, 417-419.
[18] Kleinman, D. (1965). Estimating the Parameters of Stable Probability Laws. Unpublished manuscript, University of Chicago, Graduate School of Business.
[19] Lévy, P. (1924). Bull. Soc. Math. France 52, 49-85.
[20] Lévy, P. (1937). Theorie de l'Addition des Variables Aleatoires, 2nd ed. Gauthier-Villars, Paris.
[21] Mandelbrot, B. (1963). J. Bus. 36, 349-419.
[22] Officer, R. R. (1972). J. Amer. Statist. Assoc. 67, 807-812.
[23] Paulson, A. S., Holcomb, E. W., and Leitch, R. A. (1974). The estimation of the parameters of the stable laws. Technical Report, Rensselaer Polytechnic Institute, Troy, N.Y.
[24] Praetz, P. D. (1972). J. Bus. 45, 49-55.
[25] Press, S. James. (1964). D-13075-PR, The Rand Corporation, Santa Monica, California.
[26] Press, S. James. (1967). J. Bus. 40, 317-335.
[27] Press, S. James. (1968). J. Amer. Statist. Assoc. 63, 607-613.
[28] Press, S. James. (1970). A compound Poisson process model for multiple security analysis. In Random Counts in Scientific Work, Vol. 3, edited by G. P. Patil. Penn State University Press, University Park, Pa.
[29] Press, S. James. (1972). Applied Multivariate Analysis. John Wiley and Sons, Inc., New York.

[30] Press, S. James. (1972). J. Multivariate Anal. 2, 444-462.
[31] Press, S. James. (1972). J. Amer. Statist. Assoc. 67.
[32] Press, S. James. (1972). Report No. 7227, Center for Math. Studies in Business and Economics, University of Chicago.
[33] Rvačeva, E. L. (1962). Sel. Trans. Math. Statist. and Prob. 2, 183-205. Translated by S. G. Ghurye from L'vov. Gos. Univ. Uc. Zap. Ser. Meh-Mat. 29, 5-44.
[34] Samuelson, P. (1967). Efficient Portfolio Selection for Pareto-Lévy Investments. J. Fin. Quant. Anal. 2, 107-122.
[35] Teichmoeller, J. (1971). J. Amer. Statist. Assoc. 66, 282-284.
[36] Ziemba, W. T. (1974). Mathematical Programming: Theory and Practice, P. L. Hamer and G. Zoutendijk (Eds.). North-Holland Publishing Co., 328-370.

TABLE 1 - Moment Estimator Simulation

Sample Size n = 20

Characteristic Exponent	Estimate at optimal point	Operating Point		Mean Square Error	Mean Absolute Error
α	$\hat{\alpha}$	t_1	t_2	$\|e(t_1,t_2)\|^2$	$\|e(t_1,t_2)\|$
2.0	2.00	0.05*	0.10*	2.0×10^{-6}	0.001
2.0		0.01	0.02		0.02
2.0		0.01	0.20		1.54
2.0		0.15	0.16		2.06
2.0		0.15	0.20		6.93
1.9	1.99	0.01*	0.13*	0.8×10^{-2}	0.09
1.9		0.01	0.03		0.09
1.9		0.01	0.23		1.20
1.9		0.11	0.12		2.03
1.9		0.11	0.23		14.85
1.7	1.86	0.01*	1.60*	0.05	0.20
1.7		0.01	1.50		0.22
1.7		0.01	1.70		0.28
1.7		0.11	1.50		0.23
1.7		0.11	1.70		0.25
1.5	1.62	0.11*	1.60*	0.09	0.25
1.5		0.01	1.50		0.32
1.5		0.01	1.70		0.28
1.5		0.21	1.50		0.29
1.5		0.21	1.70		0.30
1.3	1.43	0.11*	1.60*	0.13	0.29
1.3		0.01	1.50		0.40
1.3		0.01	1.70		0.37
1.3		0.21	1.50		0.32
1.3		0.21	1.70		0.33
1.1	1.19	0.24*	1.59*	0.13	0.26
1.1		0.14	1.49		0.31
1.1		0.14	1.69		0.28
1.1		0.34	1.49		0.30
1.1		0.34	1.69		0.31
1.0	1.07	0.22*	1.48	0.11	0.25
1.0		0.12	1.38		0.30
1.0		0.12	1.58		0.27
1.0		0.32	1.58		0.27
1.0		0.32	1.58		0.36

* Optimal operating point

TABLE 2 - Moment Estimator Simulation

Sample Size n = 250

Characteristic Exponent	Estimate at optimal Point	Operating Point		Mean Square Error	Mean Absolute Error
α	$\hat{\alpha}$	t_1	t_2	$\|e(t_1,t_2)\|^2$	$\|e(t_1,t_2)\|$
2.0	2.0	0.12*	0.13*	3. x 10^{-6}	0.001
2.0		0.02	0.03		0.03
2.0		0.02	0.23		1.67
2.0		0.22	0.23		2.05
2.0		0.22	0.23		0.004
1.9	1.89	0.30*	0.82*	0.01	0.05
1.9		0.20	0.72		0.06
1.9		0.20	0.92		0.32
1.9		0.40	0.72		0.06
1.9		0.40	0.92		0.79
1.7	1.71	0.17*	1.32*	0.01	0.07
1.7		0.07	1.22		0.101
1.7		0.07	1.42		0.12
1.7		0.27	1.22		0.08
1.7		0.27	1.42		0.18
1.5	1.50	0.22*	1.08*	0.01	0.08
1.5		0.12	0.98		0.09
1.5		0.12	1.18		0.13
1.5		0.32	0.98		0.09
1.5		0.32	1.18		0.25
1.3	1.30	0.19*	1.47*	0.01	0.07
1.3		0.09	1.37		0.09
1.3		0.09	1.57		0.10
1.3		0.29	1.37		0.76
1.3		0.29	1.57		0.14
1.1	1.11	0.20*	1.43*	0.01	0.07
1.1		0.10	1.33		0.08
1.1		0.10	1.53		0.09
1.1		0.30	1.33		0.09
1.1		0.30	1.53		0.13
1.0	1.01	0.14*	1.48*	0.01	0.06
1.0		0.04	1.38		0.08
1.0		0.04	1.58		0.08
1.0		0.24	1.38		0.08
1.0		0.24	1.58		0.10

* Optimal operating point

STRUCTURAL PROPERTIES AND STATISTICS OF FINITE MIXTURES

Javad Behboodian

Pahlavi University, Shiraz, Iran

SUMMARY. Some general properties of mixtures of distributions which are useful in theory and practice are surveyed. The structural properties of finite mixtures are explained, and some useful statistics are introduced. The statistical works regarding mixing distributions and the parameters of finite mixtures are mentioned.

KEY WORDS. General mixtures, finite mixtures, location mixtures, scale mixtures, identifiability, estimation.

1. INTRODUCTION. The point estimation of the parameters in a mixture of two normal distributions was studied by K. Pearson (1894); he applied the method of moments to estimate the parameters. This problem and other probabilistic and statistical problems regarding general mixtures and finite mixtures of distributions have received some attention in the literature in recent years. However, due to the structural complexity of mixtures many of the related problems have been ignored and many of them, particularly in the case of finite mixtures, have been solved but not quite satisfactorily. Mixtures of distributions often arise in various biological, psychological, and physical applications.

For example, suppose in fishery biology it is desired to measure a certain characteristic in a special population of fish. If this characteristic varies with color of fish, then its distribution is a fintie mixture; but if the characteristic varies with age of fish then its distribution is an infinite or a

G. P. Patil et al. (eds.), Statistical Distributions in Scientific Work, Vol. 1, 103-112.

continuous mixture. Theoretically, mixtures are important because of their connections with conditional distributions, the theory of sums of a random number of random variables, convolution operations, and construction of empirical Bayes decision procedures.

In this article we first survey some general properties of mixtures which might be useful for investigation of the properties of finite mixtures. Next, we explain some structural properties of finite mixtures, and we introduce some useful statistics. Finally, we look at the statistical works regarding mixing distributions and the parameters of finite distributions.

2. SOME GENERAL PROPERTIES OF MIXTURES. Let S be a fixed set of the real line, such as an interval or the set of integers, and let W be a random variable defined on S with distribution $G(w)$. Assume that $F(x;w)$ is the distribution of a random variable X_w for every w in S. If the expectation of $F(x;W)$ exists for every real value x, then it is easy to show that

$$M(x) = E[F(x;W)] \tag{1}$$

is the distribution function of a random variable X. The distribution $M(x)$ is called a mixture on the family of distribution $\mathcal{F} = \{F(x;w): w\varepsilon X\}$ with mixing distribution $G(w)$ or briefly $M(x)$ is a G-mixture on $\mathcal{F}$. If G runs in $\mathcal{G}$, the class of all possible distributions on S, then $M(x)$ runs in a class $\mathcal{M}$, which is called the induced class of mixtures on $\mathcal{F}$.

For example, let $S = [a,b]$, where a and b are two different real numbers, and let X_w be $N(w,\sigma^2)$ for every $w\varepsilon S$. If W is a discrete random variable on S with $P(W=a)=p$, $P(W=b)=q$, $0<p<1$, and $p+q=1$, then $M(x)$ is a mixture of two normal distributions $N(a,\sigma^2)$ and $N(b,\sigma^2)$ with mixing proportions p and q. If W has a uniform distribution on S, then $M(x)$ is a mixture of infinitely many normal distributions with a uniform mixing distribution. Some investigation about the properties of a general mixture $M(x)$ has been done in the literature of probability and statistics by Robbins (1948) and Teicher (1960). In the following we shall briefly explain the structure and useful properties of $M(x)$.

(a) The formula (1) can be generalized to the case in which S is a subset of the Euclidean m-space, i.e., M is an m-dimensional random vector, and also X_w is an n-dimensional random vector.

(b) If X_w is an absolutely continuous random variable with density function $f(x;w)$, then X is also absolutely continuous with density function

$$m(x) = E[f(x;W)] \cdot \tag{2}$$

(c) The distributions $F(x;w)$ and $G(w)$ determine a joint distribution of two random variables X and W. Thus, the distribution $F(x;w)$ which depends on the parameter w, is the conditional distribution of X given W, and the mixture $M(x)$ is, in fact, the marginal distribution of X obtained from the joint distribution.

(d) Let $u(x)$ be a function of the real variable x with real or complex values. If $E[u(x)]$ exists, we can use conditional expectation to obtain

$$E\,[u(X)] = E[E[u(X)|W]] \cdot \tag{3}$$

From (3) we note that the expectation of $u(X)$ is a G-mixture of the expectation of $u(X_w)$. In particular for $u(x)=x^k$ and $u(x)=e^{itx}$ we observe that the kth moment of X is a G-mixture of the kth moment of X_w, and the characteristic function of X is a G-mixture of the characteristic function of X_w.

(e) By using (1) we can easily show that the distribution of $Z=u(X)$ is a G-mixture of the distribution of $Z_w=u(X_w)$ for any continuous function $z=u(x)$. For example, if X has the mixture of two normal distributions mentioned above, then the distribution of $Z=e^X$ is a mixture of two lognormal distributions with the same mixing proportions p and q.

(f) Let $\mathcal{F}$ be a location parameter family of distributions, i.e., $F(x;w)=H(x-w)$. If S includes zero, then a distribution $H(v)$ of a random variable, say V, generates the location family. It follows from (1) that the mixture $M(x)$, which is called a location mixture, is the convolution of $H(x)$ and $G(w)$. In terms of random variables, if V and W are independent, then X is distributed as V+W. For example, a mixture of two normal distributions with density

$$m(x) = \left[pe^{-(x-a)^2/2}+qe^{(x-b)^2/2}\right]\Big/\sqrt{2\Pi} \tag{4}$$

is a location mixture, where V is $N(0,1)$ and W is a discrete random variable with probability mass p and q at a and b. Using this analysis, we may be able to find the distribution of the sample mean for a location mixture.

(g) Let $\mathcal{F}$ be a scale parameter family of distributions, i.e., $F(x;w)=K(x/w)$, where S is a subset of $(0,\infty)$. If S includes one, then a distribution $K(t)$ of a random variable, say T, generates the family. It follows from (1) that the mixture $M(x)$, which is called a scale mixture, is the same as the distribution of TW, where T and W are independent. For example, a mixture of two exponential distributions with density function

$$m(x)=(p/a)e^{-x/a}+(q/b)e^{-x/b} \tag{5}$$

is a scale mixture, where T has distribution $K(t)=1-e^{-t}$ and W is a discrete random variable with probability mass p and q at a and b. Using this analysis for a scale mixture we can easily find moments of X.

(h) When it comes to estimation problems for mixtures, a very important concept is identifiability. We say a mixture $M(x)$ is "identifiable" if for two mixing distributions $G_1(w)$ and $G_2(w)$ with

$$M(x)=E_{G_1}[F(x;W)] = E_{G_2}[F(x;W)] \tag{6}$$

we have $G_1(w)=G_2(w)$. A class of mixtures $\mathcal{M}$ is identifiable if each element of that is identifiable. Some conditions for identifiability of general mixtures are given by Teicher (1961, 1963) and Yakowitz (1968). Certain problems of identifiability of mixtures of discrete distributions have been studied by Patil (1965, 1967) and Patil and Bildikar (1966).

3. STRUCTURE OF FINITE MIXTURES. In this section we consider the special case in which the mixing distribution $G(w)$ or the random variable W is discrete with $P(W=w_i)=p_i$, $0<p_i<1$, $\sum_{i=1}^{n} p_i=1$, where $w_1,w_2,\ldots,w_n$ is a subset of S. Since $G(w)$ is a jump function the formula (1) becomes

$$M(x) = \sum_{i=1}^{n} p_i F(x;w_i) \cdot \tag{7}$$

We say $M(x)$, given by (7), is a finite mixture on the family $\mathcal{F}$ with the mixed distributions $F(x;w_1)$, $F(x;w_2),\ldots,$ $F(x;w_n)$ and mixing proportions $p_1,p_2,\ldots,p_n$. If $\mathcal{M}_o$ denotes the class of all such finite mixture on $\mathcal{F}$, then $\mathcal{M}_o$ is the class of all finite weighted averages of the elements of $\mathcal{F}$ or the convex hull of $\mathcal{F}$. It is clear that $\mathcal{M}_o$ is a subclass of $\mathcal{M}$, the class of all possible mixtures on $\mathcal{F}$. When $\mathcal{F}$ has only a finite number n of distinct elements, we denote them by $F_1, F_2, \ldots, F_n$. For this classical situation we have

$$M_o = \{M(x):M(x) = \sum_{i=1}^{n} p_i F_i \text{ with } p_i \varepsilon [0,1] \text{ and } \sum_{i=1}^{n} p_i = 1\}. \quad (8)$$

A finite mixture has all properties of a general mixture, which we referred to in Section 2, and the related expectations can be calculated easily by finite sums. However, we would like to give a probabilistic meaning of a finite mixture and to add more words regarding identifiability of a finite mixture.

Let X be a random variable whose distribution is given by (7), and let Q_i be a population with distribution $F_i(x)$, i=1,2, ...,n. Suppose D_i is the event that X comes from population Q_i with $p(D_i)=p_i$. Now assuming that Q_i's are mutually exclusive populations, we have

$$P(X \leq x) = \sum_{i=1}^{n} p(D_i) p(X \leq x | D_i), \quad (9)$$

which is another form of (7), and it says that X comes from a mixed population whose distribution is given by (7). This interpretation of a mixture might be sometimes useful for studying finite mixtures. For example, if we have a random sample from a finite mixture of normal distributions with common variance σ^2 and distinct means $\mu_1, \mu_2, \ldots, \mu_n$, we may be able to classify the sample by some suitable procedure and estimate the parameters. Of course, to obtain reliable results, the mixed distributions should be enough separated and none of the mixing proportions p_i should be too small.

A finite mixture given by (7) is identifiable if it has a unique representation as far as the mixed distributions, their number, and the mixing proportions are concerned. In other words

$$M(x) = \sum_{i=1}^{m} p_i F(x;w_i) = \sum_{j=1}^{n} p'_j F(x;w'_j) \quad (10)$$

should imply n=m and for each i, $1 \leqq i \leqq m$ there is some j, $1 \leqq j \leqq n$, such that $p_i = p'_j$ and $F(x,w_i) = F(x,w'_j)$.

The topic of identifiability for finite mixtures has been investigated first by Teicher (1963). He gives some general theorems which contain necessary and sufficient conditions for the identifiability of finite mixtures. In particular, his results imply that the mixtures of normals and gamma distributions are identifiable. In a recent paper Yakowitz and Spragins (1968) prove that the class M_o of all finite mixtures on the family F of distribution is identifiable if and only if F is linearly independent in its span over the field of real numbers.

Using the general theorems of mixtures or directly we can obtain the following useful results regarding the identifiability of finite mixtures.

(a) The class of finite location mixtures on the family $F=\{F(x;w)=H(x-w):w\varepsilon S\}$ is identifiable. This may be proved directly by using the fact that X is distributed as V+W, with the notations we had in Section 2, and characteristic function.

(b) The class of finite scale mixtures on the family $F=\{F(x;w)=K(x/w):w\varepsilon S\}$ is identifiable if the distribution K(t) has r^{th} moment for some $r>0$. This result can be proved by using transformations $x=e^y$ and $w=e^{-u}$ to change a scale distribution to a location distribution. The result is also generally true for scale mixtures as Keilson and Steutel (1974) show.

(c) If F consists of a finite number n of distinct and known distributions $F_1,F_2,\ldots,F_n$ then any mixture of these distributions is identifiable.

When F is a family of location-scale distributions, there are not simple sufficient conditions for identifiability of finite mixtures on F. Theorem 2 of Teicher (1963) answers this question and by that it is proved that finite mixtures of normal distributions and finite mixtures of gamma distributions are identifiable. However, the conditions of this theorem cannot always be checked for a finite location-scale mixture, and the problem may be treated by the theorem of Yakowitz and Spragins (1968) or otherwise.

4. A CLASS OF USEFUL STATISTICS FROM A FINITE MIXTURE. Consider the finite mixture M(x) given by (7) and a random sample $X_1,X_2,\ldots,X_N$ from this mixture. We can show that the joint distribution of this random sample is itself a mixture in multivariate sense. It is not easy to find a simple expression for this joint distribution, or for the corresponding joint density which can be viewed as the likelihood function, in general. However, one can use the probabilistic meaning of a finite mixture to show that this joint distribution for a finite mixture is a multinomial mixture with

parameters $P_1, P_2, \ldots, P_n$, N for the multinomial mixing distribution; but the mixed distributions are not usually simple.

Considering the complexity of the joint distribution, it is clear that we cannot find the distribution of a general statistics $T=u(X_1, X_2, \ldots, X_N)$ easily, although it follows from the general discussion of Section 2 that the distribution should be a finite mixture. Let us now study the special case that the function $t=u(x_1, x_2, \ldots, x_n)$ of real variables $x_1, x_2, \ldots, x_n$ is a symmetric function, i.e., a function which is unchanged by any permutation on the x_i's. It is clear that the distribution of the statistic T in this case is invariant by any permutation on X_i's; so we call T a symmetric statistic of the ramdom sample. Examples of such statistics are sample moments, sample variance, student's ratio $\overline{X}\sqrt{N}/S$, order statistics which are all important for statistical analysis. Fortunately the distribution of a symmetric T is tractable and has a compact form. For simplicity we consider a mixture of two distributions $M(x)=pF_1(x)+qF_2(x)$ with density function $m(x)=pf_1(x)+qf_2(x)$.

We use the probabilistic meaning of a finite mixture given in Section 3. Let E_k, $k=0,1,\ldots,N$, be the event that exactly k of the $X_1, X_2, \ldots, X_N$ have density $f_1(x)$ and the remaining ones have density $f_2(x)$. Using conditional density, we have

$$f_T(t) = \sum_{k=o}^{N} P(E_k) f_T(t|E_k), \quad \text{where} \quad P(E_k) = \binom{N}{k} p^k q^{n-k} \quad \text{and by}$$

the summetry of T one can show $f_T(t|E_k)=f_{T_k}(t)$, where $T_k=u(X_{k1}, X_{k2}, \ldots, X_{kn})$, $k=0,1,\ldots,N$, is a statistic for which the X_{ki}'s are independent with density $f_1(x)$ if $i \leq k$ and density $f_2(x)$ if $i>k$. Therefore, the density of T is a binomial mixture $B(N,p)$ of the densities of T_k's. More details and several examples with an application are given by Behboodian (1972).

5. STATISTICAL ANALYSIS OF FINITE MIXTURES. In this section we briefly review some statistical works regarding mixing proportions and the parameters of mixed distributions in finite mixtures. There are also some theoretical results by Choi and Bulgren (1968), Deely and Kruse (1968), and Ralph (1968) about the estimation of mixing distributions in general mixtures which may be useful for finite mixtures.

First we consider a mixture $M(x) = \sum_{i=1}^{n} p_i F_i(x)$ in which the mixed distributions $F_i(x)$'s are known and the only unknown parameters are the mixing proportions p_i's. Hill (1963) finds the Fisher information about p_1 in a mixture of two known normal distributions and also two known exponential distributions to show how estimation of p_1 requires a large sample when it is close to zero or one. Boes (1966) suggests a class of unbiased estimators for proportions which converge to the parameters with probability one, and in (1967) he gives minimax unbiased estimators for proportions. Behboodian (1972) gives Bayesian estimation for the proportions relative to a beta prior distribution.

There are several articles regarding estimation of finite mixtures with all or some parameters unknown, like mixtures of binomial, poisson, normal, exponential distributions, which can be found in the given references of this article and we do not want to list all of them here. Let us briefly look at some of the most recent investigations.

Kabir (1968) gives a procedure for estimation of the parameters of a mixture of the exponential family of distributions, and he shows that the resulting estimators are consistent and asymptotically normally distributed. He applies his method to mixtures of two binomial and two negative exponential distributions. Day (1969) considers the problem of estimating the components of a mixture of two normal distributions, multivariate or otherwise, with common but unknown covariance matrices. He mainly examines the maximum likelihood estimators, but he also discusses about the moment, minimum chi-square, and Bayes estimators. Robertson and Fryer (1970) suggest a new method for finding the biases and covariances of moment estimators to order n^{-2}, and apply this method to mixtures of normal distributions. Scheaffer (1974) uses a certain inverse sampling scheme to estimate the expected value of a mixture of two distributions.

The class of location-scale mixtures which was referred to in Section 3 is very important since it includes many mixtures like mixtures of normals and exponentials. For simplicity we consider the location-scale family generated by the density $g(u)$, namely $f(x) = \frac{1}{\sigma} g(\frac{x-\mu}{\sigma})$, where $-\infty<x<\infty$, $-\infty<\mu<\infty$, $\sigma>0$. Let us now look at a mixture $m(x) = pf_1(x)+qf_2(x)$ of the mixed densities $f_i(x) = \frac{1}{\sigma_i} g\left(\frac{x-\mu_i}{\sigma_i}\right)$, $i=1,2$.

As we mentioned in Section 3, there is no simple criterion for identifiability of m(x) and it is a problem which needs more attention. The following transformation, which reduces the number of parameters, may be useful for recognizing the structural properties of m(x) and statistical analysis regarding its parameters.

Let $\sigma_1 \leq \sigma_2$ without loss of generality. Now consider the linear transformation $y = \varepsilon(x - \bar{\mu})/\bar{\sigma}$. Assuming that $\varepsilon=1$ for $\mu_1 \leq \mu_2$ and $\varepsilon=-1$ for $\mu_1 > \mu_2$ where

$$\bar{\mu} = (\mu_1+\mu_2)/2, \quad \bar{\sigma} = \sqrt{\sigma_1\sigma_2} .$$

Take also $D = |\mu_2-\mu_1|/2\bar{\sigma}$, $r = \sigma_1/\sigma_2$, where $D \geq 0$ and $0 < r \leq 1$. One can easily show that this transformation sends the location-scale mixture m(x) to another mixture with three parameters p,D,r instead of five parameters $p, \mu_1, \mu_2, \sigma_1, \sigma_2$.

This transformation has been applied by Behboodian (1970) to answer the question of whether a mixture of two normal distributions with five parameters is unimodal or not. It has also been applied by him (1972) to find the Fisher information matrix about the five parameters of the same mixture. I believe that this approach might be useful for further investigation of the properties of a location-scale mixture.

REFERENCES

[1] Behboodian, J. (1970). Technometrics 12, 131-139.
[2] Behboodian, J. (1972). Technometrics 14, 919-923.
[3] Behboodian, J. (1972). J. Statist. Comput. Simul. 1, 295-314.
[4] Behboodian, J. (1972). Sankhyā Ser B 34, 15-20.
[5] Boes, D. C. (1966). Ann. Math. Statist. 37, 177-187.
[6] Boes, D. C. (1967). Sankhyā Ser A 29, 417-420.
[7] Choi, K. and Bulgren, W. G. (1968). J. Roy. Statist. Soc. 30, 444-460.
[8] Day, N. E. (1969). Biometrika 56, 463-474.
[9] Deely, J. J. and Kruse, R. L. (1968). Ann. Math. Statist. 39, 286-288.
[10] Feller, W. (1971). An Introduction to Probability Theory and Its Applications, Vol. II (2nd ed.). John Wiley and Sons, Inc., New York.
[11] Hill, B. M. (1963). J. Amer. Statist. Assoc. 58, 918-932.
[12] Ifram, A. F. (1970). J. Amer. Statist. Assoc. 65, 749-753.
[13] Keilson, J. and Steutel, F. W. (1974). Ann. Prob. 2, 112-129.

[14] Lutful Kabir, A. B. M. (1968). J. R. Statist. Soc. B 30, 472-482.
[15] Patil, G. P. (1965). Sankhyā Ser A 27, 259-270.
[16] Patil, G. P. (1967). Sankhyā Ser B 30, 355-366.
[17] Patil, G. P. and Bildikar, S. (1966). Proc. Camb. Philos. Soc. 62, 484-494.
[18] Pearson, K. (1894). Phill. Trans. Roy. Soc. A 185, 71-110.
[19] Robbins, H. (1948). Ann. Math. Statist. 19, 360-367.
[20] Robbins, H. and Pitman, E. J. G. (1949). Ann. Math. Statist. 20, 552-560.
[21] Robbins, H. (1964). Ann. Math. Statist. 35, 1-20.
[22] Robertson, C. A. and Fryer, J. G. (1970). Biometrika 57, 57-65.
[23] Rolph, J. E. (1968). Ann. Math. Statist. 39, 1289-1302.
[24] Scheaffer, R. L. (1974). Biometrics 30, 187-198.
[25] Teicher, H. (1960). Ann. Math. Statist. 31, 55-73.
[26] Teicher, H. (1961). Ann. Math. Statist. 32, 244-248.
[27] Teicher, H. (1961). Ann. Math. Statist. 34, 1265-1269.
[28] Yakowitz, S. and Spragins, J. (1968). Ann. Math. Statist. 39, 209-214.
[29] Yakowitz, S. (1969). Ann. Math. Statist. 40, 1728-1735.

DISTRIBUTION THEORY FOR THE VON MISES-FISHER DISTRIBUTION AND ITS APPLICATION

K. V. Mardia

Leeds University, Leeds, England

SUMMARY. The von Mises-Fisher distribution is the most important distribution in directional data analysis. We derive the sampling distributions of the sample resultant length, the sample mean direction and the component lengths. For the multi-sample case, the conditional distribution of the individual sample resultant lengths given the combined sample resultant length is derived. These results depend heavily on the corresponding distributions for the isotropic random walk on hypersphere. Using these results we investigate some optimum properties of various important tests. Most of these tests were formulated intuitively by Watson and Williams (1956). Mardia (1972) in his book concentrated on the optimum properties of the circular and spherical cases, and this paper extends and unifies some of the parametric work.

KEY WORDS. Directional data analysis, von Mises-Fisher distribution, multi-sample problems.

1. INTRODUCTION. Let $\underset{\sim}{1}' = (1_1,\ldots,1_p)$ be a unit random vector taking values on the surface of a p-dimensional hypersphere S_p of unit radius and having its centre at the origin. The unit random vector $\underset{\sim}{1}$ is said to have a p-variate von Mises-Fisher distribution, $M_p(\underset{\sim}{\mu},\kappa)$ if its probability density function (pdf) is given by

$$c_p(\kappa)e^{\kappa\underset{\sim}{\mu}'\underset{\sim}{1}}, \quad \kappa > 0, \quad \underset{\sim}{1} \in S_p , \tag{1.1}$$

where κ is the concentration parameter, $\underset{\sim}{\mu}$ is the mean-direction

G. P. Patil et al. (eds.), Statistical Distributions in Scientific Work, Vol. 1, 113-130.

vector and

$$c_p(\kappa) = \kappa^{\frac{1}{2}p-1}/\{(2\pi)^{\frac{1}{2}p} I_{\frac{1}{2}p-1}(\kappa)\}, \tag{1.2}$$

with $I_r(\kappa)$ denoting the modified Bessel function of the first kind and order r. A discussion of this distribution will be found in Watson and Williams (1956) and Mardia (1974a).

On transforming $\underset{\sim}{1}$ to the spherical polar co-ordinates $\underset{\sim}{\theta}' = (\theta_1,\ldots,\theta_{p-1})$ with the help of the transformation

$$1_j = u_j(\underset{\sim}{\theta}),\ j = 1,\ldots,p \tag{1.3}$$

$$u_j(\underset{\sim}{\theta}) = \cos\theta_j \prod_{i=0}^{j-1} \sin\theta_i,\ j=1,\ldots,p;\ \sin\theta_o = \cos\theta_p = 1, \tag{1.4}$$

it is found that the pdf. of the polar vector $\underset{\sim}{\theta}$ is

$$g(\underset{\sim}{\theta};\underset{\sim}{\mu}_o,\kappa) = c_p(\kappa)\exp\{\kappa \underset{\sim}{u}'(\underset{\sim}{\mu}_o)\underset{\sim}{u}(\underset{\sim}{\theta})\}a_p(\underset{\sim}{\theta}),$$

$$0 < \theta_i \leq \pi,\ i=1,\ldots,p-2;\ 0 < \theta_{p-1} \leq 2\pi,\ \kappa > 0, \tag{1.5}$$

with

$$\underset{\sim}{u}(\underset{\sim}{\theta}) = \{u_1(\underset{\sim}{\theta}),\ldots,u_p(\underset{\sim}{\theta})\}'$$

and

$$a_p(\underset{\sim}{\theta}) = \prod_{j=2}^{p-1} \sin^{p-j}\theta_{j-1},\ a_2(\underset{\sim}{\theta}) = 1. \tag{1.6}$$

For $\kappa = 0$, the distribution reduces to the uniform distribution on S_p and the pdf. is then

$$f(\underset{\sim}{1}) = c_p,\ \underset{\sim}{1} \in S_p,$$

where

$$c_p = c_p(0) = \Gamma(\tfrac{1}{2}p)/(2\pi)^{\frac{1}{2}p}. \tag{1.7}$$

Some other properties of the distribution are given in Appendix 1.

2. DISTRIBUTIONS

2.1 Introduction. Let $\underset{\sim}{l}_1,\ldots,\underset{\sim}{l}_n$ be a random sample from $M_p(\underset{\sim}{\mu},\kappa)$. For inference problems, the following statistics are of importance. Let R be the sample resultant length and $\bar{\underset{\sim}{x}}_o$ be the sample mean-direction vector. That is,

$$\sum_{i=1}^{n} \underset{\sim}{l}_i' = (R_{x_1},\ldots,R_{x_p}) = \underset{\sim}{R}_x' = R\underset{\sim}{l}_o' = R\underset{\sim}{u}'(\bar{\underset{\sim}{x}}_o). \tag{2.1}$$

The statistics $R_{x_1},\ldots,R_{x_p}$ are described as the $x_1,\ldots,x_p$ - components of $\underset{\sim}{l}_i$, $i=1,\ldots,n$. The roles of $\bar{\underset{\sim}{x}}_o$ and R in the statistics of directional data are as important as those of the sample mean vector and covariance matrix in normal theory data analysis.

In most cases, the sampling distribution of R and $\bar{\underset{\sim}{x}}_o$ from the von Mises-Risher population can be derived with the help of the corresponding distributions for the uniform case (see §§ 2.2-2.3). In fact, the distribution theory for the uniform case is comparatively well-developed as it is related to the problem of the isotropic random walk on a hypersphere.

The distribution of $\bar{\underset{\sim}{x}}_o$ has not received sufficient attention although it plays an important role (Mardia, 1972). We derive one of the most significant results in Section 3.4.1, namely that the conditional distribution of $\bar{\underset{\sim}{x}}_o$ given R for the von Mises-Fisher population is again $M_p(\underset{\sim}{\mu},\kappa R)$. A conjecture of Watson and Williams (1956) regarding a distributional problem is resolved in Section 2.4.2. The sampling distributions derived in Section 3.4 are of importance in inference problems, and are used in Section 4 to study optimality properties of various tests. The section concludes with an outline of some useful approximations.

2.2 The Characteristic Function Method and the Isotropic Case. Let l be a random vector having a density with the characteristic function (cf.) $\phi(\underset{\sim}{t})$. Then the pdf. of R_x can be obtained from the inversion theorem. Further, the pdf. of $(R,\bar{\underset{\sim}{x}}_o)$ is given by

$$f(R,\bar{\underset{\sim}{x}}_o) = (2\pi)^{-p}R^{p-1}a_p(\bar{\underset{\sim}{x}}_o)\int_0^{\infty}\int \Psi^n(\rho,\underset{\sim}{\Phi})e^{-i\rho R\underset{\sim}{u}'(\underset{\sim}{\Phi})\underset{\sim}{u}(\bar{\underset{\sim}{x}}_o)}\rho^{p-1}a_p(\underset{\sim}{\Phi})\,d\rho d\underset{\sim}{\Phi} \tag{2.2}$$

where

$$\Psi(\rho,\underset{\sim}{\Phi}) \equiv [\phi(\underset{\sim}{t})]_{\underset{\sim}{t}} = \rho \underset{\sim}{u}(\underset{\sim}{\Phi}),$$

and the second integral is taken over $0 < \Phi_i \leq \pi$, $i=1,\ldots,p-2$; $0 < \Phi_{p-1} \leq 2\pi$. Also, the density of R is found to be

$$f(R) = (2\pi)^{-p}R^{p-1}\int_0^\infty \int \Psi^n(\rho,\underset{\sim}{\Phi})\rho^{p-1}\{c_p(i\rho R)\}^{-1}a_p(\underset{\sim}{\Phi})\,d\rho\, d\underset{\sim}{\Phi}. \tag{2.3}$$

The following results can be deduced for the isotropic random walk with n equal steps, starting from the origin, in p-dimensions. Let $\underset{\sim}{l}_1,\ldots,\underset{\sim}{l}_n$ be the direction-vectors of the n From equation (A1.1) of Appendix 1, the c.f. of $\underset{\sim}{l}$ simplifies to

$$\Psi(\rho,\underset{\sim}{\Phi}) = c_p/c_p(i\rho), \tag{2.4}$$

where

$$c_p(i\rho) = \rho^{\frac{1}{2}p-1}/\{(2\pi)^{\frac{1}{2}p} J_{\frac{1}{2}p-1}(\rho)\}. \tag{2.5}$$

Substituting (2.5) in (2.2) and integrating with respect to $\underset{\sim}{\Phi}$, we deduce that

(i) R and $\bar{\underset{\sim}{x}}_o$ are independently distributed,

(ii) $\bar{\underset{\sim}{x}}_o$ is uniformly distributed on S_p,

and

(iii) the pdf. of R is given by

$$h_n(R) = (2\pi)^{-p}c_p^{n-1}R^{\frac{1}{2}p}\int_0^\infty \rho^{p-1}\{c_p(i\rho R)c_p^n(i\rho)\}^{-1}d\rho. \tag{2.6}$$

Further, the pdf. of $\underset{\sim}{R}_x$ is

$$h_n(\underset{\sim}{R}_x'\underset{\sim}{R}_x). \tag{2.7}$$

If the successive steps are of different length $\beta_1,\ldots,\beta_n$, then by precisely the same argument, the pdf. of $\bar{\underset{\sim}{x}}_o$ is again uniform whereas the pdf. of R is

$$h_n(R;\beta_1,\ldots,\beta_n) = (2\pi)^{-p} c_p^{n-1} R^{p-1} \int_0^\infty \rho^{p-1} \{c_p(i\rho R) \prod_{j=1}^{n} c_p(i\rho\beta_j)\}^{-1} d\rho. \tag{2.8}$$

The distributions (2.6) and (2.8) for $p = 2$ and $p = 3$ have received the attention of various prominent workers including S. Chandrasekhar, W. Feller, R. A. Fisher, A. A. Markov, Karl Pearson, M. H. Quenouille and Lord Rayleigh. A detailed historical account for these special cases will be found in Mardia (1972, p. 96, p. 240). An approximation to the distribution of R is given in Section 2.4.

2.3 Distributional Problems for the von Mises-Fisher Case.

2.3.1. Single Sample Problems. Let us assume that $\underset{\sim}{l}_1,\ldots,\underset{\sim}{l}_n$ is a random sample from $M_p(\underset{\sim}{\mu},\kappa)$. We consider three sampling distributions.

(i) Distribution of R and $\bar{\underset{\sim}{x}}_o$. On integrating the likelihood function

$$c_p^n(\kappa)e^{\kappa\underset{\sim}{\mu}'\underset{\sim}{R}_x}$$

over constant values of $\underset{\sim}{R}_x$, we find that the pdf of $\underset{\sim}{R}_x$ is

$$f_\kappa(\underset{\sim}{R}_x,\underset{\sim}{\mu}) = \{c_p^n(\kappa)/c_p^n\}e^{\kappa\underset{\sim}{\mu}'\underset{\sim}{R}_x} f_o(\underset{\sim}{R}_x), \tag{2.9}$$

where f_o is the pdf. of $\underset{\sim}{R}_x$ under uniformity and is given by (2.7). Hence, the pdf. of $(R,\bar{\underset{\sim}{x}}_o)$ is

$$g_\kappa(R;\bar{\underset{\sim}{x}}_o;\underset{\sim}{\mu}_o) = \{c_p^n(\kappa)/c_p^n\}e^{\kappa R u'(\underset{\sim}{\mu}_o)\underset{\sim}{u}(\bar{\underset{\sim}{x}}_o)} h_n(R)a_p(\bar{\underset{\sim}{x}}_o), \tag{2.10}$$

where $h_n(R)$ is the pdf. of R for the uniform case given by (2.6). On integrating over $\bar{\underset{\sim}{x}}_o$, the pdf. of R is given by

$$h_n^{(\kappa)}(R) = c_p^{-n}c_p^n(\kappa)h_n(R)/c_p(\kappa R). \tag{2.11}$$

The result (2.11) is due to Watson and Williams (1956). From (2.10) and (2.11), it follows that

$$\bar{\mathbf{x}}_0 | R \sim M_p(\boldsymbol{\mu}, \kappa R). \tag{2.12}$$

Further, the distribution of $\bar{\mathbf{x}}_0$ depends on R unless $\kappa = 0$ which is in contrast to the normal case where the mean vector and the covariance matrix are independently distributed.

(ii) Distribution of R_{x_1}. From equation (A1.1) of Appendix 1, the cf. of R_{x_1} is

$$c_p^n(\kappa)/c_p^n(i\omega),$$

where

$$\omega = \{(t-i\kappa\mu)^2 - \kappa^2\nu^2\}^{\frac{1}{2}}, \ \mu=\cos\mu_{0,1}, \ \nu=\sin\mu_{0,1}.$$

On applying the inversion theorem the pdf. of R_{x_1} is

$$f(R_{x_1}) = (2\pi)^{-1} c_p^n(\kappa) \int_{-\infty}^{\infty} e^{-itR_{x_1}} c_p^{-n}(i\omega)\,dt.$$

By contour integration around a rectangle, it is found that

$$f(R_{x_1}) = \pi^{-1} c_p^n(\kappa) e^{\kappa\mu R_{x_1}} b(R_{x_1};\lambda), \ -\infty < R_{x_1} < \infty, \tag{2.13}$$

where

$$b(R_{x_1};\lambda) = \int_0^{\infty} \cos tR_{x_1}\, c_p^{-n}\{i(t^2-\lambda^2)^{\frac{1}{2}}\}dt, \ \lambda=\kappa\nu. \tag{2.14}$$

For $\boldsymbol{\mu}_0 = \mathbf{0}$, we have

$$f(R_{x_1}) = \pi^{-1} c_p^n(\kappa) \int_0^{\infty} (\cos tR_{x_1}) c_p^{-n}(it)\,dt. \tag{2.15}$$

(iii) Distribution of $R|R_{x_1}$. First, consider the pdf. of $(R, \bar{x}_{0,1})$. Let us write $\mathbf{u}_p(\boldsymbol{\theta})$ for $\mathbf{u}(\boldsymbol{\theta})$ given by (1.4). We then have

$$\underset{\sim}{u}_p'(\underset{\sim}{\mu}_o)\underset{\sim}{u}_p(\bar{\underset{\sim}{x}}_o) = \cos \bar{x}_{o,1}\cos\mu_{o,1} + \sin \bar{x}_{o,1}\sin \mu_{o,1}\underset{\sim}{u}_{p-1}'(\underset{\sim}{\mu}_{o,2})\underset{\sim}{u}_{p-1}(\bar{\underset{\sim}{x}}_{o,2}) ,$$

where

$$\bar{\underset{\sim}{x}}_o' = (x_{o,1} \vdots \bar{\underset{\sim}{x}}_{o,2}'), \quad \underset{\sim}{\mu}_o' = (\mu_{o,1} \vdots \underset{\sim}{\mu}_{o,2}').$$

Further,

$$a_p(\bar{\underset{\sim}{x}}_o) = \sin^{p-2}\bar{x}_{o,1}a_{p-1}(\bar{\underset{\sim}{x}}_{o,2}).$$

Substituting these results in (2.10) and integrating over $\bar{\underset{\sim}{x}}_{o,2}$, we obtain the pdf. of $(R,\bar{x}_{o,1})$. Since $R_{x_1} = R \cos \bar{x}_{o,1}$, we then find from (2.13) that the conditional pdf. of $R|R_{x_1}$ for $p > 2$ is

$$\pi c_p^{-n}[c_{p-1}\{\lambda(R^2-R_{x_1}^2)^{\frac{1}{2}}\}]^{-1}h_n(R)(R^2-R_{x_1}^2)^{\frac{1}{2}(p-3)}/b(R_{x_1};\lambda), \qquad (2.16)$$

where $0 < R_{x_1}^2 < R^2$. Hence the distribution of R given R_{x_1} depends only on λ but there does not seem to be an intuitive explanation for this behaviour. For $\mu_{o,1} = 0$ or $\underset{\sim}{\mu}_o = \underset{\sim}{0}$, (2.16) reduces to

$$\pi(2\pi)^{-\frac{1}{2}np}c_p^{-n}c_{p-1}^{-1}h_n(R)(R^2-R_{x_1}^2)^{\frac{1}{2}(p-3)}/\int_o^\infty(\cos tR_{x_1})c_p^{-n}(it)dt. \qquad (2.17)$$

Hence for $\mu_{o,1} = 0$, $R|R_{x_1}$ does not depend on κ. The last result also follows on noting that for this case R_{x_1} is a sufficient estimate of κ. For $\underset{\sim}{\mu} = \underset{\sim}{0}$, the distributions (2.15) and (2.17) were obtained by Stephens (1962). The result is not true for p=2; for this particular case, we refer to Mardia (1972, p. 142). The above results can obviously be expressed in terms of the Bessel functions on using (1.2) and (2.5).

2.3.2. Multi-sample Problems. Let $R_1,\dots,R_q$ denote the resultants for q-independent random samples of sizes $n_1,\dots,n_q$ drawn from $M_p(\underset{\sim}{\mu},\kappa)$. Let $n = \Sigma n_i$. Suppose that R and $\underset{\sim}{\bar{x}}_o$ are the resultant and the mean-direction vector for the combined sample. For inference problems, the distribution of $\underset{\sim}{R}^* = (R_1,\dots,R_q)'$ given R is required. It is easily seen that the pdf. of $(R,\underset{\sim}{R}^*,\underset{\sim}{\bar{x}}_o)$ is given by

$$f_\kappa(R,\underset{\sim}{R}^*,\underset{\sim}{\bar{x}}_o) = \{c_p(\kappa)/c_p\}^n e^{\kappa R \underset{\sim}{u}'(\underset{\sim}{\mu}_o)\underset{\sim}{u}(\underset{\sim}{\bar{x}}_o)} f_o(R,\underset{\sim}{R}^*,\underset{\sim}{\bar{x}}_o), \qquad (2.18)$$

where f_o denotes the corresponding pdf. for the uniform case which can be obtained as follows. Obviously, the pdf. of $\underset{\sim}{R}^*$ is

$$f_o(\underset{\sim}{R}^*) = \prod_{i=1}^{q} h_{n_i}(R_i). \qquad (2.19)$$

Further, the distribution of $(R,\underset{\sim}{\bar{x}}_o)|\underset{\sim}{R}^*$ is equivalent to the distribution for the isotropic random walk consisting of q steps when the ith step is of fixed length R_i, $i=1,\dots,q$. Hence, from Section 2.2, (i) $\underset{\sim}{\bar{x}}_o$ is uniformly distributed (ii) $\underset{\sim}{\bar{x}}_o$ and R are independently distributed and (iii) the pdf. of $R|\underset{\sim}{R}^*$ is given by $h_q(R;\underset{\sim}{R}^*)$ which is given at (2.8). Consequently, we have

$$f_o(R,\underset{\sim}{R}^*,\underset{\sim}{\bar{x}}_o) = c_p h_q(R;\underset{\sim}{R}^*)\{\prod_{i=1}^{q} h_{n_i}(R_i)\} a_p(\underset{\sim}{\bar{x}}_o). \qquad (2.20)$$

On substituting (2.20) into (2.18) and integrating over $\underset{\sim}{\bar{x}}_o$, we obtain the pdf. of $(R,\underset{\sim}{R}^*)$ which when divided by the pdf. of $\underset{\sim}{R}^*$ given at (2.11) gives

$$f_\kappa(\underset{\sim}{R}^*|R) = h_q(R;\underset{\sim}{R}^*) \prod_{i=1}^{q} h_{n_i}(R_i)/h_n(R). \qquad (2.21)$$

Hence, the density of $\underset{\sim}{R}^*|R$ does not depend on κ. This fact leads to its applications in inference on $\underset{\sim}{\mu}$ when κ is unknown.

For a history of this result for particular cases of p = 2 and p = 3, we again refer to Mardia (1972, p. 103, p. 244). It should be noted that the result (3.22) was derived for p = 3 and q = 2 by Fisher (1953). For p = 2, the result is due to J. S. Rao (1969). It was conjectured by Watson and Williams (1956) that for p = 3 , (2.21) might be of the form

$$R\{\prod_{i=1}^{q} h_{n_i}(R_i)/R_i\}/h_n(R)$$

which was shown to be false by Stephens (1967); Mardia (1972, pp. 242-244) provided the exact solution to the problem. In brief, earlier attempts to obtain (2.21) for $p = 3$ were unsuccessful. For $p = 2$, Mardia (1972, pp. 155-156) examines the reason for this independnce on κ. This question will be discussed further in Section 3.

We now further simplify (2.21) for $q = 2$. From Watson G. N. (1948, p. 411), we have for $m > 0$,

$$\int_0^\infty J_m(xR)J_m(xR_1)J_m(xR_2)x^{1-m}dx = \frac{(R_1R_2)^{m-1}(\sin\lambda)^{2m-1}}{(2R)^m\Gamma(\frac{1}{2})\Gamma(m+\frac{1}{2})},$$

where

$$\sin\lambda = [\{(R_1+R_2)^2-R^2\}\{R^2-(R_1-R_2)^2\}]^{\frac{1}{2}}/(2R_1R_2). \tag{2.22}$$

Using this result in (2.21) and (2.8), we find that

$$f_\kappa(R_1,R_2|R) = \{\beta(\tfrac{1}{2},\tfrac{1}{2}(p-1))\}^{-1}(\sin\lambda)^{p-3}Rh_{n_1}(R_1)h_{n_2}(R_2)/\{R_1R_2h_n(R)\} \tag{2.23}$$

where

$$|R_1-R_2| < R < R_1 + R_2,\ 0 < R_i < n_i,\ i = 1,2.$$

Watson and Williams (1956) obtained this result for $p = 2$ whereas Fisher (1953) derived it for $p = 3$.

2.4 Approximations. The distributions derived above have no obvious connection with standard normal theory sampling distributions. For large κ or for large n, the situation is different.

For the uniform case, we have from the central limit theorem that $\underset{\sim}{R}_x$ is distributed asymptotically as $N_p(\underset{\sim}{0},(n/p)\underset{\sim}{I})$. Hence

$$pR^2/n \simeq \chi_p^2. \tag{2.24}$$

For $M_p(\underset{\sim}{0},\kappa)$, we have for large κ,

$$2\kappa(1-\cos\theta_1) \simeq \chi^2_{p-1}.$$

Hence, we have the following approximation (Watson and Williams, 1956):

$$2\kappa(n-R_{x_1}) \simeq \chi^2_{(p-1)n}, 2\kappa(n-R) \simeq \chi^2_{(p-1)(n-1)}, \tag{2.25}$$

$$2\kappa(n-\Sigma R_i) \simeq \chi^2_{(p-1)(n-q)}, 2\kappa(\Sigma R_i - R) \simeq \chi^2_{(p-1)(q-1)}, \tag{2.26}$$

where the last two random variables are independently distributed. These approximations can be improved by replacing the multiplier κ by γ. For the two important particular cases, the values of γ are

$$p = 2,\ \gamma^{-1} = \kappa^{-1} + \tfrac{3}{8}\kappa^{-2},\ \text{(Stephens, 1969a)} \tag{2.27}$$

$$p = 3,\ \gamma^{-1} = \kappa^{-1} - \tfrac{1}{5}\kappa^{-3};\ \text{(Mardia, 1972, p. 247).} \tag{2.28}$$

Such approximations have obvious advantages since they preserve the analysis of variance type decomposition as the distribution theory of the components is identical to normal theory. We now consider an extension. Define γ^* by

$$E\{2\gamma^{*-1}(n-R_{x_1})\} = p \tag{2.29}$$

so that we can have approximately

$$2\gamma^{*-1}(n-R_{x_1}) \sim \chi^2_{np}\ . \tag{2.30}$$

It is found that

$$\gamma^{*-1} = 2\{1-A(\kappa)\}/(p-1)$$

which leads to

$$\gamma^{*-1} \doteqdot \frac{1}{\kappa} - \frac{(p-3)}{4\kappa^2} - \frac{(p-3)}{4\kappa^3} + \frac{(p-3)(p-7)(p+3)}{64\kappa^4} = \delta^{-1},\ \text{say.} \tag{2.31}$$

For $p > 3$, we can replace γ^* by δ in (2.26). For $p = 2$, Upton (1974) has examined the adequacy of the approximation by taking the first three terms in δ^{-1}. However, for $p > 2$, further investigation is necessary. For moderately large κ, we expect improved approximations on replacing κ by δ in (2.25) and (2.26).

Let θ be distributed as $M_p(\underset{\sim}{0},\kappa)$. Numerical work for p = 2 and p = 3 shows that for $\kappa > 0$, $(\kappa-\frac{1}{2})\theta_1 \sim \chi^2_{p_2}$ provides a better approximation to the tails of θ_1 than $\kappa\theta_1 \sim \chi^2_p$.

If n is large and $\kappa > 0$, we can obtain approximations by the usual method. For example,

$$\bar{R} \sim N(\rho, A'(\kappa)/n), \tag{2.32}$$

where

$$\rho = A(\kappa) = \{E(\underset{\sim}{1}')E(\underset{\sim}{1})\}^{\frac{1}{2}}. \tag{2.33}$$

A proof of (2.33) is given in equation (A1.3) of Appendix 1 which also defines the notation $A(\kappa)$. We can write

$$A'(\kappa) = \kappa/\{n(\kappa-\rho-\kappa\rho^2)\}.$$

Therefore, it is possible to obtain a variance stabilizing transformation for $\bar{R}$ which may be applicable for small n after some modification. Mardia (1972, pp. 159-161, pp. 265-266) obtains such transformations for p = 2 and p = 3 by considering different intervals of the concentration parameter. For p > 3, the topic is yet unexplored.

3. HYPOTHESES TESTING

3.1 Introduction. Before examining the various tests, we briefly mention a few estimation properties. Following the notation of Section 2, we note that R_{x_i} , i = 1,...,p are jointly sufficient and complete estimators of (κ,μ). If κ is given then $R_{\underset{\sim}{x}}$ is a vector of minimal sufficient statistics for $\underset{\sim}{\mu}$ which implies that $\bar{\underset{\sim}{x}}_o$ itself does not contain all the information about $\underset{\sim}{\mu}_o$. This fact results in different inference procedures from those arising from a standard exponential family. If $\underset{\sim}{\mu}_o$ is given, by taking $\underset{\sim}{\mu}_o = \underset{\sim}{0}$, we see from (2.13) that R_{x_1} is a complete sufficient statistic for κ. Let

$$\hat{\kappa}_o = A^{-1}(\bar{R}_{x_1}), \tag{3.1}$$

where $\bar{R}_{x_1} = R_{x_1}/n$. Since $E\{A(\hat{\kappa}_o)\} = A(\kappa)$, a simple application of the Rao-Blackwell theorem shows that $\bar{R}_{x_1}$ is the best unbiased estimate of $A(\kappa)$.

It is seen that the maximum likelihood estimator (m.l.e.) of $\underset{\sim}{\mu}_o$ is $\bar{\underset{\sim}{x}}_o$. The m.l.e. of κ is given by

$$\hat{\kappa} = A^{-1}(\bar{R}), \ \bar{R} = R/n. \tag{3.2}$$

These estimators are biased. For large n, it is found that $\hat{\kappa}$ and $\bar{\underset{\sim}{x}}_o$ are asymptotically independent normal with

$$\text{var}(\hat{\kappa}) = \kappa/[n\{\kappa - A(\kappa) - \kappa A^2(\kappa)\}] = 1/nA'(\kappa),$$

and

$$\text{var}(\bar{x}_{o,i}) = \frac{1}{n}\{\kappa A(\kappa) \prod_{j=1}^{i-1} \sin^2 \mu_{o,j}\}^{-1}, \ i = 1,\ldots,p-1.$$

Tabulated values of $A(\kappa)$ and its inverse are available for $p = 2$ and $p = 3$. Mardia and Zemroch (1974a,b) give algorithms to calculate these estimators and other relevant statistics for $p = 2$ and $p = 3$.

3.2 Tests of Uniformity.

3.2.1 Mean Vector Known.

Consider the problem of testing

$$H_o : \underset{\sim}{\mu}_o = \underset{\sim}{0}, \ \kappa = 0 \text{ against } H_1 : \underset{\sim}{\mu}_o = \underset{\sim}{0}, \ \kappa = \kappa_1 \neq 0.$$

The Neyman-Pearson lemma leads to the best critical region (BCR)

$$R_{x_1} > K, \tag{3.3}$$

where K will be used as a generic constant in similar contexts.

3.2.2. Mean Vector Unknown (The Rayleigh Test).

Consider

$$H_o : \kappa = 0 \text{ against } H_1 : \kappa \neq 0$$

where $\underset{\sim}{\mu}_o$ is unknown.

We first show that the likelihood ratio (LR) principle leads to the critical region

$$R > K. \tag{3.4}$$

Let λ be the LR statistic. On using

$$c_p'(\kappa)/c_p(\kappa) = -A'(\kappa) \tag{3.5}$$

it is found that

$$\partial(\log\lambda)/\partial\hat{\kappa} = -n\hat{\kappa}A'(\hat{\kappa}),$$

which is non-negative since

$$A'(\kappa) = \mathrm{var}(\cos\theta_1) \geq 0. \tag{3.6}$$

Hence, λ is a monotonically decreasing function of $\hat{\kappa}$ and consequently the critical region is given by (3.4).

Next, we show that this test is a uniformly most powerful (UMP) invariant test under rotations. It is found from Beran (1968) that the most powerful invariant test is

$$\{c_p(\kappa)\}^n \int \prod_{i=1}^{n} \{\exp(\kappa \underset{\sim}{l}' \underset{\sim}{l}_i)\, d\underset{\sim}{l} > K,$$

where the region of integration is given by $\underset{\sim}{l} \in S_p$. Hence, the test is defined by

$$\{c_p(\kappa)\}^n / c_p(\kappa R) > K. \tag{3.7}$$

Since

$$A(\kappa) \geq 0,\ c_p(\kappa) \geq 0 \tag{3.8}$$

we have

$$c_p'(\kappa) = -A(\kappa)c_p(\kappa) \leq 0.$$

Therefore, $c_p(\kappa R)$ is a decreasing function of κ, and consequently (3.8) leads to the desired result.

3.3 Tests for the Mean Directions.

We now assume that $\kappa \neq 0$ so that the uniform case is ruled out.

3.3.1. Concentration Parameter Known.

We wish to test

$$H_o : \underset{\sim}{\mu}_o = \underset{\sim}{0} \text{ against } H_1 : \underset{\sim}{\mu}_o = \underset{\sim}{\mu}_o^* \neq \underset{\sim}{0}. \tag{3.9}$$

In this case, there is no UMP test. A discussion for $p = 2$ on this point will be found in Mardia (1972, p. 137).

The Fisher Ancillary Principle. Denote the pdf. of $\bar{\underset{\sim}{x}}_o | R$ given at (2.12) by g_1 and the pdf. of R at (2.11) by g_2. We find that the likelihood function can be expressed as

$$L = g_1(\bar{\underset{\sim}{x}}_o | R; \underset{\sim}{\mu}_o, \kappa) g_2(R;\kappa) h(\underset{\sim}{1}_1, \ldots, \underset{\sim}{1}_n),$$

where $h(\cdot)$ does not involve $\underset{\sim}{\mu}_o$ and κ. For this situation, the Fisher ancillary principle implies that the vector $\bar{\underset{\sim}{x}}_o | R$ should be used to test hypotheses about $\underset{\sim}{\mu}_o$. The principle is conceptually appealing because it takes into account the actual precision achieved. However, again there is no conditional BCR. For $p = 2$, $|\bar{x}_o| > K$, $-\pi < \bar{x}_o < \pi$ provides a conditional unbiased test (see Mardia 1972, p. 140).

Let δ be the angle between the directions $\underset{\sim}{\mu}$ under H_o and H_1. It is of some interest to test

$$H_o : \delta = 0 \text{ against } H_1 : \delta = \delta_1 \ (\neq 0).$$

By selecting the axes appropriately, we can reformulate the above hypotheses as

$$H_o : \mu_{o,1} = 0 \text{ against } H_1 : \mu_{o,1} = \delta_1, \tag{3.10}$$

where $\mu_{o,j} = 0$, $j > 1$. For this case, the best critical region (BCR) is found to be

$$\bar{x}_{o,1} | R > K. \tag{3.11}$$

From (3.12), the pdf. of $\bar{x}_{o,1} | R$ under H_o is

$$\{c_p(\kappa R)/c_{p-1}\} e^{\kappa R \cos \bar{x}_{o,1}} \sin^{p-2} \bar{x}_{o,1}, \quad 0 < \bar{x}_{o,1} < \pi, \tag{3.12}$$

which is of the form of the marginal distribution of θ_1 when $\underset{\sim}{\theta}$ is $M_p(\underset{\sim}{0},\kappa)$. For $p = 2$ and $p = 3$, some selected critical values of (3.11) are given in Mardia (1972). A $(1-\alpha)100\%$ confidence cone around $\underset{\sim}{\mu}$ can be constructed as follows. Let $\bar{x}_{o,1}$ be distributed as (3.12). Suppose that

$$P(\bar{x}_{o,1} > \delta \mid R) = \alpha.$$

Then, for given R, the probability that the true mean direction lies within a cone with vertex at the origin, axis as the sample mean direction and a semi-vertical angle δ is $1-\alpha$.

It may be noted that the LR test amounts to rejecting H_o for

$$R - R_{x_1} > K$$

which for given R reduces to (3.11). The power of the test can be written down with the help of (2.13) and (2.16).

2.3.2. Concentration Parameter Unknown. Consider the problem of testing

$$H_o : \underset{\sim}{\mu}_o = \underset{\sim}{0} \text{ against } H_1 : \underset{\sim}{\mu}_o \neq \underset{\sim}{0}. \tag{3.13}$$

Using (2.14) and (2.16), we find that a UMP invariant test is given by

$$R > K \mid R_{x_1} . \tag{3.14}$$

For a critical study of this test and the corresponding confidence cone, we refer to Mardia (1974b).

For large κ, we can use Watson & Williams' approximation

$$\frac{(n-1)(n-R_{x_1})}{n(n-R)} \simeq F_{(p-1)n,(p-1)(n-1)} . \tag{3.15}$$

For $p = 2$ and 3, its adequacy has been fully investigated.

3.4. Tests for Concentration Parameters.

3.4.1. Mean Direction Known. Let $\underset{\sim}{\mu}_o = 0$. To test

$$H_o : \kappa = \kappa_o \text{ against } H_1 : \kappa = \kappa_1 > \kappa_o,$$

the BCR is given by $R_{x_1} > K$. The test is UMP for testing H_o against the alternatives $\kappa > \kappa_o$. A UMP unbiased test exists for the two-sided alternative $\kappa \neq \kappa_o$ since R_{x_1} is sufficient for κ and its pdf. belongs to the standard exponential family. In this case, the CR can be written down formally from (2.13), but the resulting expression is complicated (see Mardia, 1972, p. 149). For large κ, the approximations given in Section 2.4 can be used.

3.4.2. Mean Direction Unknown. Consider the problem of testing

$$H_0 : \kappa = \kappa_0 \text{ against } H_1 : \kappa > \kappa_0.$$

The problem remains invariant under rotations. Now, $\underset{\sim}{R}^*_x$ is sufficient and complete for $(\underset{\sim}{\mu}_0, \kappa)$ and it is seen that $R^2 = \underset{\sim}{R}^{*\prime}_x \underset{\sim}{R}^*_x$ is a maximal invariant for the problem. Hence, invariant tests for this situation depend only on R. We now show that the LR test defined by CR

$$R > K \tag{3.16}$$

is a UMP invariant test. Using the pdf. $h_n^{(\kappa)}(R)$ of R given by (2.11), it is found that

$$\partial^2 h_n^{(\kappa)}(R)/\partial R \partial \kappa = \partial\{RA(\kappa R)\}/\partial R = A(\kappa R) + R\kappa A'(\kappa R)$$

which is non-negative in view of (3.6) and (3.8). Hence, the density $h_n^{(\kappa)}(R)$ has a montone likelihood ratio in κ (see Lehmann, 1959, p. 111). Hence, the test defined by (3.16) is a UMP invariant test.

For two-sided alternatives, an expression for a locally unbiased test can be written down but does not lead to any simplification (cf. Mardia 1972, p. 150, for p = 2). Again, for large κ approximations can be used to obtain the critical values. For large n, the normal approximation given by (2.32) can be used.

3.5 Multi-sample Tests. Optimum tests for multi-sample problems do not exist. We illustrate the point by considering a two-sample test for the equality of mean directions. Suppose that random samples of sizes n_1 and n_2 are drawn from $M_p(\underset{\sim}{\mu}_{0,1}, \kappa)$ and $M_p(\underset{\sim}{\mu}_{0,2}, \kappa)$ respectively. Following Section 2.3.2, let the corresponding resultant lengths be R_1 and R_2. Further, let $\bar{\underset{\sim}{x}}_0$ and R be the mean direction and the resultant length for the combined sample respectively.

We are interested in testing $H_o : \underset{\sim}{\mu}_{o,1} = \underset{\sim}{\mu}_{o,2} = \underset{\sim}{\mu}_o$ against $H_1 : \mu_{o,1} \neq \mu_{o,2}$, where $\underset{\sim}{\mu}_o$ and κ are unknown.

Since $(\overline{\underset{\sim}{x}}_o, R)$ are jointly complete sufficient statistics for $(\underset{\sim}{\mu}_o, \kappa)$, all similar tests should be conditional on $(\overline{\underset{\sim}{x}}_o, R)$. Even for p = 2, there is no UMP similar test (Mardia, 1972, pp. 152-153).

We can look for an invariant test under rotations. Such a test should depend on (R_1, R_2) given $\underset{\sim}{R}$. Indeed, we have seen in (2.23) that the null distribution of (R_1, R_2) given R does not involve κ.

However, the question arises as to which function of $(R_1, R_2)|R$ should be used. On intuitive grounds, Watson and Williams (1956) have suggested rejecting H_o for large values of $R_1 + R_2|R$.

REFERENCES

[1] Beran, R. J. (1968). J. Appl. Prob. 5, 177-195.
[2] Fisher, R. A. (1953). Proc. Roy. Soc. Lond. A 217, 295-305.
[3] Lehmann, E. L. (1959). Testing Statistical Hypotheses. Wiley, New York.
[4] Mardia, K. V. (1972). Statistics of Directional Data. Academic Press, London.
[5] Mardia, K. V. (1974a). Characterizations of directional distributions. Presented to the First International Conference on Characterizations. Calgary, Canada.
[6] Mardia, K. V. (1974b). Statistics of directional data. To be presented to the Roy. Statist. Soc.
[7] Mardia, K. V. and Zemroch, P. J. (1974a). Circular statistics. Algorithm to appear in J. Roy. Statist. Soc. Ser. C.
[8] Mardia, K. V. and Zemroch, P. J. (1974b). Spherical statistics. Algorithm to appear in J. Roy. Statist. Soc. Ser. C.
[9] Rao, J. S. (1969). Some contributions to the analysis of circular data. Ph.D. Thesis, Indian Statistical Institute, Calcutta.
[10] Stephens, M. A. (1962). The statistics of directions. Ph.D. Thesis, Toronto University, Canada.
[11] Stephens, M. A. (1967). Biometrika 54, 211-223.

[12] Upton, G. J. G. (1974). Biometrika, 61, 369-374.
[13] Watson, G. N. (1948). A Treatise on the Theory of Bessel Functions. 2nd. edn. Cambridge Univ. Press.
[14] Watson, G. S. and Williams, E. J. (1956). Biometrika, 43, 344-352.

APPENDIX 1

Some Properties of the von Mises-Fisher Distribution

(i) The 'characteristic function' of $\underset{\sim}{l}$ is

$$\phi(\underset{\sim}{t}) = E(e^{i\underset{\sim}{t}'\underset{\sim}{l}}) = c_p(\kappa)/c_p(\omega), \tag{A1.1}$$

where

$$\omega = (\kappa^2 - \underset{\sim}{t}'\underset{\sim}{t} + 2i\kappa\underset{\sim}{t}'\underset{\sim}{\mu})^{\frac{1}{2}}.$$

A proof follows on verifying that

$$\int_{S_p} \exp(i\underset{\sim}{t}'\underset{\sim}{l} + \kappa\underset{\sim}{\mu}'\underset{\sim}{l})\, ds_p = c_p(\omega), \tag{A1.2}$$

where ds_p is the surface element corresponding to S_p.

(ii) From (A1.1), we have

$$E(\underset{\sim}{l}) = A(\kappa)\underset{\sim}{\mu}, \tag{A1.3}$$

and

$$\mathrm{cov}(\underset{\sim}{l}) = \underset{\sim}{\mu}\underset{\sim}{\mu}'\{1 - p\kappa^{-1}A(\kappa) - A^2(\kappa)\} + \kappa^{-1}A(\kappa)\underset{\sim}{I} \tag{A1.4}$$

where $\underset{\sim}{I}$ is the identity matrix and

$$A(\kappa) = I_{\frac{1}{2}p}(\kappa)/I_{\frac{1}{2}p-1}(\kappa). \tag{A1.5}$$

(iii) $M_p(\underset{\sim}{\mu},\kappa)$ is invariant under orthogonal transformations.

CERTAIN STATISTICAL DISTRIBUTIONS INVOLVING SPECIAL FUNCTIONS AND THEIR APPLICATIONS

Frank McNolty, J. Richard Huynen and Eldon Hansen

Lockheed Palo Alto Research Laboratory, Palo Alto, California

SUMMARY. The objective of the paper is to show that a certain generalized Bessel distribution is a useful theoretical tool because, through mixing procedures, specialization of parameter values and integral representation considerations, it unifies the theory of a broad class of special distributions; and that it is a useful tool because it has applications in radio communication. The random sine wave problem and distributions for fluctuating radar targets are studied in terms of this distribution with particular emphasis on amplitude, phase and component distributions. The marginal pdf and characteristic function corresponding to Bennett's (Rice's) distribution of a random sine wave plus stationary Gaussian noise are obtained when the sine wave amplitude is assigned a generalized Bessel prior distribution. The Bennett problem is also reduced to a randomly phased sine wave without noise and the corresponding marginal pdf and characteristic function are obtained from the noise corrupted case. Non-uniform phase distributions are also treated in terms of generalized distributions and the corresponding amplitude and component pdf^s are provided. The distribution of a useful quadratif form in which the random variates are the squares of a generalized component corrupted by Gaussian noise is also provided. The quadratic form distribution is then used as a basic model for fluctuating radar cross section (RCS) and includes the Swerling RCS models and the Nakagami amplitude distributions as special cases. The authors also provide the corresponding pulse-train probability distributions for two pulse integration schemes using either scan-to-scan or pulse-to-pulse amplitude independence. Many of the pdf^s and characteristic functions provided in the paper are expressed in terms of both closed form expressions and mixture representations.

G. P. Patil et al. (eds.), Statistical Distributions in Scientific Work, Vol. 1, 131-160.

KEY WORDS. Generalized Bessel distribution, mixture representations, characteristic functions, random sine wave, fluctuating radar cross section.

1. RANDOM SINE WAVE PLUS GAUSSIAN NOISE.

The random quantity of interest is

$$y(t) = Q\sin(\omega_o t + \phi) + n(t) \tag{1}$$

where ω_o and Q are positive constants, ϕ is uniformly distributed over $(0,2\pi)$ and $n(t)$ is mean zero stationary Gaussian noise with variance σ_n^2.

In 1944 W. R. Bennett provided the probability density function (pdf) of the instantaneous value of $y(t)$,

$$f(\alpha)d\alpha = \text{Prob}[\alpha < y(t) \leq \alpha + d\alpha]$$

$$= \frac{d\alpha}{\pi\sqrt{2\pi}\,\sigma_n}\int_0^{\pi} \exp\left[-\frac{(\alpha - Q\cos\theta)^2}{2\sigma_n^2}\right]d\theta, \quad -\infty < \alpha < \infty \tag{2}$$

where the characteristic function corresponding to (2) is

$$\theta(t) = J_o(tQ)\cdot\exp(-t^2\sigma_n^2/2), \quad -\infty < t < \infty. \tag{3}$$

1.1. The Generalized Prior Distribution.

A prior distribution for the amplitude Q in (1) is now introduced

$$g(Q)dQ = 2^{\frac{a}{4}-\frac{1}{2}}\cdot\frac{Q^{\frac{a}{2}}}{\gamma^{\frac{a}{2}-1}}\cdot\beta^{\frac{a}{2}}\exp[-\gamma^2/4\beta - \beta Q^2/2]I_{\frac{a}{2}-1}\left(\gamma\frac{Q}{\sqrt{2}}\right)dQ,$$

$$Q \geq 0 \tag{4}$$

where $a > 0$; $\gamma, \beta > 0$ and $I_\nu(x)$ is the modified Bessel function of the first kind of order ν.

Expression (4) is a generalized Bessel distribution which has been formulated so that for simple choices of the parameters it yields several familiar pdfs as special cases. For instance:

(a) for $\gamma = 0$, $a = 1$ and $\beta = 1/\sigma^2$ we obtain the one-sided Gaussian distribution

$$g(Q)\,dQ = \frac{1}{\sigma}\sqrt{\frac{2}{\pi}} \cdot \exp(-Q^2/2\sigma^2)\,dQ, \quad Q \geq 0 \tag{5}$$

(b) for $\gamma = 0$, $a = 2$ and $\beta = 1/\sigma^2$ expression (4) becomes the Rayleigh distribution

$$g(Q)\,dQ = \frac{Q}{\sigma^2} \cdot \exp(-Q^2/2\sigma^2)\,dQ, \quad Q \geq 0 \tag{6}$$

(c) for $\gamma = 0$, $a = 3$ and $\beta = 1/\sigma^2$ we obtain the Maxwell-Boltzmann pdf

$$g(Q)\,dQ = \frac{1}{\sigma^3}\sqrt{\frac{2}{\pi}}\; Q^2 \exp(-Q^2/2\sigma^2)\,dQ, \quad Q \geq 0 \tag{7}$$

(d) for $\gamma = 0$, $a = 2\lambda$ and $\beta = 2\lambda/\overline{Q^2}$ we have Swerling's (1957) generalized amplitude distribution

$$g(Q)\,dQ = \frac{2}{\Gamma(\lambda)}\left(\frac{\lambda}{\overline{Q^2}}\right)^{\lambda} \cdot Q^{2\lambda-1} \cdot \exp(-\lambda Q^2/\overline{Q^2})\,dQ, \quad Q \geq 0 \tag{8}$$

$\overline{Q^2} = E(Q^2)$ and letting $\lambda = m$ in (8) gives Minoru Nakagami's (1960) m-distribution for fast fading statistics where $m \geq 1/2$, $m = \overline{Q^2}^2/E[(Q^2 - \overline{Q^2})^2]$ and $E(Q^4) = \frac{m+1}{m} \cdot \overline{Q^2}^2$.

(e) for $Q = R$, $\gamma = Q\sqrt{2}/\sigma_n^2$, $a = 2$ and $\beta = 1/\sigma_n^2$ expression (4) yields the Rice distribution for the envelope of (1) when $n(t)$ is narrowband Gaussian noise with variance σ_n^2, i.e. the noncentral 'chi' distribution for two degrees of freedom

$$g(R)\,dR = \frac{R}{\sigma_n^2}\exp(-R^2/2\sigma_n^2 - Q^2/2\sigma_n^2) \cdot I_o(QR/\sigma_n^2)\,dR, \quad R \geq 0 \tag{9}$$

Additional special cases may easily be obtained from (4). For instance:

(f) letting $a = 2N$, $\beta = 2$, $\gamma = 2\sqrt{NR_p}$ and $Q^2 = y$ gives Marcum's (1960) pdf for the sum y of N noncoherently integrated, randomly (uniformly distributed) phased signal-plus-noise pulses of constant signal amplitude out of a square-law detector where R_p = ratio of the ma imum single-pulse instantaneous signal power to average noise power out of the matched filter (preceding the square-law device),

$$f(y)dy = \left(\frac{2y}{NR_p}\right)^{\frac{N-1}{2}} \cdot e^{-y} \cdot \exp(-NR_p/2) \cdot I_{N-1}(\sqrt{2NR_p y})dy,$$

$$= \sum_{m=0}^{\infty} P(m;NR_p/2) \cdot g_s(y;1,N+m)dy, \quad y \geq 0 \tag{10}$$

where P(m;a) is the discrete Poisson distribution

$$P(m;a) = \frac{a^m}{m!} e^{-a}$$

and $g_s(x;\alpha,\beta)$ is Swerling's pdf (11) below.

(g) For $\gamma = 0$, $a = 2\lambda$, $\beta = 2\lambda/\bar{\sigma}_c$ and $Q^2 = \sigma_c$ expression (4) becomes Swerling's generalized distribution (1957) for fluctuating radar cross section,

$$g_s(\sigma_c)d\sigma_c = g_s(\sigma_c;\alpha,\beta)d\sigma_c = \frac{1}{\Gamma(\lambda)} \cdot \frac{\lambda}{\bar{\sigma}_c}\left(\frac{\lambda\sigma_c}{\bar{\sigma}_c}\right)^{\lambda-1} \cdot \exp\left(-\frac{\lambda\sigma_c}{\bar{\sigma}_c}\right)d\sigma_c,$$

$$0 \leq \sigma_c < \infty \tag{11}$$

where $\bar{\sigma}_c = E(\sigma_c)$, $\mathrm{Var}(\sigma_c) = \bar{\sigma}_c^2/\lambda$, $\lambda > 0$ and the gamma distribution parameters corresponding to (11) are $\alpha = \lambda/\bar{\sigma}_c$ and $\beta = \lambda$.

2. THE UNCONDITIONAL DISTRIBUTION AND CHARACTERISTIC FUNCTION

2.1 Uniform Phase (ϕ) Distribution, $\gamma \neq 0$.

Using (4) as a prior distribution for Q in expression (2) gives the unconditional (marginal) distribution for (1),

$$h(\alpha)d\alpha = d\alpha \int_0^\infty g(Q)f(\alpha|Q)dQ$$

$$= \frac{\exp(-\alpha^2/4\sigma_n^2-\gamma^2/4\beta)}{\sqrt{2\pi}\,\sigma_n\Gamma(a/2)} \sum_{m=0}^{\infty} \frac{(m+a/2-1)!}{(m!)^2(2\beta\sigma_n^2)^m} D_{2m}(\alpha/\sigma_n)$$

$$\cdot {}_1F_1(m+a/2;a/2;\gamma^2/4\beta)\,d\alpha, \quad -\infty < \alpha < \infty \qquad (12)$$

where ${}_1F_1(a;c;x)$ is the confluent hypergeometric function and $D_p(u)$ is the parabolic cylinder function defined by

$$D_p(u) = 2^{P/2}\cdot\exp(-u^2/4)\left[\frac{\sqrt{\pi}}{\Gamma(\frac{1-P}{2})}\cdot{}_1F_1(-P/2;1/2;u^2/2) - \frac{u\sqrt{2\pi}}{\Gamma(-P/2)}\right.$$

$$\left.\cdot\,{}_1F_1(\tfrac{1-P}{2};3/2;u^2/2)\right].$$

Similarly, the characteristic function corresponding to (12) is given by

$$\phi(t) = \exp(-\gamma^2/4\beta)\cdot\exp(-t^2\sigma_n^2/2)\sum_{m=0}^{\infty}\frac{(\gamma^2/4\beta)^m}{m!}$$

$$\cdot\; {}_2F_1(-m,-m-a/2+1;1;-2t^2/\gamma^2) \qquad (13)$$

where ${}_2F_1(a,b;c;x)$ is the hypergeometric function.

Some alternative forms of expressions (12) and (13) are given in Appendix I.

2.2. Uniform Phase (ϕ) Distribution, $\gamma = 0$.

When $\gamma = 0$ in (4) the prior pdf for Q takes the same general form as Swerling's amplitude pdf (8) so that expressions (12) and (13) may be written more simply. In this case (12) becomes

$$h_1(\alpha)d\alpha = \frac{d\alpha}{\sqrt{2\pi}\,\sigma_n\Gamma(a/2)}\sum_{k=0}^{\infty}\sum_{m=0}^{\infty}\frac{\Gamma(K+m+1/2)\Gamma(m+a/2)(-1)^{m+K}\alpha^{2K}}{K!\Gamma(K+1/2)(m!)^2\beta^m\sigma_n^{2m+2K}2^K},$$

$$-\infty < \alpha < \infty \qquad (14)$$

while (13) reduces to

$$\phi_1(t) = \exp(-t^2\sigma_n^2/2)\cdot\exp(-t^2/2\beta)\,{}_1F_1(1-a/2;1;t^2/2\beta) \qquad (15)$$

where ${}_1F_1(a;c;x)$ is the confluent hypergeometric function.

Appendix II provides some interesting alternative forms of the marginal distribution (14).

Recalling from (6) that $\gamma = 0$, $a = 2$ and $\beta = 1/\sigma^2$ in (4) yields a Rayleigh distributed amplitude Q, and since (throughout this section) ϕ is assumed to be uniformly distributed, it follows that for these parameter values (1) is Gaussian. Thus, setting $a = 2$ and $\beta = 1/\sigma^2$ in (15) we obtain $\exp(-t^2\sigma_n^2/2)\cdot\exp(-t^2\sigma^2/2)$ which does indeed correspond to a Gaussian distribution.

3. NOISE TERM ABSENT. In the absence of the noise term n(t) in (1), only the random sine wave remains,

$$x(t) = Q\sin(\omega_o t + \phi) \qquad (16)$$

which has a characteristic function [4] $J_o(tQ)$ and pdf $\frac{1}{\pi}(Q^2-x^2)^{-1/2}$, $-Q < x < Q$ where here Q is a positive real constant.

3.1 Uniform Phase (ϕ) Distribution, $\gamma \neq 0$. When n(t) is not present ($\sigma_n^2 = 0$) in expression (1) the prior distribution (4) yields the following unconditional characteristic function from (3),

$$\tau(t) = \sum_{m=0}^{\infty} P(m;\gamma^2/4\beta)\cdot{}_2F_1(-m,-m-a/2+1;1;-2t^2/\gamma^2)$$

$$= \frac{1}{\Gamma(a/2)}\exp(-\gamma^2/4\beta)\sum_{m=0}^{\infty}\frac{\Gamma(m+a/2)}{(m!)^2}(-t^2/2\beta)^m$$

$$\cdot{}_2F_1(-m,-m;a/2;-\gamma^2/2t^2) \qquad (17)$$

where P(m;a) is again the discrete Poisson distribution and ${}_2F_1(a,b;c;x)$ is the hypergeometric function.

3.2 Uniform Phase (ϕ) Distribution, $\gamma = 0$. For $\gamma = 0$ the unconditional characteristic function (15) reduces to

$$\tau_1(t) = \exp(-t^2/2\beta)\cdot {}_1F_1(1-a/2;1;t^2/2\beta) \tag{18}$$

and the inversion of (18) yields the marginal pdf for the random sine wave (16) when Q is distributed according to (4) with $\gamma = 0$, i.e.,

$$w(\alpha)\,d\alpha = \sqrt{\frac{\beta}{2\pi}}\cdot\exp(-\beta\alpha^2/4)\cdot\frac{d\alpha}{\Gamma(1-a/2)}\cdot\sum_{m=0}^{\infty}\frac{(-1)^m\Gamma(m+1-a/2)}{2^m(m!)^2}\cdot D_{2m}(\alpha\sqrt{\beta}),$$

$$-\infty < \alpha < \infty \tag{19}$$

where again $D_\nu(x)$ is the parabolic cylinder function defined below expression (12).

4. NON-UNIFORM PHASE DISTRIBUTIONS. In order to consider non-uniform phase and non-Gaussian component distributions we write

$$\begin{aligned} z(t) &= Q(t)\cos[\omega_o t + \phi(t)] \\ &= x(t)\cos\omega_o t - y(t)\sin\omega_o t \end{aligned}$$

where

$$x(t) = Q(t)\cos\phi(t) \tag{20}$$

$$y(t) = Q(t)\sin\phi(t). \tag{21}$$

Expression (4) is now assumed to be the amplitude (radial) distribution for the quantity $Q = \sqrt{x^2+y^2}$ where the independent variates x and y are given by (20) and (21) and need not have identical distributions. We define the distribution of the x-component as

$$f(x)\,dx = f(x;\beta,\gamma_1,a_1)\,dx = 2^{\frac{a_1}{4}-\frac{3}{2}}\cdot\gamma_1^{-\frac{a_1}{2}+1}|x|^{\frac{a_1}{2}}\beta^{\frac{a_1}{2}}$$

$$\cdot\exp\left[-\frac{1}{2}\left(\frac{\gamma_1^2}{2\beta}+\beta x^2\right)\right]\cdot I_{\frac{a_1}{2}-1}\left(\gamma_1\frac{|x|}{\sqrt{2}}\right)dx,\quad -\infty < x < \infty \tag{22}$$

where $I_\nu(x)$ is the modified Bessel function of the first kind of order ν and similarly the distribution of the y-component has the same general form as (22) but now only the β parameter need remain the same, i.e.,

$$f(y)dy = f(y;\beta,\gamma_2,a_2)dy \tag{23}$$

where $\beta > 0$; $\gamma_1,\gamma_2 \geq 0$ and $a_1,a_2 > 0$. Since the corresponding parameters γ_1,γ_2 and a_1,a_2 need not be the same the actual forms of (22) and (23) may be quite different in terms of special cases. It is not difficult to show that (4) is indeed the pdf for $Q = \sqrt{x^2 + y^2}$ where in (4) we would have $a = a_1 + a_2$, $\beta = \beta$ and $\gamma^2 = \gamma_1^2 + \gamma_2^2$.

The distribution of $\phi = \arctan \frac{y}{x}$ is obtained from (22) and (23) in a straight-forward manner,

$$p(\phi)d\phi = \frac{1}{2}\sum_{i=0}^{\infty}\sum_{j=0}^{\infty} H(i,j;\gamma_1,\gamma_2;a_1,a_2)|\sin\phi|^{a_2+2j-1}\cdot|\cos\phi|^{a_1+2i-1}d\phi,$$

$$0 \leq \phi \leq 2\pi \tag{24}$$

where

$$h(i,j;\gamma_1,\gamma_2;a_1,a_2) = \exp\left[-\frac{1}{2\beta}\left(\frac{\gamma_1^2+\gamma_2^2}{2}\right)\right]\cdot\left(\frac{\gamma_1}{2}\right)^{2i}\cdot\left(\frac{\gamma_2}{2}\right)^{2j}$$

$$\cdot\frac{1}{\beta^{i+j}i!j!B\left(\frac{a_1}{2}+i,\frac{a_2}{2}+j\right)}$$

and $B(x,y)$ is the beta function.

When expression (22) and (23) happen to be identical (i.e., $a_1 = a_2$, $\gamma_1 = \gamma_2$) then (4) may be written in a somewhat more symmetric form by virtue of the fact that in this case $a = a_1 + a_2 = 2a_1$ and $\gamma^2 = \gamma_1^2 + \gamma_2^2 = 2\gamma_1^2$. Thus, writing a for a_1 and γ for γ_1 the pdf of $\sqrt{x^2 + y^2}$ is given by

$$g(Q)dQ = \left[(\beta Q)^a/\gamma^{a-1}\right]\cdot\exp\left[-(\gamma^2/2\beta)-(\beta/2)Q^2\right]\cdot I_{a-1}(\gamma Q)dQ \tag{25}$$

where $a, \beta > 0$; $\gamma, Q \geq 0$ and again $I_\nu(x)$ is the modified Bessel function of the first kind of order ν.

In specializing (4), (22), (23) and (24) we will assume, for brevity, that the independent x,y components are identically distributed. Then

(a) corresponding to the Rayleigh amplitude (6),

the x and y components are each distributed according to

$$f(x)dx = \frac{1}{\sqrt{2\pi}\,\sigma} \exp(-x^2/2\sigma^2)dx, \quad -\infty < x < \infty \tag{26}$$

while the phase distribution is uniform

$$p(\phi)d\phi = \frac{d\phi}{2\pi}, \quad 0 \leq \phi \leq 2\pi \tag{27}$$

where in (4), (22), (23) and (24) we have let $a_1 = a_2 = 1$, $\gamma_1 = \gamma_2 = 0$ and $\beta = 1/\sigma^2$ to obtain (6), (26) and (27).

(b) corresponding to the Maxwell-Boltzmann amplitude distribution (7),

the x, y components each have the pdf

$$f(x)dx = \frac{1}{2^{3/4}\Gamma(3/4)\sigma^{3/2}}|x|^{1/2}\cdot\exp(-x^2/2\sigma^2)dx, \quad -\infty < x < \infty \tag{28}$$

and the corresponding phase distribution is

$$p(\phi)d\phi = \frac{|\sin 2\phi|^{1/2}\sqrt{\pi}}{4\sqrt{2}\,\Gamma^2(3/4)}, \quad 0 \leq \phi \leq 2\pi \tag{29}$$

where in arriving at (7), (28) and (29) we have let $a_1=a_2=3/2$, $\gamma_1=\gamma_2=0$ and $\beta = 1/\sigma^2$ in (4), (22), (23) and (24).

(c) corresponding to the Swerling amplitude pdf (8)

the x, y component distributions are each

$$f(x)dx = \left(\frac{\lambda}{Q^2}\right)^{\lambda/2}\cdot\frac{|x|^{\lambda-1}}{\Gamma(\lambda/2)}\cdot\exp\left(-\frac{\lambda x^2}{Q^2}\right)dx, \quad -\infty < x < \infty \tag{30}$$

and the phase distribution is given by

$$p(\phi)d\phi = \frac{\Gamma(\lambda)}{\Gamma^2(\lambda/2)} \cdot \frac{|\sin 2\phi|^{\lambda-1}}{2^\lambda} d\phi, \quad 0 \leq \phi \leq 2\pi \tag{31}$$

where $a_1 = a_2 = \lambda$, $\gamma_1 = \gamma_2 = 0$ and $\beta = 2\lambda/\overline{Q^2}$ in (4), (22), (23) and (24).

(d) for the amplitude distribution (9) the corresponding x, y component and phase distributions are, respectively,

$$f(x)dx = \frac{1}{\sqrt{2\pi}\,\sigma_n} \cdot \exp\left[-\frac{1}{2\sigma_n^2}\left(\frac{Q^2}{2} + x^2\right)\right] \cdot \cosh\left(\frac{Qx}{\sqrt{2}\,\sigma_n^2}\right)dx,$$

$$-\infty < x < \infty \tag{32}$$

and

$$p(\phi)d\phi = \exp(-Q^2/2\sigma_n^2) \sum_{i=0}^{\infty} \sum_{j=0}^{\infty} \left[\frac{Q}{2\sigma_n}\right]^{2i+2j} \cdot \frac{(i+j)!\sin^{2j}\phi \cdot \cos^{2i}\phi}{2i!j!\Gamma(i+1/2)\Gamma(j+1/2)},$$

$$0 \leq \phi \leq 2\pi \tag{33}$$

{recall the change of notation from Q to R in writing (9) from expression (4)}.

If from (22) we write the distribution of $u = x^2$ as a mixture representation it becomes,

$$g(u)du = \sum_{m=0}^{\infty} P\left(m; \frac{\gamma_1^2}{4\beta}\right) \cdot g_s(u; \beta/2, a_1/2+m)du, \quad u \geq 0 \tag{34}$$

where $P(m;a)$ is again the discrete Poisson distribution and $g_s(u;\alpha,\beta)$ is a distribution of the Swerling (gamma) form, i.e.,

$$g_s(u;\alpha,\tau) = \frac{\alpha^\tau}{\Gamma(\tau)} u^{\tau-1} e^{-\alpha u}, \quad u \geq 0. \tag{34a}$$

The characteristic function for (34) is

$$\theta_u(t) = \left[1 - \frac{2it}{\beta}\right]^{-a_1/2} \cdot \exp\left[\frac{it\gamma_1^2}{2\beta^2(1-\frac{2it}{\beta})}\right] = \sum_{m=0}^{\infty} P(m; \frac{\gamma_1^2}{4\beta})$$

$$\cdot \left[1 - \frac{2it}{\beta}\right]^{-a_1/2-m} . \qquad (35)$$

An extensive treatment of mixture representations, such as in (10), (17), (34) and (35), is given in McNolty (1972a). Also see Chapter IX of Luke (1969).

If we now regard $z(t) = Q(t)\cos[\omega_o t + \phi(t)]$ as a stationary narrowband Gaussian process with mean zero and variance σ_n^2 in which $Q(t)$ and $\phi(t)$ vary slowly in comparison with $\cos \omega_o t$ then a quantity of interest is the distribution of

$$w(t) = \{x(t) + Q(t)\cos[\omega_o t + \phi(t)]\}^2 \qquad (36)$$

where Q is Rayleigh, x is distributed according to (22) and ϕ is uniformly distributed over $(0,2\pi)$ where x, Q and ϕ are all statistically independent.

More generally one can write the quadratic form expression

$$S_k = \sum_{i=1}^{k} w_i = \sum_{i=1}^{k} (x_{oi} + x_{1i})^2 \qquad (37)$$

where x_{oi} is distributed according to (22) with parameters a_1, γ_1, β and x_{1i} is a mean zero Gaussian variate with variance σ_n^2. All x_{oi} and x_{1i} are statistically independent. Actually, the distribution of S_k may be obtained provided only that β and σ_n^2 remain fixed, while the a_1 and γ_1 parameters may vary from one w_i to the next. For brevity, however, it will be assumed that the x_{oi} are identically distributed according to (22) and in order to simplify the notation we will write b in place of a_1 and τ in place of γ_1. In the discussion of fluctuating radar cross section which is to follow it will also be preferable to use the symbol A to denote amplitude, rather than Q as was used in previous sections.

It may be shown that the characteristic function of any single term $(x_{oi} + x_{1i})^2$ in (37) is

$$\theta(t) = \frac{\beta^{b/2}(1-2it\sigma_n^2)^{\frac{b-1}{2}}}{[\beta - 2it(1+\beta\sigma_n^2)]^{b/2}} \cdot \exp\left\{\frac{it\gamma^2}{2\beta[\beta-2it(1+\beta\sigma_n^2)]}\right\} \tag{38}$$

and that the distribution of (37) is given by

$$f(s)ds = \frac{1}{2\pi}\int_{-\infty}^{\infty} \exp(-its) \cdot [\theta(t)]^k \, dt \tag{39a}$$

$$f(s)ds = (\sigma_n^2)^{K(b-1)/2} \cdot \beta^{Kb/2}[2^{K/2}\Gamma(K/2)(1+\beta\sigma_n^2)^{Kb/2}]^{-1}$$

$$\cdot \exp(-s/2\sigma_n^2 - K\tau^2/4\beta) \cdot \sum_{m=0}^{\infty} \frac{(K\tau^2)^m \sigma^{2m}}{4^m m!(1+\beta\sigma_n^2)^m}$$

$$\cdot {}_1F_1[Kb/2 + m; K/2; \frac{s}{2\sigma_n^2(1+\beta\sigma_n^2)}], \quad 0 \leq s < \infty \tag{39b}$$

where again ${}_1F_1(a; c; x)$ is the confluent hypergeometric function.

For the simple two dimensional case

$$S_2 = A^2 = (x_o + x_1)^2 + (y_o + y_1)^2 \tag{40}$$

the pdf (39b) reduces to

$$f(s)ds = \frac{\beta^b(\sigma_n^2)^{b-1}ds}{2(1+\beta\sigma_n^2)^b} \cdot \exp(-s/2\sigma_n^2 - \tau^2/2\beta) \cdot \sum_{m=0}^{\infty} \frac{\tau^{2m}(\sigma_n^2)^m}{2^m m!(1+\beta\sigma_n^2)^m} \cdot$$

$$\cdot {}_1F_1\left[b + m; 1; \frac{s}{2\sigma_n^2(1+\beta\sigma_n^2)}\right], \quad 0 \leq s < \infty \tag{41}$$

The cumulative distribution corresponding to (41) becomes,

$$F(A_1^2) = \text{Prob}(s \leq A_1^2) = \int_0^{A_1^2} f_2(s)\,ds$$

$$= \left[(\beta\sigma_n^2)/(1+\beta\sigma_n^2)\right]^b \cdot \exp(-\tau^2/2\beta) \sum_{m=0}^{\infty} \left[\gamma(m+1), A_1^2/2\sigma_n^2)/m(1+\beta\sigma_n^2)^m \cdot m!\right] \cdot$$

$$\cdot \left[1/B(m,b)\right] \cdot {}_1F_1\left\{b+m; b; \left[\tau^2\sigma_n^2/2(1+\beta\sigma_n^2)\right]\right\} \tag{42}$$

where $\gamma(a,x)$ is the incomplete gamma function and $B(x,y)$ is the beta function.

Expression (42) is readily reduced to any of the special cases for x_o and y_o described in the preceding section (where x_1, y_1 always remain as mean zero Gaussian variates with variance σ_n^2). For instance, corresponding to Gaussian x_o and y_o components as in (26) we have

$$F(A_1^2) = \frac{\sigma_n^2}{\sigma^2+\sigma_n^2} \sum_{m=0}^{\infty} \frac{\gamma(m+1, A_1^2/2\sigma_n^2)(\sigma^2)^m}{m!(\sigma^2+\sigma_n^2)^m} = 1 - \exp\left[-A_1^2/2(\sigma^2+\sigma_n^2)\right] \tag{43}$$

and for the Maxwell-Boltzmann amplitude case in which x_o and y_o are distributed according to (28)

$$F(A_1^2) = \frac{2\sigma_n^3}{(\sigma^2+\sigma_n^2)^{3/2}} \sum_{m=0}^{\infty} \frac{(2m+1)!}{(m!)^3 2^{2m+1}} \left(\frac{\sigma^2}{\sigma^2+\sigma_n^2}\right)^m \cdot \gamma(m+1, A_1^2/2\sigma_n^2) \tag{44}$$

where, of course, $F(0) = 0$ and for $A_1 \to \infty$,

$$\lim_{A_1\to\infty} F(A_1^2) = \frac{\sigma_n^3}{(\sigma^2+\sigma_n^2)^{3/2}} \sum_{m=0}^{\infty} \frac{(2m+1)!}{(m!)^2 2^{2m}} \left(\frac{\sigma^2}{\sigma^2+\sigma_n^2}\right)^m = 1.$$

When the amplitude A is distributed according to the non-central "chi" distribution for N degrees of freedom we simply set $a = N/2$ in (25), i.e., recalling the change of notation described near the beginning of this section we restrict the values of b to 1/2, 1, 3/2, 2, 5/2, 3,... in (42).

Corresponding to the case for Swerling (Nakagami) amplitude (8), the cumulative distribution (42) becomes,

$$F(A_1^2) = \frac{(2\lambda\sigma_n^2)^\lambda}{\Gamma(\lambda)(\overline{A^2}+2\lambda\sigma_n^2)^\lambda} \sum_{m=0}^{\infty} \frac{\Gamma(m+\lambda)}{(m!)^2} \left[\frac{\overline{A^2}}{\overline{A^2}+2\lambda\sigma_n^2}\right]^m \cdot\gamma(m+1, A_1^2/2\sigma_n^2) \tag{45}$$

where again $\gamma(a,x)$ denotes the incomplete gamma function.

5. FLUCTUATING RADAR CROSS SECTION. The statistical model, (22) and (40), for fluctuating radar cross section (RCS) presented in this paper was formulated so that the existing RCS models due to Swerling (1954), (1957) and the related amplitude models of Nakagami (1960) are obtained as special cases. The model is especially adapted to fit a format of general target decompositions developed in a thesis by J. R. Huynen (1970). A primary result of that work is that the electromagnetic target return voltage $V_T(t)$ observed during an interval (t_1,t_2) may be decomposed into scattering due to an "effective target" voltage $V_{To}(t)$ and a "residue target" voltage $V_{T1}(t)$ such that $V_T(t) = V_{To}(t) + V_{T1}(t)$. The voltages $V_{To}(t)$ and $V_{T1}(t)$ have the property that corresponding powers are additive in the average, i.e.,

$$\langle P_T \rangle = \langle P_{To} \rangle + \langle P_{T1} \rangle \ . \tag{46}$$

The effective target return voltage $V_{To}(t)$ may be quite general, whereas, the residue component $V_{T1}(t)$ can be viewed as an additive clutter term for which the most common example is the Gaussian model. The electromagnetic model also accounts for the polarization used during transmission and reception, Huynen (1970). In McNolty (1974b) it is shown that (40) is indeed an appropriate mathematical representation of the squared amplitude of the radar echo voltage; thus, the analysis of fluctuating radar cross section which follows will be based upon the two dimensional model (40) and (41) in which the components x_o and y_o are distributed according to (22). An important property of the authors' generalized Bessel distribution (22) is that independent x_o and y_o components distributed according to (22) and (23), respectively, transform to separable phase (ϕ) and amplitude (A) variates for the important special case

$\gamma_1 = \gamma_2 = 0$. Also, the phase distribution (24) associated with (22) and (23) through $\phi = \text{arc tan } (y/x)$ permits one to define a wide-sense stationary process $s(t) = A\cos(\omega_o t + \phi)$, since $E(\cos 2\phi) = E(\sin 2\phi) = E(\cos \phi) = E(\sin \phi) = 0$. Also, the structure of the generalized Bessel distribution (22) is such that one may readily construct mixture representations [see (10, (53) and (54)] as alternative and sometimes more convenient, descriptors of the corresponding closed form distributions.

We remark that the radar cross section σ_c is obtained from (40) by the simple transformation $\sigma_c = \frac{1}{K} A^2 = kA^2$ where K is an appropriate constant. Also, it is worthwhile to reiterate that the parameters a_1 and γ_1 in (22) have been changed to b and τ, respectively, in the following discussion.

Using the usual notation E( ) to denote an expected value one can write the following simple relationships from (40)

$$E(A^2) = E(x_o^2 + y_o^2) + E(x_1^2 + y_1^2) = E(x_o^2 + y_o^2) + 2\sigma_n^2 \,. \tag{47}$$

In order to obtain an expression for $E(x_o^2 + y_o^2)$ in terms of the parameters of (22) it is recalled that, for identically distributed components, (25) is the pdf of $(x_o^2 + y_o^2)^{1/2}$. Thus, letting $u = A^2$ in (25) we obtain

$$h(u)du = \beta^b [2\tau^{b-1}]^{-1} \cdot u^{\frac{b-1}{2}} \cdot \exp(-\tau^2/2\beta - \beta u/2) \cdot I_{b-1}(\tau\sqrt{u})\,du, \quad 0 \leq u < \infty \tag{48}$$

which has a characteristic function,

$$\theta_u(t) = (1 - 2it/\beta)^{-b} \cdot \exp\left[\frac{it\tau^2}{\beta^2(1-2it/\beta)}\right] \tag{49}$$

and from either (48) or (49) we readily find that

$$E(u) = E(x_o^2 + y_o^2) = 2b/\beta + \tau^2/\beta^2 \tag{50}$$

and, thus, from (47)

$$E(A^2) = 2b/\beta + \tau^2/\beta^2 + 2\sigma_n^2 \,. \tag{51}$$

By defining $2k\sigma_n^2$ as $\overline{\sigma}_{c_1}$ expression (51) yields,

$$\text{ave(RCS)} = \overline{\sigma}_c = kE(A^2) = k\overline{A^2} = 2kb/\beta + k\tau^2/\beta^2 + \overline{\sigma}_{c_1}$$

$$= \overline{\sigma}_{c_o} + \overline{\sigma}_{c_1} \tag{52}$$

Using the relationships in (52), expression (41) may be rewritten in terms of σ_c via the substitution $s = K\sigma_c$.

As a matter of interest (48) and (49) will be written in mixture-representation form [recall (10)],

$$h(u)du = \sum_{m=0}^{\infty} P(m;\tau^2/2\beta)g_s(u;\ \beta/2,\ b+m)du, \quad u \geq 0 \tag{53}$$

and

$$\theta_u(t) = \sum_{m=0}^{\infty} P(m;\ \tau^2/2\beta)\cdot(1-2it/\beta)^{-b-m} \tag{54}$$

where again $P(m;a)$ is the discrete Poisson distribution

$$P(m;a) = \frac{a^m}{m!}e^{-a}$$

and $g_s(u;\ \alpha,\ \lambda)$ is Swerling's (gamma) pdf

$$g_s(u;\alpha,\lambda) = \frac{\alpha^\lambda}{\Gamma(\lambda)}u^{\lambda-1}e^{-\alpha u}.$$

An extensive treatment of mixture representations is given in McNolty (1972a). In some cases the mixture representation permits more simplified manipulations than does the corresponding closed form.

For $\tau = 0$ expressions (41) becomes

$$f_2(s)ds = \frac{\beta^b(\sigma_n^2)^{b-1}}{2(1+\beta\sigma_n^2)^b}\cdot\exp(-s/2\sigma_n^2)\,{}_1F_1\left[b;1;\frac{s}{2\sigma_n^2(1+\beta\sigma_n^2)}\right]ds,$$

$$s \geq 0 \tag{55}$$

and letting $\beta = 2\lambda/\overline{A^2}$ and $b = \lambda$ in (55) yields the modified Swerling distribution McNolty (1974a) for fluctuating radar cross section σ_c,

$$g_m(\sigma_c)d\sigma_c = \frac{\lambda^\lambda d\sigma_c}{(\bar{\sigma}_{c_1})^{1-\lambda}(\bar{\sigma}_{c_o}+\lambda\bar{\sigma}_{c_1})^\lambda}\cdot \exp(-\sigma_c/\bar{\sigma}_{c_1}) \cdot$$

$$\cdot {}_1F_1\left[\lambda;1;\frac{\bar{\sigma}_{c_o}\sigma_c}{\bar{\sigma}_{c_1}\cdot(\lambda\bar{\sigma}_{c_1}+\bar{\sigma}_{c_o})}\right], \quad \sigma_c \geq 0 \tag{56}$$

where now expression (52) has become,

$$\bar{\sigma}_c = k\overline{A^2} = k\overline{A_o^2} + 2k\sigma_n^2 = \bar{\sigma}_{c_o} + \bar{\sigma}_{c_1} .$$

When $\sigma_n = 0$ expression (56) readily reduces to

$$g_s(\sigma_c)d\sigma_c = \frac{1}{\Gamma(\lambda)} \cdot \frac{\lambda^\lambda}{\bar{\sigma}_{c_o}}\left(\frac{\sigma_c}{\bar{\sigma}_{c_o}}\right)^{\lambda-1}\cdot \exp(-\lambda\sigma_c/\bar{\sigma}_{c_o})d\sigma_c, \quad \sigma_c \geq 0 \tag{57}$$

and (57) is identical to Swerling's distribution (11), since $\bar{\sigma}_c = \bar{\sigma}_{c_o}$ for $\sigma_n = 0$. Of course, (57) and (11) include Swerling's (1954) exponential and one-dominant-plus-Rayleigh distributions (58 and (59), respectively:

$$h(\sigma_c)d\sigma_c = \frac{1}{\bar{\sigma}_c}\exp(-\sigma_c/\bar{\sigma}_c)d\sigma_c, \quad 0 \leq \sigma_c < \infty \tag{58}$$

$$h(\sigma_c)d\sigma_c = \frac{4\sigma_c}{\bar{\sigma}_c^2} \cdot \exp(-2\sigma_c/\bar{\sigma}_c)d\sigma_c, \quad 0 \leq \sigma_c < \infty \tag{59}$$

where again $\bar{\sigma}_c = E(\sigma_c)$, $\overline{A^2} = K\bar{\sigma}_c$.

6. PULSE INTEGRATION.

6.1. Scan-to-Scan Amplitude Independence.

Expression (10) is the pdf for the sum of N noncoherently

integrated, randomly phased (uniformly distributed) signal-plus-noise pulses all of constant signal amplitude out of a square-law detector. The parameter R_p is the ratio of the maximum single pulse instantaneous signal-power to average noise power out of the matched filter preceding the square-law device. Thus, (10) is specifically the distribution of the quantity $y = \sum_{i=1}^{N} \frac{1}{2} r_i^2$ where r_i is the envelope of the ith signal-plus-noise (narrowband Gaussian) pulse. From we have

$$R_p = 2E_p/N_o = 2A^2\varepsilon_p/N_o = A^2 \tag{60}$$

where in (60): the energy ε_p of the rf quadrature components has been normalized via $2\varepsilon_p/N_o = 1$; = energy of a single pulse = $A^2\varepsilon_p$; $N_o/2$ = two-sided noise power spectrum and A = signal amplitude. The amplitude of all N signal pulses is assumed to remain constant in (10).

Since the characteristic function (cf) for a single pulse out of the square-law detector is

$$\phi_1(t) = (1-it)^{-1}\cdot\exp\left[\frac{itR_p}{2(1-it)}\right] = (1-it)^{-1}\cdot\exp\left[\frac{itA^2}{2(1-it)}\right] \tag{61}$$

it follows that the cf corresponding to (10) is ($R_p = A^2$)

$$\phi_N(t) = \left[\phi_1(t)\right]^N = (1-it)^{-N}\cdot\exp\left[\frac{itNA^2}{2(1-it)}\right] . \tag{62}$$

In expression (41) the simple substitution $s = A^2$ provides the pdf for the fluctuating amplitude A, i.e.,

$$g(A)\,dA = \frac{\beta^b(\sigma_n^2)^{b-1}A\,dA}{(1+\beta\sigma_n^2)^b}\cdot\exp(-A^2/2\sigma_n^2 - \tau^2/2\beta)\cdot\sum_{m=0}^{\infty}\frac{\tau^{2m}(\sigma_n^2)^m}{2^m m!(1+\beta\sigma_n^2)^m}\cdot$$

$$\cdot{}_1F_1\left[b+m;\ 1;\ \frac{A^2}{2\sigma_n^2(1+\beta\sigma_n^2)}\right],\quad A \geq 0 \tag{63}$$

The unconditional characteristic function for the scan-to-scan case is obtained from (62) and (63),

$$\int_0^\infty \phi_N(t|A)g(A)\,dA = \frac{\beta^b[1-it(1+N\sigma_n^2)]^{b-1}}{(1-it)^{N-1}\{\beta-it[N+\beta(1+N\sigma_n^2)]\}^b}$$

$$\cdot\exp\left\{\frac{itN\tau^2}{2\beta\{\beta-it[N+\beta(1+N\sigma_n^2)]\}}\right\} \tag{64}$$

With $\beta = 2\lambda/\overline{A^2}$, $\tau = 0$ and $b = \lambda$ in (64) we obtain the characteristic function for the scan-to-scan pulse train pdf when the fluctuating signal amplitude has a modified Swerling distribution, i.e., let $\sigma_c = kA^2$ in (56),

$$\alpha(t) = \frac{(2\lambda)^\lambda[1-it(1+N\sigma_n^2)]^{\lambda-1}}{(1-it)^{N-1}\{2\lambda-it[2\lambda(1+N\sigma_n^2)+N\overline{A^2}]\}^\lambda} \tag{65}$$

The unconditional pulse train characteristic function (65) corresponds to the case where x_o and y_o in (40) are each distributed according to (30), with Q_o replaced by A, rather than (22).

One can show that the distribution corresponding to (64) is given by

$$f(y) = \frac{\beta^b(\sigma_n^2)^{b-1}\exp(-\tau^2/2\beta)y^{N-1}e^{-y}}{\Gamma(N)(1+\beta\sigma_n^2)^{b-1}[N(1+\beta\sigma_n^2)+\beta]}\cdot\sum_{m=0}^\infty\sum_{n=0}^\infty\frac{(-1)^n\tau^{2m}(\sigma_n^2)^{m-n}\Gamma(1-b-m+n)}{\Gamma(1-b-m)n!m!2^m(1+\beta\sigma_n^2)^m}\cdot$$

$$\cdot\frac{1}{[N(1+\beta\sigma_n^2)+\beta]^n}\cdot{}_1F_1\left\{n+1;\,N;\,\frac{N(1+\beta\sigma_n^2)y}{[N(1+\beta\sigma_n^2)+\beta]}\right\},\quad y\geq 0 \tag{66}$$

Letting $\tau = 0$ in (66) gives

$$f(y) = \frac{\beta^b(\sigma_n^2)^{b-1} y^{N-1} e^{-y}}{\Gamma(N)[N(1+\beta\sigma_n^2)+\beta](1+\beta\sigma_n^2)^{b-1}} \cdot \sum_{n=0}^{\infty} \frac{(-1)^n\Gamma(1-b+n)}{\Gamma(1-b)n![N(1+\beta\sigma_n^2)+\beta]^n(\sigma_n^2)^n} \cdot$$

$$\cdot {}_1F_1\left\{n+1;\ N;\ \frac{N(1+\beta\sigma_n^2)y}{[N(1+\beta\sigma_n^2)+\beta]}\right\},\ y \geq 0 \qquad (67)$$

and (67) is the distribution for the scan-to-scan case when the fluctuating amplitude has a modified Swerling distribution. When $\beta = 2\lambda/\overline{A^2}$ and $b = \lambda$ then (67) corresponds exactly to the characteristic function (65).

The case $\sigma_n = 0$ in (67) is also interesting and yields (see Appendix) the expression,

$$f(y) = \frac{\beta^b y^{N-1} e^{-y}}{\Gamma(N)(N+\beta)^b} \cdot {}_1F_1(b;\ N;\ \frac{Ny}{N+\beta}),\ y \geq 0. \qquad (68)$$

By letting $\beta = 2\lambda/\overline{A^2}$ and $b = \lambda$ in (68) we obtain

$$f(y) = \frac{(2\lambda)^\lambda y^{N-1} e^{-y}}{\Gamma(N)(N\overline{A^2}+2\lambda)^\lambda}\ {}_1F_1(\lambda;N;\ \frac{Ny\overline{A^2}}{N\overline{A^2}+2\lambda}),\ y \geq 0 \qquad (69)$$

and (69) is the scan-to-scan pdf when the fluctuating amplitude is distributed according to Swerling's distribution (8).

The term scan-to-scan denotes that the signal amplitude A is assumed to remain constant for all N pulses of the scan, but may be assumed as statistically independent from scan-to-scan.

6.2 Pulse-to-Pulse Amplitude Independence

In this case the signal amplitude A fluctuates independently from pulse-to-pulse, so that we will first obtain the unconditional characteristic function $\psi_1(t)$ for a single pulse out of the square law detector by letting $N = 1$ in (64). For the sum of N independent pulses the characteristic function is,

$$[\psi_1(t)]^N = \Psi_N(t) = \frac{\beta^{Nb}[1-it(1+\sigma_n^2)]^{N(b-1)}}{\{\beta-it[1+\beta(1+\sigma_n^2)]\}^{Nb}}$$

$$\cdot \exp\left\{\frac{itN\tau^2}{2\beta\{\beta-it[1+\beta(1+\sigma_n^2)]\}}\right\} \tag{71}$$

From McNolty (1973) one can show that the pdf corresponding to (71) is,

$$f(y) = \frac{\beta^{Nb}(1+\sigma_n^2)^{N(b-1)}y^{N-1}}{[1+\beta(1+\sigma_n^2)]^{Nb}} \cdot \exp\left\{-\frac{N\tau^2}{2\beta[1+\beta(1+\sigma_n^2)]}\right\} \cdot \exp\left(-\frac{y}{1+\sigma_n^2}\right)$$

$$\cdot \sum_{m=0}^{\infty} \frac{1}{m!\Gamma(m+N)} \left\{\frac{N\tau^2 y}{2[1+\beta(1+\sigma_n^2)]^2}\right\}^m \cdot {}_1F_1\left\{m+Nb;m+N;\frac{y}{(1+\sigma_n^2)[1+\beta(1+\sigma_n^2)]}\right\} \tag{72}$$

and again (72) may be specialized to familiar pulse train distributions.

In the three cases which follow, the distribution of (40) will be given when the x_o and y_o components are assigned pdfs other than (22) and (23). Here we will not require x_o and y_o to have identically the same distribution and again x_o, y_o and the zero mean Gaussian n(t) are all assumed to be statistically independent.

Case I

The distribution of the x-component is defined as

$$f(x)dx = f(x;a,b;c;\beta,\gamma)dx$$

$$= \frac{a^{\gamma-\beta}(a-b)^{\beta}}{\Gamma(\gamma)} \exp\left(-\frac{c}{a} - ax^2\right) \cdot |x|^{2\gamma-1} \cdot \Phi_3(\beta,\gamma;bx^2,cx^2)dx,$$

$$-\infty < x < \infty \tag{73}$$

where $a > b \geq 0$, $c \geq 0$, $\gamma > 0$, $\beta \geq 0$ and Φ_3 is defined by

$$\Phi_3(\beta;\gamma;\, bu, cu) = \sum_{m=0}^{\infty} \sum_{n=0}^{\infty} \frac{(\beta)_m}{(\gamma)_{m+n} m!n!} b^m c^n u^{m+n} .$$

Similarly, the independent y-component has a pdf of the same form

$$f(y)dy = f(y;a,b;d;\alpha,\lambda)dy, \quad -\infty < y < \infty \tag{74}$$

where $a > b \geq 0$, $d \geq 0$, $\lambda > 0$, $\alpha \geq 0$ and in (73) and (74) only the parameters a and b need be the same.

For this case one can show that the distribution of (40) is (writing u for S_2),

$$g(u)du = a^{\gamma+\lambda-\beta-\alpha}(a-b)^{\beta+\alpha} \cdot \exp\left(-\frac{c+d}{a}\right)$$

$$\cdot \sum_{m=0}^{\infty} \sum_{n=0}^{\infty} \sum_{r=0}^{\infty} \sum_{K=0}^{\infty} \frac{(\beta)_m (\alpha)_r b^{m+r} c^n d^k}{m!n!r!k!} \cdot G(M,a,\sigma_n,u)du, \quad u \geq 0 \tag{75}$$

where

$$G(m,a,\sigma_n,u) = \frac{2^{M-1}\sigma_n^{2M}}{\sigma_n^2(1+2a\sigma_n^2)^M} \cdot \exp(-u/2\sigma_n^2) \cdot {}_1F_1\left[M;1;\frac{u}{2\sigma_n^2(1+2a\sigma_n^2)}\right] \tag{76}$$

and $M = \gamma + \lambda + m + n + r + K$.

In this case the non-uniform phase distribution for $\phi = \arctan \frac{y}{x}$ is

$$p(\phi)d\phi = \sum_{m_1=0}^{\infty} \sum_{n_1=0}^{\infty} \sum_{m_2=0}^{\infty} \sum_{n_2=0}^{\infty} P(n_1;\tfrac{c}{a})P(n_2;\tfrac{d}{a}) \cdot NB(m_1;\beta,1-\tfrac{b}{a}) \cdot$$

$$\cdot NB(m_2;\alpha,1-\tfrac{b}{a}) \frac{\Gamma(\gamma+\lambda+m_1+m_2+n_1+n_2)}{\Gamma(\gamma+m_1+n_1)\Gamma(\lambda+m_2+n_2)}$$

$$\cdot \frac{|\cos\phi|^{2(\gamma+m_1+n_1)} \cdot |\sin\phi|^{2(\lambda+m_2+n_2)}}{|\sin 2\phi|} d\phi \quad 0 \leq \phi \leq 2\pi \tag{77}$$

where again P(m;a) is the discrete Poisson distribution and now NB(m;s,1-x) is the negative binomial distribution,

$$NB(m;s,1-x) = C_{s-1}^{s+m-1} (x)^m (1-x)^s .$$

Case II

The distribution of the x-component is now defined as

$$f(x)dx = f(x;a;b,c;\gamma;\lambda,\tau)dx$$

$$= \frac{(a-\lambda)^b (a-\tau)^c}{\Gamma(\gamma)\, a^{b+c-\gamma}} \cdot \exp(-ax^2)\cdot|x|^{2\gamma-1}\Phi_2(b,c;\gamma;\lambda x^2,\tau x^2)dx,$$

$$-\infty < x < \infty \qquad (78)$$

where $a > \lambda$, $a > \tau$, $a > 0$; $\lambda,\tau \geq 0$; $\gamma > 0$ and $b,\ c \geq 0$,

$$\Phi_2(b,c;\gamma;\lambda z,\tau z) = \sum_{m=0}^{\infty} \sum_{n=0}^{\infty} \frac{(b)_m (c)_n}{(\gamma)_{m+n} m!n!} \lambda^m \tau^n z^{m+n} .$$

The distribution of the independent y-component has the same form as (78); however, three corresponding parameters may be different

$$f(y)dy = f(y;a;B,d;\alpha;\lambda,\tau)dy \qquad (79)$$

with the same restrictions on a, λ, τ where B, $d \geq 0$ and $\alpha > 0$.

In this case the distribution of (40) becomes (again writing u for S_2),

$$g(u)du = \frac{(a-\lambda)^{b+B}\cdot(a-\tau)^{c+d}du}{a^{B+d+b+c-\gamma-\alpha}} \sum_{m=0}^{\infty} \sum_{n=0}^{\infty} \sum_{r=0}^{\infty} \sum_{K=0}^{\infty} \cdot$$

$$\cdot \frac{(b)_m (c)_n (B)_r (d)_K \lambda^{m+r} \tau^{n+K}}{m!n!r!K!} \cdot G(M,a,\sigma_n,u),\ U \geq 0 \qquad (80)$$

where G( ) is the same function shown in (76) above except that now $M = \gamma + \alpha + m + n + r + K$.

The non-uniform phase distribution for $\phi = \arctan \frac{y}{x}$ is now,

$$p(\phi)\,d\phi = \sum_{m_1=0}^{\infty}\sum_{n_1=0}^{\infty}\sum_{m_2=0}^{\infty}\sum_{n_2=0}^{\infty} NB(m_1;b,1-\frac{\lambda}{a})\cdot NB(m_2;B,1-\frac{\lambda}{a})$$

$$\cdot NB(n_1;c,1-\frac{\tau}{a})\cdot NB(n_2;d,1-\frac{\tau}{a})\cdot\frac{\Gamma(m_1+m_2+n_1+n_2+\gamma+\alpha)}{\Gamma(m_1+n_1+\gamma)\Gamma(m_2+n_2+\alpha)\cdot|\sin 2\phi|}$$

$$\cdot|\cos\phi|^{2(\gamma+m_1+n_1)}\cdot|\sin\phi|^{2(\alpha+m_2+n_2)}\cdot d\phi,\quad 0\le\phi\le 2\pi \qquad (81)$$

where again $NB(m;s,1-x)$ is the discrete negative binomial distribution.

Case III

The distribution of the x-component is

$$f(x)\,dx = f(x;a,b;\nu,\mu)\,dx$$

$$= \frac{a^{\nu}b^{\nu}}{\Gamma(\nu+\mu)}\cdot\exp(-bx^2)\cdot|x|^{2\mu+2\nu-1}\cdot{}_1F_1\left[\nu;\nu+\mu;(b-a)x^2\right]dx,$$

$$-\infty < x < \infty \qquad (82)$$

where $b > a > 0$, $\nu > 0$, $\nu + \mu > 0$ (μ may be negative) and again ${}_1F_1(a;c;x)$ is the confluent hypergeometric function.

The y-component pdf has the same form as (82) except that the ν,μ parameters may be different,

$$f(y)\,dy = f(y;a,b;\gamma,\beta)\,dy,\quad -\infty < y < \infty \qquad (83)$$

where $\gamma > 0$, $\gamma + \beta > 0$ (β may be negative) or, alternatively, for $a > b$, $\beta > 0$, $\gamma + \beta > 0$ and γ may be negative.

The distribution of S_2 is now

$$g(u)\,du = a^{\nu+\gamma}\cdot b^{\mu+\beta}\sum_{m=0}^{\infty}\sum_{n=0}^{\infty}\frac{(\nu)_m(\gamma)_n(b-a)^{m+n}}{m!n!}\cdot G(M,b,\sigma_n,u)\,du,\quad u\ge 0 \qquad (84)$$

where $G(\ )$ is the same function defined in (76) and now $M = \nu + \mu + \beta + \gamma + m + n$.

The non-uniform phase distribution for $\phi = \arctan\frac{y}{x}$ is given by,

$$p(\phi)d\phi = \frac{a^{\nu+\gamma}\, d\phi}{\Gamma(\nu)\Gamma(\gamma)b^{\nu+\gamma}} \sum_{i=0}^{\infty}\sum_{j=0}^{\infty} \frac{\Gamma(\nu+i)\Gamma(\gamma+j)(b-a)^{i+j}}{i!j!b^{i+j}B(\mu+\nu+i,\beta+\gamma+j)} \cdot$$

$$\cdot \frac{|\sin\phi|^{2\beta+2\gamma+2j} \cdot |\cos\phi|^{2\mu+2\nu+2i}\, d\phi}{|\sin 2\phi|}$$

$$= (\frac{a}{b})^{\nu+\gamma} \cdot \frac{|\cos\phi|^{2(\beta+\gamma)} \cdot |\sin\phi|^{2(\mu+\nu)}}{|\sin 2\phi|} \sum_{n=0}^{\infty} B(\mu+\nu+n,\beta+\gamma) \cdot C_{\nu-1}^{\nu-n+1} \cdot$$

$$\cdot (1-\frac{a}{b})^n \cos^{2n}\phi \cdot {}_2F_1\left[\gamma,\nu+\mu+\beta+\gamma;\beta+\gamma;(1-\frac{a}{b})\cos^2\phi\right] d\phi, \quad 0 \leq \phi \leq 2\pi. \tag{85}$$

APPENDIX I

In this appendix the authors provide some interesting alternative forms for expressions (12) and (13). Expression (12) may be written as

$$h(\alpha) = M(\alpha)\sum_{K=0}^{\infty}\sum_{m=0}^{\infty} \frac{(-1)^K \varepsilon_K \alpha^{2K+2m}\Gamma(m+a/2)\, {}_2F_1[-m,-m-K;a/2;\gamma^2\sigma_n^4/2\alpha^2]}{m!(m+2K)!\sigma_n^{2m+2K}(b^2-1)^{c/2}} \cdot$$

$$\cdot P_{c-1}^{K}\left(\frac{b}{\sqrt{b^2-1}}\right) \tag{86}$$

$$= \frac{\exp(-\gamma^2/4\beta)}{\sqrt{2\pi}\,\sigma_n\Gamma(a/2)} \sum_{K=0}^{\infty}\sum_{m=0}^{\infty} \frac{\Gamma(K+m+1/2)\Gamma(m+a/2)(-1)^{m+K}\alpha^{2K}}{K!\Gamma(K+1/2)(m!)^2\beta^m\sigma_n^{2m+2K}2^K}$$

$$\cdot {}_1F_1(m+a/2;a/2;\gamma^2/4\beta) \tag{87}$$

$$= \frac{\exp(-\gamma^2/4\beta)}{\pi\sqrt{2}\,\sigma_n} \sum_{m=0}^{\infty} \frac{(m-1/2)!(m+a/2)}{(\beta\sigma_n^2)^m} \lambda_{2m}(m+a/2,1,i\alpha\sqrt{\beta/2})$$

$$\cdot {}_1F_1(m+a/2;a/2;\gamma^2/4\beta) \tag{88}$$

$$= \frac{\exp(-\alpha^2/2\sigma_n^2-\gamma^2/4\beta)}{\sqrt{2\pi}\,\sigma_n\Gamma(a/2)} \sum_{m=0}^{\infty} \frac{(m+a/2-1)!}{(m!)^2(2\beta\sigma_n^2)^m 2^m} \cdot H_{2m}(\alpha/\sigma_n\sqrt{2})$$

$$\cdot {}_1F_1(m+a/2;a/2;\gamma^2/4\beta) \tag{89}$$

where in (86) $P_\mu^\nu(x)$ is the associated Legendre function, ε_K is Neumann's factor (=1 for K=0 and =2 for $K \neq 0$), ${}_2F_1(a,b;c;x)$ is the hypergeometric function, $c = m + a/2 + K$, $b = 1 + 2\beta\sigma_n^2$, and $M(\alpha) = \beta^{a/2}2^{\frac{a}{2}-\frac{1}{2}}\sigma_n^{a-1}\exp(-\alpha^2/2\sigma_n^2 - \gamma^2/4\beta) \cdot [\sqrt{\pi}\Gamma(a/2)]^{-1}$; in (87) ${}_1F_1(a;c;x)$ is the confluent hypergeometric function; in (88) $\lambda_n(\alpha,\beta,x)$ is Rainville's (1960) special polynomial; in (89) $H_n(x)$ is a Hermite polynomial and in (86)-(89), $-\infty < \alpha < \infty$.

Some alternative forms for (13) are:

$$\phi(t) = \gamma^{-a}\exp(-\gamma^2/4\beta - t^2\sigma_n^2/2) \cdot \sum_{m=0}^{\infty} \frac{(\gamma^2+2t^2)^{2m+a/2}}{m!(4\beta\gamma^2)^m}$$

$$\cdot {}_2F_1(m+1,m+a/2;1;-2t^2/\gamma^2) \tag{90}$$

$$= \exp(-t^2\sigma_n^2/2) \sum_{m=0}^{\infty} P(m;\gamma^2/4\beta) \cdot {}_2F_1[-m,-m-a/2+1;1;-2t^2/\gamma^2] \tag{91}$$

$$= \frac{1}{\Gamma(a/2)}\exp(-t^2\sigma_n^2/2-\gamma^2/4\beta) \sum_{m=0}^{\infty} \frac{\Gamma(m+a/2)}{(m!)^2}\left(-\frac{t^2}{2\beta}\right)^m \cdot$$

$$\cdot {}_2F_1(-m,-m;a/2;-\gamma^2/2t^2) \tag{92}$$

where in (90), (91) and (92) ${}_2F_1(a,b;c;x)$ is again the hypergeometric function; and, in (91) $P(m;a)$ is the discrete Poisson Distribution.

APPENDIX II

Alternative forms for expression (14) are:

$$h_1(\alpha) = \frac{1}{\sqrt{2\pi}\,\sigma_n\Gamma(a/2)}\cdot\exp(-\alpha^2/4\sigma_n^2)\sum_{m=0}^{\infty}\frac{(m+a/2-1)!}{(m!)^2(2\beta\sigma_n^2)^m}\cdot D_{2m}(\alpha/\sigma_n) \quad (93)$$

$$= \frac{1}{\pi\sqrt{2}\,\sigma_n\Gamma(a/2)}\sum_{K=0}^{\infty}\frac{(a/2-1)!(K-1/2)!}{(2K)!}(-2\alpha^2/\sigma_n^2)$$

$$\cdot {}_2F_1(a/2,K+1/2;1;-\frac{1}{\beta\sigma_n^2}) \quad (94)$$

$$= \frac{\exp(-\alpha^2/2\sigma_n^2)}{\sqrt{2\pi}\,\sigma_n\Gamma(a/2)}\sum_{m=0}^{\infty}\frac{(m+a/2-1)!}{(m!)^2(2\beta\sigma_n^2)^m 2^m}\cdot H_{2m}(\frac{\alpha}{\sigma_n\sqrt{2}}) \quad (95)$$

$$= \frac{1}{\pi\sqrt{2}\,\sigma_n}\sum_{m=0}^{\infty}\frac{(m-1/2)!}{(\beta\sigma_n^2)^m}(m+a/2)\cdot\lambda_{2m}(m+a/2,1,i\alpha\sqrt{\beta/2}) \quad (96)$$

$$= \frac{1}{\pi\sqrt{2}\,\sigma_n\Gamma(a/2)}\sum_{m=0}^{\infty}\frac{(m+a/2-1)!(-1)^m(m-1/2)!}{(m!)^2\beta^m\sigma_n^{2m}}\cdot {}_1F_1(m+\tfrac{1}{2};\tfrac{1}{2};-\alpha^2/2\sigma_n^2) \quad (97)$$

$$= G(\alpha)\sum_{K=0}^{\infty}\sum_{m=0}^{\infty}(-1)^K\varepsilon_K x^{2K}\frac{\Gamma(m+a/2)z^m}{m!(m+2K)!\sigma_n^{2m}}\cdot P^K_{a/2+m+K-1}(t) \quad (98)$$

where in (98) $t = b/(b^2-1)^{1/2}$, $x = \alpha/\sigma_n(b^2-1)^{1/4}$, $z = \alpha^2/(b^2-1)^{1/2}$, $G(\alpha) = 2^{(a/2)-(1/2)}\sigma_n^{a-1}\beta^{a/2}\cdot\exp(-\alpha^2/2\sigma_n^2)\cdot[\sqrt{\pi}\,\Gamma(a/2)(b^2-1)^{a/4}]^{-1}$, and $b = 1+2\beta\sigma_n^2$ while all other special functions in (93)-(98) have previously been defined. A comprehensive discussion of these functions is given in Luke (1962), (1969a), (1969b).

APPENDIX III

This appendix provides some distributions and characteristic functions which are related to expressions (73), (78) and (82).

Expression (73).

The pdf for the random variable, $u = x^2$ is

$$h(u;a,b;c;\beta,\gamma)du = \sum_{m=0}^{\infty} \sum_{n=0}^{\infty} P(n;\frac{c}{a}) \cdot NB(m;\beta,1-\frac{b}{a}) \cdot g_s(u;a,\gamma+m+n)du,$$

$$u \geq 0 \qquad (99)$$

where again $P(m;a)$ and $NB(m;s,1-x)$ are the discrete Poisson and negative binomial distributions, respectively, and $g_s(u;\alpha,\tau)$ is the gamma distribution defined in (34a).

The characteristic function corresponding to (99) is

$$\theta(t) = \frac{(a-b)^{\beta}(a-it)^{\beta-\gamma}}{a^{\beta-\gamma}(a-b-it)^{\beta}} \cdot \exp\left[\frac{itc}{a(a-it)}\right] . \qquad (100)$$

Expression (78).

The distribution of the random variable $u = x^2$ is

$$h(u;a;b,c;\gamma;\lambda,\tau)du = \sum_{m=0}^{\infty} \sum_{n=0}^{\infty} NB(m;b,1-\frac{\lambda}{a}) \cdot NB(n;c,1-\frac{\tau}{a}) \cdot$$

$$\cdot g_s(u;a,m+n+\gamma)du, \quad u \geq 0 \qquad (101)$$

and the characteristic function for (101) is

$$\theta(t) = \frac{(a-\lambda)^{b}(a-\tau)^{c}(a-it)^{b+c-\gamma}}{(a-it-\lambda)^{b}(a-it-\tau)^{c}a^{b+c-\gamma}} . \qquad (102)$$

Expression (82).

The distribution of $u = x^2$ is

$$h(u;a,b;\nu,\mu)du = \sum_{n=0}^{\infty} NB(n;\nu,\frac{a}{b}) \cdot g_s(u;b,\mu+\nu+n)du, \quad u \geq 0 \qquad (103)$$

and the characteristic function corresponding to (103) is given by

$$\theta(t) = \frac{a^{\nu}b^{\mu}}{(a-it)^{\nu}(b-it)^{\mu}} = \sum_{n=0}^{\infty} NB(n;\nu,a/b) \cdot (1-it/b)^{-n-\mu-\nu} \qquad (104)$$

APPENDIX IV.

The purpose of this appendix is to show that (68) is the limit of (67) as $\sigma_n \to 0$.

$$\lim_{\sigma_n \to 0} (67) = \lim_{\sigma_n \to 0} \frac{\beta^b (\sigma_n^2)^{b-1} y^{N-1} e^{-y}}{\Gamma(N)(N+\beta)\Gamma(1-b)} \sum_{m=0}^{\infty} \frac{(-1)^m \Gamma(1-b+m)}{m!(N+\beta)^m (\sigma_n^2)^m} \cdot$$

$$\cdot {}_1F_1 \left[m+1; N; \frac{Ny}{N+\beta} \right]$$

$$= \lim_{\sigma_n \to 0} \frac{\beta^b y^{N-1} e^{-y}}{\Gamma(N)(N+\beta)\Gamma(1-b)} \sigma^{2b-2} \cdot \sum_{K=0}^{\infty} \sum_{m=0}^{\infty} \left(\frac{Ny}{N+\beta}\right)^K \frac{(m-b)!(-1)^m (m+K)!}{K!(N)_K (m!)^2 (N+\beta)^m \sigma^{2m}}$$

$$= \lim_{\sigma_n \to 0} \frac{\beta^b y^{N-1} e^{-y}}{\Gamma(N)(N+\beta)\Gamma(1-b)} \sigma^{2b-2} \sum_{K=0}^{\infty} \frac{(-b)!}{(N)_K} \left(\frac{Ny}{N+\beta}\right)^K \cdot {}_2F_1 \left[1-b, K+1; 1; \frac{-1}{(N+\beta)\sigma_n^2} \right]$$

$$= \frac{\beta^b y^{N-1} e^{-y}}{\Gamma(N)(N+\beta)\Gamma(1-b)} \cdot \frac{(-b)!}{(N+\beta)^{b-1}} \cdot {}_1F_1\left(b; N; \frac{Ny}{N+\beta}\right) = (68).$$

REFERENCES

[1] Bennett, W. R. (1944). J. Acoust. Soc. of Amer. 15, 164-172.

[2] Huynen, J. R. (1970). Phenomenological Theory of Radar Targets. Drukkerij Bronder-Offset N.V., Rotterdam.

[3] Johnson, N. L. and Kotz, S. (1970). Continuous Univariate Distributions, Vol. I. John Wiley and Sons, New York.

[4] Johnson, N. L. and Kotz, S. (1970). Continuous Univariate Distributions, Vol. II. John Wiley and Sons, New York.

[5] Johnson, N. L. and Kotz, S. (1972). Continuous Multivariate Distributions, Vol. III. John Wiley and Sons, New York.

[6] Luke, Y. L. (1962). Integrals of Bessel Functions. McGraw-Hill Book Co., Inc., New York.

[7] Luke, Y. L. (1969a). The Special Functions and Their Approximations, Vol. I. Academic Press, New York.

[8] Luke, Y. L. (1969b). The Special Functions and Their Approximations, Vol. II. Academic Press, New York.

[9] Marcum, J. I. (1960). IRE Trans. on Infor. Theory. IT-6, 59-267.

[10] McNolty, F. (1962). Ann. Math. Statist. 33, 796-801.
[11] McNolty, F. (1968). J. Oper. Res. Soc. 16, 1027-1040.
[12] McNolty, F. and Tomsky, J. (1972a). Sankhȳa Ser. B, 34, 251-264.
[13] McNolty, F. (1972b). Sankhȳa, 34 Ser. B, 21-26.
[14] McNolty, F. (1973). Math. Comput. 27, 495-504.
[15] McNolty, F. and Hansen, E. (1974a). IEEE Trans. on Aerospace and Electronic Systems AES-10, 281-285.
[16] McNolty, F.; Huynen, J. R. and Hansen, E. (1974b). Component distributions for fluctuating radar targets. Submitted for publication.
[17] Nakagami, M. (1960). In Statistical Methods in Radio Wave Propagation. Pergamon Press, New York, 3-36.
[18] Rainville, E. D. (1960). Special Functions. The MacMillan Co., New York.
[19]. Rice, S. O. (1944). Bell System Tech. J. 23, 282-332.
[20] Rice, S. O. (1945). Bell System Tech. J. 24, 46-156.
[21] Swerling, P. (1954). Probability of Detection for Fluctuating Targets. Rand Corp. Research Memo, RM-1217. Rand Corp., Santa Monica, Ca.
[22] Swerling, P. (1957). IRE Trans. on Infor. Theory, IT-3, 175-178.
[23] Swerling, P. (1960). IRE Trans. on Infor. Theory IT-6, 269-308.

TAILWEIGHT, STATISTICAL INFERENCE AND FAMILIES OF DISTRIBUTIONS - A BRIEF SURVEY

Thomas P. Hettmansperger and Michael A. Keenan

Department of Statistics, The Pennsylvania State University, University Park, Pa., U.S.A.

SUMMARY. There are surprisingly many concepts and definitions of tailweight. For symmetric distributions F and G the main results are based on the convexity or starshapedness of $G^{-1}F$. This has essentially replaced kurtosis as measure of tailweight. Recently in research on the location problem there has been interest in families of distributions ordered by tailweight and in statistical procedures for assessing tailweight. This problem of assessing tailweight has long been a problem of interest to researchers developing life models. There is a rich variety of results for testing the exponential family against an alternative with lighter tails. Hopefully, there will soon be an equally rich variety of results for testing the tails of a symmetric distribution in the location model.

KEY WORDS. Tailweight, symmetric distributions, life models, convex and starshaped functions.

1. INTRODUCTION. The possibility of ordering distribution functions according to the weight contained in the tails of the distribution is of considerable interest to researchers studying robustness properties of statistical procedures and to model builders attempting to select the appropriate underlying distribution. In the latter case researchers, selecting a life model, have been concerned about tailweight for some time. In the former case robustness studies have used families of distributions ordered by tailweight in power and efficiency studies and more recently, in adaptive inference, research has concentrated on allowing the sample to select the underlying distribution model from a specified family.

G. P. Patil et al. (eds.), Statistical Distributions in Scientific Work, Vol. 1, 161-172.

In surveying the literature we find that each author has developed his own ordering of distributions to fit his particular needs. There are, for example, six definitions for families of symmetric distributions alone. In this survey we will try to present the material along the following lines: definitions, examples, relationships among the various orderings, results relating the ordering to properties of statistical procedures, statistical inference for the selection of a distribution from an ordered family, and some open questions.

There are two main headings: symmetric distributions and life distributions. We will give no proofs here since in most cases they can be found in the references. We do not claim this survey to be exhaustive and apologize to those researchers who have contributed material that we have missed.

2. FAMILIES OF SYMMETRIC DISTRIBUTIONS.

One of the earliest partial orderings of symmetric, continuous distributions is given in:

Definition 2.1. van Zwet (1964, p. 65). F is lighter tailed than G, denoted $F \underset{s}{<} G$, if $G^{-1}F(x)$ is convex for $x > x_0$, x in the support of F, and x_0 the point of symmetry of F.

Without loss of generality we will suppose the distributions are all centered at 0. Let $G^{-1}F(x) = \phi(x)$, convex. Then since ϕ is non-decreasing, for $x > 0$, $\phi(x) \geq x$ and $G(\phi(x)) = F(x)$. Hence $G(x) \leq F(x)$ and G has more probability in the tail than F.

Example 2.1. van Zwet (1964). The following distributions have been used by many researchers in the investigation of the efficiency properties of tests.

Uniform $\underset{s}{<}$ normal $\underset{s}{<}$ logistic $\underset{s}{<}$ double exponential $\underset{s}{<}$ Cauchy

Typically the uniform is referred to as "light" tailed, the normal and logistic as "medium" tailed and the double exponential and Cauchy as "heavy" tailed.

Equivalent statements of this ordering are given in the proposition below and will be helpful in the later discussion.

Proposition 2.1. van Zwet (1964, p. 66). $F \underset{s}{<} G$ if there is a non-decreasing, non-constant function $a(u)$, $\frac{1}{2} < u < 1$, such that $[G^{-1}(u)]' = a(u)[F^{-1}(u)]'$. This can also be written $f(F^{-1}(u)) = a(u)\, g(G^{-1}(u))$.

A weaker form of partial ordering is given in:

Definition 2.2. Hajek (1969, p. 150). F is lighter tailed than G, denoted $F \underset{h}{<} G$, if there is a non-decreasing, non-constant function $a(u)$, $1/2 < u < 1$, such that $F^{-1}(u) = a(u)\, G^{-1}(u)$.

Hence Hajek places a condition on F^{-1} rather than fF^{-1}. In his proof of Theorem 34A, p. 150, he establishes:

Proposition 2.2. Hajek (1969, p. 150). If $F \underset{s}{<} G$ then $F \underset{h}{<} G$.

Hence the examples for $\underset{s}{<}$ also apply to $\underset{h}{<}$. In fact Hajek's partial ordering is the weakest considered here and hence results that require $\underset{h}{<}$ will then hold for the other partial orderings. In Proposition 2.9 we point out that Hajek's condition is essentially equivalent to $G^{-1}F$ being starshaped rather than convex.

Hajek (1969) and Gastwirth (1969, 1970) relate the concepts of partial ordering of families of distributions to the weight function which defines the asymptotically most powerful rank test (amprt). If F is the underlying distribution the amprt is defined by the weight function:

$$\phi_F(u) = -f'(F^{-1}(u))/f(F^{-1}(u)).$$

Hajek places conditions on ϕ_F and ϕ_G to establish $F \underset{h}{<} G$. Gastwirth, on the other hand, suggests these conditions on ϕ_F and ϕ_G as defining partial orderings on F and G. Gastwirth (1970) also suggests conditions on ϕ_F' and ϕ_G' which define a further partial ordering. Note that these conditions on ϕ and ϕ' are essentially defined on $f'F^{-1}$ and $f''F^{-1}$ and will induce stronger orderings than the two previous definitions.

Definition 2.3. Hajek (1969, p. 150). ϕ_F increases more rapidly than ϕ_G, denoted ϕ_F imr ϕ_G, if there is a non-decreasing, non-constant function $a(u)$, $1/2 < u < 1$, such that $\phi_F(u) = a(u)\, \phi_G(u)$.

Definition 2.4. Gastwirth (1969). F is lighter tailed than G, denoted $F \underset{g}{<} G$, if ϕ_F imr ϕ_G.

Proposition 2.3. Hajek (1969, p. 150). $F \underset{g}{<} G$ implies $F \underset{s}{<} G$ in turn implies $F \underset{h}{<} G$.

Definition 2.5. Gastwirth (1970). ϕ_F emphasizes the middle less than ϕ_G, denoted ϕ_F emℓ ϕ_G, if there is a non-decreasing, non-constant, function a(u), 1/2 < u < 1, such that $\phi_F'(u) = a(u)\, \phi_G'(u)$.

Definition 2.6. F is lighter tailed than G, denoted $F \underset{k}{<} G$, if ϕ_F emℓ ϕ_G. An equivalent form of this definition is $F \underset{k}{<} G$ if $\phi_G^{-1}\phi_F(u)$ is concave for $1/2 < u < 1$.

Proposition 2.4. $F \underset{k}{<} G$ implies $F \underset{g}{<} G$ implies $F \underset{s}{<} G$ implies $F \underset{h}{<} G$.

It might be noted again that the above implication reflects the fact that conditions are placed on $f''F^{-1}$, $f'F^{-1}$, fF^{-1} and F^{-1}, respectively. A further insight on the relation between the amprt score function and the distribution generating it is provided by Puri (1975). Although he shows under certain conditions a one to one correspondence between the distribution and the score function generating the locally most powerful rank test the result will probably hold for amprts.

Example 2.2. Gastwirth (1969). By constructing a censored family from a symmetric, strongly unimodal density and displaying the weight function of the amprt, Gastwirth shows the family is ordered under $\underset{s}{<}$ but not $\underset{g}{<}$. In general the implications in Proposition 2.4 cannot be reversed.

Example 2.3. Gastwirth (1969). Consider the family of distributions

$$f_n(x) = \frac{1}{2}\,\frac{n-1}{n}\,\frac{1}{\left(1 + \frac{|x|}{n}\right)^n},\quad n = 2, 3, \ldots;\quad f_\infty(x) = \frac{1}{2}\, e^{-|x|}.$$

The ϕ function generating the amprt is given by

$$\phi_n(u) = \begin{cases} \left(\frac{n+1}{n-1}\right)^{\frac{1}{2}} [2(1-u)]^{\frac{1}{n-1}} & u > \frac{1}{2} \\ -\left(\frac{n+1}{n-1}\right)^{\frac{1}{2}} [2u]^{\frac{1}{n-1}} . & u < \frac{1}{2} . \end{cases}$$

For $n' > n$ we have $F_{n'} \underset{g}{<} F_n$. In fact $F_{n'} \underset{k}{<} F_n$ so the family is ordered by all partial orders considered so far.

Example 2.4. Gastwirth (1969). Consider the t family

$$f_k(x) = \frac{1}{\sqrt{k\pi}} \frac{\Gamma(\frac{1}{2}k + \frac{1}{2})}{\Gamma(\frac{1}{2}k)} (1 + \frac{x^2}{k})^{-\frac{(k+1)}{2}} \quad , k = 1,2,\ldots;$$

$$f_\infty(x) = (2\pi)^{-1/2} \exp(-x^2/2).$$

This family ranges from the heavy-tailed Cauchy distribution to the normal distribution. Intuitively, we would expect that for $k > k'$, $F_{k'} \underset{g}{<} F_k$ but this has not been proved. (See Example 2.7)

Example 2.5. Mielke (1972). For $r > -1/2$ let f_r be the density characterized by $f_r(x(p)) = (1 - |2p-1|^{r+1})/2(r+1)$ where $F(x(p)) = p$, $0 < p < 1$. If $r = 0$ or 1, F is the double exponential or logistic distribution. The amprt for this family is defined by

$$\varphi_r(u) = \begin{cases} |u - 1/2|^r & \text{if } u > 1/2 \\ -|u - 1/2|^r & \text{if } u < 1/2 \end{cases}$$

For $u > 1/2$ and $r < s$ we have $\varphi_r^{-1} \varphi_s(u) = (u - 1/2)^{r/s} + 1/2$ which is convex. Hence if $r < s$, $F_s \underset{k}{<} F_r$.

The partial orderings considered give rise to results on the Pitman asymptotic relative efficiency of tests.

Proposition 2.5. van Zwet (1964, p. 104). Let W and N denote the Wilcoxon and normal scores tests and $e(W,N|F)$ the Pitman efficiency of W with respect to N when sampling from F. If $F \underset{s}{<} G$ then $e(W,N|F) \leq e(W,N|G)$.

This result is generalized in

Proposition 2.6. Gastwirth (1970). Suppose ϕ_1, ϕ_2 are certain types of amprts. If ϕ_1 emℓ ϕ_2, and $F \underset{s}{<} G$ then $e(\phi_2, \phi_1|F) \leq e(\phi_2, \phi_1|G)$.

Intuitively this says that as the tails of the underlying distribution become heavier, tests which place less emphasis on the middle ranks, become less efficient. Gastwirth (1969) states another version of this proposition which actually only uses $\underset{g}{<}$ of Definition 2.4.

Proposition 2.7. Gastwirth (1969). If ϕ_1 imr ϕ_2, and $F \underset{g}{<} H$ then $e(\phi_2, \phi_1|F) \leq e(\phi_2, \phi_1|G)$.

We next consider the relationship between the power of monotone rank tests [see Doksum (1969)] and partial orderings.

Definition 2.7. Doksum (1969). F is lighter tailed than G, denoted $F \leq_t G$, if $G^{-1}F(x) - x$ is increasing for $x \in I$, the support of F.

Doksum then proves that when monotone rank tests are used in testing for location the probabilities of type I and II error increase as the tails of the underlying distribution become heavier according to Definition 2.7.

Definition 2.8. Lawrence (1966). F is lighter tailed than G, denoted $F \leq_r G$, if $G^{-1}F(x)$ is starshaped on $\{x: 1/2 < F(x) < 1\}$ and $G^{-1}F(-x)$ is starshaped on $\{x: 0 < F(-x) < 1/2\}$. Note that h is starshaped if $h(\lambda x) \leq \lambda h(x)$, $0 \leq \lambda \leq 1$.

Proposition 2.8. Doksum (1969). If f,g are continuous at 0 ang $g(0) \leq f(0)$ then $F \leq_r G$ implies $F \leq_t G$.

Proposition 2.9. Doksum (1969). If $f(0), g(0) > 0$, $F \leq_h G$ iff $F \leq_r G$.

Hence we have a characterization of Hajek's ordering in terms of starshapedness which corresponds to the equivalences for van Zwet's ordering pointed out in Proposition 2.1.

Doksum also relates the orderings $\leq_t$ and $\leq_r$ to the variances of F and G and to $P(a \leq X \leq b)$ and $P(a \leq Y \leq b)$.

Finally, Doksum shows that as the tails of the underlying distribution get heavier according to Definition 2.8 the probabilities of type I and II errors of a monotone rank test of scale increase.

It was recently pointed out by Ali (1974) that for symmetric distributions kurtosis is a measure of tailedness. He further discusses the relationship between stochastic ordering and kurtosis. Kurtosis is more than a partial ordering since it provides a numerical value for tail weight. A formal connection between kurtosis and tail weight is given by:

Proposition 2.10. van Zwet (1964, p. 20). If $F <_s G$ then $K(F) \leq K(G)$ where $K(\cdot)$ is the kurtosis. Two examples of families ordered by kurtosis are:

Example 2.6. Hogg (1972). Take $f(x; \theta) = \{2\Gamma[(\theta+1)/\theta]\}^{-1}$ $\exp(-|x|^{\theta})$ for $-\infty < x < \infty$. Now $f(x; 1)$ is a double exponential distribution, $f(x; 2)$ is a normal distribution and the limiting distribution as $\theta \to \infty$ is uniform. In this case the kurtosis $E(x^4)/[E(x^2)]^2 = \Gamma(\frac{5}{\theta})\,\Gamma(\frac{1}{\theta})/[\Gamma(\frac{3}{\theta})]^2 = K(\theta)$. For $\theta \geq 1$, $K(\theta)$ is decreasing and orders the family. This family has played an important role in Hogg's work on adaptive inference, first in adaptive estimation of the center of a symmetric distribution and later in the construction of a statistic to assess tail weight (see Example 2.9.).

Example 2.7. The kurtosis of the t-distribution with k degrees of freedom given in Example 2.4 is $K(k) = 3 + b/(k-4)$. Hence for $k \geq 5$, $K(k)$ decreases and orders the family.

When making statistical inferences about a location parameter one strategy is to find a procedure that is relatively insensitive to changes in the tailweight of the underlying distribution. The search for such procedures has spawned a large body of literature in robustness.

Recently interest has centered on the problem of adaptive inference. In this case a family of underlying distribution containing a range from light or medium to heavy tails is assumed and the sample indicates which the appropriate model is. Once the model has been selected optimal rank methods or some other techniques appropriate to that model are used. Hence the initial inference problem is to use the observations to assess the tail weight.

We will give three examples of this approach without detail. The pertinent references contain full discussions of these methods.

Example 2.8. Jaeckel (1970). For estimating the center of symmetry he considers the family of α trimmed means. For $0 < \alpha_0 < \alpha < \alpha_1 < 1/2$ he determines that α which minimizes an estimate of the asymptotic variance of the trimmed mean. The underlying idea is that the heavier the tails of the distribution samples the more we should trim.

Example 2.9. Randles, Ramberg and Hogg (1973) and Randles and Hogg (1973). Statistical procedures for light, medium and heavy tailed distributions are suggested. A statistic Q = (one half the range)/(mean absolute deviation from the median) is used to classify the tails of the underlying distribution. Here Q is the ratio of scale estimates from a uniform (light tail) and a double exponential (heavy tail) distribution and is derived from a scale invariant test for H_0: uniform against H_1: double exponential. The statistic Q appears to be superior to the sample kurtosis in assessing the weight in the tails.

Example 2.10. Policello and Hettmansperger (1974). A family of rank scores is considered which consist of integer and constant scores. For light tailed distributions integer scores are assigned to extreme ranks and constants to the interior ranks, and just the opposite for heavy tailed distributions. By maximizing an estimate of the Pitman efficacy the sample determines the scores to be used. The motivation for this approach is given by Proposition 2.6 above.

Techniques for drawing inferences about the tail weight of the underlying model have been much more extensively developed and studied in the life testing literature. We now turn to a discussion of some of these methods and some further definitions of tail weight.

3. LIFE MODELS. By restricting our attention to distribution satisfying $F(0) = 0$, two of the partial orderings previously defined give rise to two important classes of life model distributions. Without the symmetry assumption we say F has shorter tails than G, $F \underset{c}{<} G$, if $G^{-1}F$ is convex on the support of F. [van Zwet (1969)]. If G is the exponential distribution then we have from [2]

1) $F \underset{c}{<} G$ if and only if F has an increasing failure rate (IFR); i.e., $\frac{f(t)}{1 - F(t)}$ is an increasing function,

2) $F \underset{r}{<} G$ if and only if F has an increasing failure rate average (IFRA); i.e., $\frac{1}{t} \int_0^t \frac{f(u)}{1 - F(u)} du$ is an increasing function.

Example 3.1. Consider the Weibull distributions $F_\theta(t) = 1 - e^{-\lambda t^\theta}$, $t \geq 0$, $\lambda \geq 0$. Now $G^{-1}F_\theta(t) = t^\theta$ which is convex for $\theta \geq 1$ so $F_\theta \underset{c}{<} G$ if $\theta \geq 1$ and thus $F_\theta(t)$ has IFR for $\theta \geq 1$. Here G is the exponential.

Example 3.2. Consider the distribution $F(x) = (1 - e^{-x})(1 - e^{-kx})$, $x \geq 0$, $k > 1$. F has IFRA but not IFR.

Three tests have been proposed for testing

$$H_0: F(x) = 1 - e^{-\lambda x} \qquad x \geq 0, \lambda > 0$$

against H_1: $F(x)$ has IFR and is not exponential.

Define the normalized spacing $S_i = (n - i + 1)(X_{(i)} - X_{(i-1)})$ where $X_{(1)} \leq \cdots \leq X_{(n)}$ with $X_{(0)}$ defined to be 0.

i) Proschan and Pyke (1967). Reject H_0 in favor of H_1 if

$$V_n = \sum_{i<j}^{n} V_{ij} \text{ is large,}$$

$$\text{where } V_{ij} = \begin{cases} 1 \text{ if } S_i \geq S_j \\ 0 \text{ otherwise} \end{cases}$$

ii) Gastwirth (1969). Reject H_0 if

$$T_n = \sum_{i=1}^{n-1} \sum_{j=1}^{i} S_j / \sum_{k=1}^{n} S_k \text{ is large.}$$

iii) Bickel and Doksum (1969). Reject H_0 if

$$W_n = \sum_{i=1}^{n} i \ell n[1 - \frac{R_i}{n+1}] \text{ is large}$$

where R_i is the rank of S_i among $S_{(1)} \leq \cdots \leq S_{(n)}$.

Hollander and Proschan (1972) have made

Definition 3.1. A life distribution is new better than used (NBU) if $\overline{F}(x+y) \leq \overline{F}(x) \cdot \overline{F}(y)$, for all $x,y \geq 0$ where $F(z) = 1 - F(z)$.

In fact, from the following we see that NBU distributions arise naturally from another partial ordering. Note that every IFR distribution is an NBU distribution.

Definition 3.2. F is superadditive with respect to G if $G^{-1}F$ is superadditive, i.e., $G^{-1}F(x_1 + x_2) \geq G^{-1}F(x_1) + G^{-1}F(x_2)$, $x_1, x_2 \geq 0$.

Definition 3.3. We define $F \underset{a}{<} G$ if F is superadditive with respect to G.

Proposition 3.1. $\underset{a}{<}$ is a partial order.

Proposition 3.2. F is NBU if and only if $\bar{F} \underset{a}{<} \bar{G}$ where G is exponential distribution.

A test of

$$H_0: F(x) = 1 - e^{-\lambda x} \qquad x \geq 0, \lambda > 0$$

against H_1: F(x) is NBU (not exponential) is given by:

iv) Hollander and Proschan (1972). Reject H_0 for small values of $J_n = 2[n(n-1)(n-2)]^{-1} \Sigma' \chi(X_i, X_j + X_k)$

where $\chi(x,y) = \begin{matrix} 1 & \text{if } x > y \\ 0 & \text{if } x \leq y \end{matrix}$

and Σ' is over all n(n-1)(n-2)/2 triples (i, j, k) such that $1 \leq i, j, k \leq n$ $i \neq j$, $i \neq k$, $j < k$.

Bickel and Doksum (1969) show the efficiency of Vn with respect to Wn is 3/4 for all alternatives and hence Vn is asymptotically inadmissible. It would be interesting to know if J_n is inadmissible in a similar sense.

Another measure of tail weight is the following:

Definition 3.4. Bryson (1974). The conditional mean exceedance for a random variable X with distribution F(x) is defined as

$$CME_x = E(X-x|X \geq x) = \frac{1}{\bar{F}(x)} \int_x^\infty \bar{F}(t)\, dt.$$

Definition 3.5. A distribution is called heavy-tailed if CME_x is an increasing function of x and light-tailed if CME_x is decreasing. If CME_x is constant then F(x) must be the exponential distribution function.

Example 3.3. Bryson (1974). Consider the Pareto distribution given by

$$\bar{F}(x) = \left(\frac{a}{a+bx}\right)^{1+\frac{1}{b}} \qquad x \geq 0,\ a,b > 0.$$

This distribution has increasing CME_x; in fact $CME_x = a + bx$. However, it is easy to see that $r(x) = (1 + b)/(a + bx)$ is the failure rate for the Pareto and thus it has decreasing failure rate. Also $G^{-1}F(x)$ is concave on $[0, \infty)$ where G is the exponential distribution and so $G \underset{c}{<} F$. But this implies that G has lighter tails than the Pareto.

Bryson then constructs a test of H_0: CME_x is constant against H_1: $CME_x = a + bx$. The procedure is given by

v) Bryson (1974). Reject H_0 if $T = \bar{x}\, X_{(n)}/(n-1)\bar{x}_{GA}^{\,2}$ is too large where $\bar{x}_{GA} = \{\pi(x_i + \hat{A})\}^{1/n}$ and $\hat{A} = X_{(n)}/(n-1)$.

ACKNOWLEDGMENTS. The authors wish to thank Professor G. P. Patil for his initial suggestion to write the paper and for his continuous interest.

REFERENCES

[1] Ali, M. M. (1974). Jour. Amer. Statist. Assoc. 69, 543-545.
[2] Barlow, R.E. and Frank Proschan (1966). Ann. Math. Statist. 37, 1574-1591.
[3] Bickel, P. J. and K. Doksum (1969). Ann. Math. Statist. 40, 1216-1235.
[4] Bryson, M. C. (1974). Technometrics 16, 61-68.
[5] Doksum, K. (1969). Ann. Math. Statist. 40, 1167-1176.
[6] Gastwirth, J. L. (1969). Non-Parametric Techniques in Statistical Inference, M. L. Puri (ed.). Cambridge Univ. Press, Cambridge, England, 89-109.
[7] Gastwirth, J. L. (1970). Jour. Royal Statist. Soc. (B) 32, 227-232.
[8] Hajek, J. (1969). A Course in Nonparametric Statistics. Holden-Day, San Francisco.
[9] Hollander, M. and F. Proschan (1972). Ann. Math. Statist. 43, 1136-1146.
[10] Hogg, R. V. (1972). Jour. Amer. Statist. Assoc. 67, 422-424.
[11] Jaeckel, L. A. (1971). Ann. Math. Statist. 42, 1540-1552.
[12] Lawrence, M. J. (1966). Inequalities and Tolerance Limits for s-ordered Distributions. Technical Report ORC-66-37, Operations Research Center, University of California, Berkeley.
[13] Mielke, P. W. (1972). Jour. Amer. Statist. Assoc. 67, 850-854.
[14] Proschan, F. and R. Pyke (1967). Tests for Monotone Failure Rate. Proc. Fifth Berkeley Symp. Math. Stat. Prob. 3, 293-312.
[15] Policello, G. E. II and T. P. Hettmansperger (1974). Adaptive Robust Procedures for the One Sample Location Problem. Submitted to Jour. Amer. Statist. Assoc.
[16] Puri, P. S. (1975). In Statistical Distributions in Scientific Work, Vol. III, Characterizations and Applications, Patil, Kotz, and Ord (eds.). D. Reidel, Dordrecht and Boston, pp. 103-112.

[17] Randles, R. H., Ramberg, J. S., and Hogg, R. V. (1973). Technometrics 15, 769-778.
[18] Randles, R. H. and Hogg, R. V. (1973). Comm. in Statist. 2, 337-356.
[19] van Zwet, W. R. (1964). Convex Transformations of Random Variables. Math. Centre, Amsterdam.

THE FAMILIES WITH A "UNIVERSAL" LOCATION ESTIMATOR

A. L. Rukhin

Mathematical Institute, Leningrad, U.S.S.R.

SUMMARY. The problem of estimating the multidimensional location parameter is considered. The new family of multivariate distribution is characterized by the property of the independence of the best equivariant estimator on the even loss function choice. Some generalizations of known functional equations are introduced and solved for this aim.

KEY WORDS. Multidimensional location parameter, loss function, equivariant estimators, completely symmetrical families, D'Alembert's functional equation.

1. INTRODUCTION. Let P be some absolutely continuous probability measure on σ-algebra $\mathcal{B}$ of Borelian Sets of the Euclidean space R^m. We shall deal with the problem of estimating the location parameter θ on the base of a random sample $x_1,\ldots,x_n$ $n \geq 2$ from the family $\{P_\theta, \theta \in R^m, P_\theta(E) = P(E-\theta), E \in \mathcal{B}\}$. The best equivariant estimator $\tilde{f}$ corresponding to the invariant loss function $W(f,\theta) = W(f-\theta)$ satisfies the following relation:

$$\int W(\tilde{f}(x_1,\ldots,x_n)-\theta)\prod_1^n p(x_j-\theta)\,d\theta = \inf_{d\in R^m} \int W(d-\theta)\prod_1^n p(x_j-\theta)\,d\theta. \quad (1)$$

Here p denotes the density of P.

The practical statistician hardly knows exactly the precise form of the loss function W which he has to use. Therefore it

G. P. Patil et al. (eds.), Statistical Distributions in Scientific Work, Vol. 1, 173-184.

is natural to be interested in description of families $\{P_\theta\}$ with the best equivariant estimator $\tilde{f}$ independent of the choice of the loss function W from (possibly large) set L of loss functions. Rukhin (1970) [cf. also Kagan, Linnik, Rao (1973)] solved this problem for the case m = 1 and L consisting of all even convex functions, density p being positive and continuous. Corresponding measures are either Gaussian

$$p(u) = \frac{1}{\sqrt{2\pi}\,\sigma} \exp -\{\frac{1}{2\sigma^2}(u-\theta_0)^2\}$$

or have the form

$$p(u) = \frac{\beta}{2K_0(\alpha)} \exp\{-\alpha \operatorname{ch} \beta(u-\theta_0)\}, \tag{2}$$

where $K_0(\alpha) = \int_0^\infty e^{-\operatorname{ch} u} du$. We shall call these distributions as well as their forthcoming multidimensional generalizations completely (or strongly) symmetrical.

If we let is (2) $\beta \to 0$, $\alpha \to \infty$, so that $\alpha\beta^2 \to \sigma^{-2}$, then $K_0(\alpha) \sim \sqrt{\frac{\pi}{2\alpha}}\, e^{-\alpha}$ and

$$\frac{\beta}{2K_0(\alpha)} e^{-\alpha \operatorname{ch}\beta u} \to \frac{1}{\sqrt{2\pi}\,\sigma} e^{-u^2/2\sigma^2}.$$

If $\beta \to \infty$ and $-\frac{1}{\beta}\log\alpha \to \delta$, then $K_0(\alpha) \sim \log\frac{2}{\gamma\alpha}$ and

$$\frac{\beta}{2K_0(\alpha)} e^{-\alpha \operatorname{ch}\beta u} \to \begin{cases} \frac{1}{2\delta} & |u| < \delta \\ 0 & |u| > \delta \end{cases}.$$

Thus the normal and rectangular distributions may be regarded as extremal points of densities (2).

The optimal equivariant estimator of a location parameter for completely symmetrical distributions has the form

$$\tilde{f}_\beta(x_1,\ldots,x_n) = \frac{1}{2\beta} \log \frac{\sum_1^n e^{\beta x_j}}{\sum_1^n e^{-\beta x_j}}.$$

If $\beta \to o$ then

$$\tilde{f}_\beta(x_1,\ldots,x_n) \to \bar{x} = \frac{\sum_1^n x_j}{n} ;$$

if $\beta \to \infty$, then

$$\tilde{f}_\beta(x_1,\ldots,x_n) \to \frac{x_{max} + x_{min}}{2} ,$$

so that continuity in β holds also for "universal" estimators.

In this article we generalize the earlier results of the author to the case of arbitrary dimension m.

2. MAIN RESULTS. Let $L = \{W_\alpha, \alpha \in A\}$ be the set of differentiable loss functions W_α, $W_\alpha(-\xi) = W_\alpha(\xi)$. From (1) we see that if the best equivariant estimator of the parameter θ is independent of W_α from L then for all $\alpha \in A$

$$\int \frac{\partial}{\partial d_k} W_\alpha(\tilde{f}(x_1,\ldots,x_n)-\xi)\prod_1^n p(x_j-\xi)\,d\xi = 0,$$

where $\frac{\partial}{\partial d_k} W_\alpha(\tilde{f}-\xi)$ denotes $\frac{\partial}{\partial d_k} W_\alpha(\tilde{f}-\xi-d)\,|\,d_k = 0$. We assume that the integral $\int W(d+\xi)\prod_1^n p(x_j-\xi)\,d\xi$ may be differentiated under the sign of the integral and that all integrals $\frac{\partial^m}{\partial d_1 \ldots \partial d_m} \int W_\alpha(\xi+d)\prod_1^n p(x_j-\xi)\,d\xi$ are convergent for $\alpha \in A$ and almost all $x_1,\ldots,x_n$. It is clear that for every point $(x_1,\ldots,x_m)$ from the set $\Lambda = \{(x_1,\ldots,x_m) : \tilde{f}(x_1,\ldots,x_m) = 0\}$

$$\int_{L_j^\alpha>0} L_j^\alpha(\xi) \left[\prod_1^n p(x_j+\xi) - \prod_1^n p(x_j-\xi)\right] d\xi = 0. \tag{4}$$

If the functions $L_{Tj}(\xi) = \frac{\partial}{\partial \eta_j} W_\alpha(\xi+\eta)\big|_{\eta=0}$ are such that the condition

$$\int_{L_j^\alpha > 0} L_j^\alpha(\xi)\, h(\xi)\, d\xi = 0, \qquad \alpha \in A$$

fulfilled for all $j = 1,\ldots,m$ implies $h(\xi) = 0$ almost everywhere, then we get from (4)

$$\prod_1^n p(x_j+\xi) = \prod_1^n p(x_j-\xi) \tag{5}$$

for almost all $\xi \in R^m$. If L consists of all even convex loss functions for which the integral in (4) is absolutely convergent then (5) holds automatically. Because of assumed continuity of p (5) is valid for all $\xi \in R^m$.

Theorem. If p is continuous positive density on R^m, $m > 1$ satisfying (5) for all ξ and some $n \geq 3$ then

$$p(x) = C\exp -\{(A\,\mathrm{sh}\,\frac{(x-\theta_0)}{2}\,Tb,\ \mathrm{sh}\,\frac{(x-\theta_0)}{2}\,Tb)\}$$

where C is a positive constant, $\theta_0 \in R^m$, $b \in C^m$

$$\mathrm{shx}\ T = \frac{\exp\{x_1T_1+\ldots+x_mT_m\}-\exp\{-x_1T_1-\ldots-x_mT_m\}}{2}$$

$A, T_1,\ldots,T_m$ denote matrices of order m, where A is real, non-degenerate and symmetric, T_j are complex and $T_jT_k=T_kT_j$; T_j^2 is real, $A\,T_j^2 = T_j^2\,A$ and $[T_j]^T = T_j$; $j, k = 1,\ldots,m$.

Proof.1. Let us note at first that from the complete symmetry of p follows its symmetry. In fact the set Λ of zeroes of the continuous equivariant estimator may be characterized by the following properties:

1°. $\bigcup_c (\Lambda + c) = R^m$, $(\Lambda + c = \Lambda + (c,\ldots,c))$,

2°. $(\Lambda + c_1) \cap (\Lambda + c_2) = \phi$ for $c_1 \neq c_2$.

Thus Λ contains unique point with equal coordinates, i.e.

$$p^n(\theta_0 - \xi) = p^n(\theta_0 + \xi)$$

for all ξ and some θ_0.

By shifting the initial measure P we can achieve the equality $\theta_0 = 0$. Then $p(-\xi) = p(\xi)$ for all $\xi \in R^m$.

2. Further we prove that symmetry of p is equivalent to its complete symmetry for n = 2.

Indeed the set

$$\Lambda = \{(x_1,x_2) : p(x_1-\xi)p(x_2-\xi) = p(x_1+\xi)p(x_2+\xi) \quad \xi \in R^m\}$$

contains the set $\{(x_1,x_2) : x_1 + x_2 = 0\}$. Hence to prove that Λ is the set of zeroes of the equivariant estimator (coinciding of course with $(x_1 + x_2)/2$) it is sufficient to show that the relation $(\Lambda + c) \cap \Lambda \neq \phi$ implies $c = 0$. The condition $(\Lambda+c) \cap \Lambda \neq 0$ is tantamount to the following

$$p(x_1-\xi)p(x_2-\xi) = p(x_1+\xi)p(x_2+\xi)$$

for all ξ and some x_1 and x_2 and

$$p(x_1+c-\xi)p(x_2+c-\xi) = p(x_1+c+\xi)p(x_2+c+\xi)$$

i.e.

$$p(x_1+\xi)p(x_2+\xi) = p(x_1+2c+\xi)p(x_2+2c+\xi).$$

Thus the function $p(x_1+\xi)p(x_2+\xi)$ integrable in ξ has period equal to 2c so that $c = 0$.

We have shown that the condition

$$p(x_1-\xi)p(x_2-\xi) = p(x_1+\xi)p(x_2+\xi)$$

fulfilled for all ξ implies $x_1+x_2 = 0$.

3. Let now $\varphi(\xi) = \log p(0)/p(\xi)$. Then

$$\sum_1^n [\varphi(x_j-\xi) - \varphi(x_j+\xi)] = 0$$

for all ξ and $x_1,\ldots,x_n$ under condition $\tilde{f}(x_1,\ldots,x_n) = 0$.

Let us assume that n = 3. (The general case is treated analogously). We shall denote $\psi(u,v)$ a function for which $\tilde{f}(u,v,-\psi(u,v)) = 0$.

The value $\psi(u,v)$ is uniquely determined. Indeed if for some u and v there exist two quantities ψ_1 and ψ_2 such that

$$p(u-\xi)p(v-\xi)p(\psi_1+\xi) = p(u+\xi)p(v+\xi)p(\psi_1-\xi)$$

and

$$p(u-\xi)p(v-\xi)p(\psi_2+\xi) = p(u+\xi)p(v+\xi)p(\psi_2-\xi)$$

then

$$p(\psi_1+\xi)p(-\psi_2+\xi) = p(\psi_1-\xi)p(-\psi_2-\xi)$$

and it was proved in 2 $\psi_1 = \psi_2$.

Further let us note that the function $\psi(u,v)$ is defined for all u,v. If for some v function $R(u,v) = \varphi(u+v) - \varphi(u-v)$ takes all real values, then $\psi(u,v)$ is defined for all u,v. If for all u $|R(u,v)| < C(v)$ i.e. $\frac{p(u+v)}{p(u-v)} < e^{c(v)}$ and $p(0)e^{-c(v)} < p(2v)$, then we can assume that $c(\nu) \to \infty$ for $||\nu|| \to \infty$ and $\psi(u,v)$ is defined everywhere.

4. We have now

$$R(u,\xi) + R(v,\xi) = R(\psi(u,v),\xi) \tag{6}$$

and

$$R(u,\xi) + R(v,\xi) + R(w,\xi) = R(\psi(\psi(u,v),w),\xi).$$

On the other hand

$$R(u,\xi) + R(v,\xi) + R(w,\xi) = R(\psi(u,\psi(v,w)),\xi)$$

so that for all $\xi \in R^m$

$$R(\psi(\psi(u,v),w),\xi) = R(\psi(u,\psi(v,w)),\xi).$$

By 2 we deduce

$$\psi(\psi(u,v),w) = \psi(u,\psi(v,w)).$$

Since $\tilde{f}(u,-u,0) = 0$ we get $\psi(u,-u) = 0, \psi(0,u) = u$.

In other terms ψ defines in R^m the group operation. Corresponding group G is locally euclidean since transformation $\psi(u,v) \to u$ is measurable. Thus G is homomorphic to R^m, i.e. there exists the one to one transformation h of space R^m onto itself such that

$$h(\psi(u,v)) = h(u) + h(v).$$

5. Returning to (6) we see that the condition $\tilde{f}(x_1,x_2,x_3)=0$ is equivalent to $h(x_1) + h(x_2) + h(x_3) = 0$, so that

$$R(h^{-1}(u),\xi) + R(h^{-1}(v),\xi) = R(h^{-1}(u+v),\xi)$$

and

$$R(h^{-1}(u),\xi) = (u,a(\xi))$$

for all u, $\xi \in R^m$. Since

$$R(u,\xi) = R(\xi,u)$$

we have

$$(h(u),a(\xi)) = (h(\xi), a(u)).$$

Putting $t_j = h^{-1}(e_j), e_j = (0,\ldots,1,\ldots,0)$, $j = 1,\ldots,m$ we get

$$(h(t_j), a(\xi)) = \sum_k h_K(t_j)a_K(\xi) = a_j(\xi)$$

and

$$(h(\xi), a(t_j)) = \sum_k h_K(\xi)a_K(t_j) \quad ,$$

so that

$$a_j(\xi) = \sum_k h_K(\xi) \; a_K(t_j)$$

and

$$(h(u), a(\xi)) = \sum_{K,\ell} a_K(t_\ell)h_K(u)h_K(\xi) = (Ah(u),h(\xi)).$$

It is clear that the matrix $A = \{a_K(t_\ell)\}$ is a symmetrical one. Thus we have obtained an equation:

$$\varphi(u+v) - \varphi(u,v) = (Ah(u), h(v)) \qquad (7)$$

with $A^T = A$ and h being continuous one-to-one transformation of R^m onto itself. From (7) we see that $h(-u) = -h(u)$ and

$$p(u) = p(0)e^{-(Ah(\frac{u}{2}),\, h(\frac{u}{2}))} \quad ,$$

so that the matrix A is non-singular.

6. For arbitrary $u,v,z \in R^m$ we get from (7)

$$\varphi(u+v+z) - \varphi(u+v-z) = (Ah(u+v), h(z))$$

and

$$\varphi(u-v+z) - \varphi(u-v-z) = (Ah(u-v), h(z)),$$

so that

$$(A[h(u+v)+h(u-v)], h(z)) = \varphi(u+v-z) - \varphi(u-v-z) + $$
$$+ \varphi(u-v+z) - \varphi(u+v-z) = (Ah(u), h(z+v) + h(z-v)). \qquad (8)$$

If the matrices $A(v)$ are defined as

$$A(v)e_j = \frac{1}{2}[h(z_j+v) + h(z_j-v)],$$

where $Ah(z_j) = e_j$, $j = 1,\dots,m$ then we come to a vector equation

$$h(u+v) + h(u-v) = 2\,A(v)h(u). \qquad (9)$$

From (8) we see that for all u,v,z

$$(AA(v)h(u), h(z)) = (Ah(u), A(v)h(z)),$$

so that

$$A^T(v)A = AA(v). \qquad (10)$$

From the definition of $A(v)$ we conclude $A(-v) = A(v)$, $A^T(v) = A(v)$. Further we deduce

$$2\,A(v_2)A(v_1)h(u) = A(v_2)[h(u+v_1)+h(u-v_1)] =$$
$$= \frac{1}{2}\,[h(u+v_1+v_2) + h(u+v_1-v_2) + h(u-v_1+v_2) + h(u-v_1-v_2)] =$$
$$= A(v_1+v_2)h(u) + A(v_1-v_2)h(u)$$

and

$$A(v_1+v_2) + A(v_1-v_2) = 2A(v_2)A(v_1) = 2A(v_1)A(v_2) \qquad (11)$$

The equation (11) is multivariate generalization of D'Alembert's functional equation [cf. Aczel, (1966), p. 117].

7. If $K = \{v: A(u+v) = A(u) \text{ for all } u\}$, then K is a subgroup of R^m and $K = \{v: A(v) = I\}$.

Indeed it is not difficult to verify the following formulae [cf. for example Rejto (1972)]

$$[A(u+v) - A(u)\, A(v)]^2 = [A^2(u) - I]\, [A^2(v) - I].$$

Thus if $A(v) = I$, then $A(u+v) = A(u)A(v) = A(v)$. On the other side if $A(u+v) = A(u)$ for all u; then $A(-u+v) = A(u)$ and $A(u) = A(u)\, A(v)$ i.e. $A(v) = I$.

Because of (10) we have $A(u) = M^{-1}(u)D(u)M(u)$ with diagonal matrix $D(u)$ and an unitary matrix $M(u)$. If for some u_0 all elements of $D(u)$ are different from 1 there exists nondegenerate complex matrix B with the property

$$B^2 = A^2(u_0) - I = \frac{1}{2}[A(2u_0) - I].$$

We can also to choose the matrix B commuting with all matrices $A(u)$. If now

$$G(u) = B^{-1}[A(u)(B-A(u_0)) + A(u+u_0)] =$$
$$= A(u) + B^{-1}[A(u+u_0) - A(u)A(u_0)],$$

then [cf. Retjo (1972)]

$$G(u+v) = G(u)\, G(v)$$

and

$$A(u) = \frac{G(u) + G^{-1}(u)}{2}.$$

8. If all matrices $D(u)$ do contain the identity block:

$$D(u) = \begin{pmatrix} I_s & 0 \\ 0 & \overline{D}(u) \end{pmatrix},$$

where $\overline{D}(u)$ is a diagonal matrix without elements equal to 1, and I_s - is a identity matrix of dimension $S = S(u)$. If u_0 is some point for which the number $S = S(u_0) \geq 1$ is minimal, then we can choose the matrix $M(u_0)$ having the form

$$M(u_0 = \begin{pmatrix} M_1(u_0) & 0 \\ 0 & M_2(u_0) \end{pmatrix}$$

with $M_1(u)$ being square matrix of the dimension s. Then

$$A(u_0) = \begin{pmatrix} I_s & 0 \\ 0 & M_2^{-1}(u_0)\overline{D}(u_0)M_2(u_0) \end{pmatrix}$$

and because of the relation $A(u)\ A(v) = A(v)\ A(u)$ all matrices $A(u)$ do have the form

$$A(u) = \begin{pmatrix} I & 0 \\ 0 & \overline{A}(u) \end{pmatrix}$$

where the matrix $\overline{A}^2(u) - I$ is nonsingular. Since $\overline{A}(u)$ satisfies (11) we get

$$\overline{A}(u) = \frac{\overline{G}(u) + \overline{G}^{-1}(u)}{2}$$

where $\overline{G}$ is a homomorphism of the group R^m into the group of non-degenerate complex matrices of dimension equal to m-s. If

$$G(u) = \begin{pmatrix} I & 0 \\ 0 & \overline{G}(u) \end{pmatrix},$$

then

$$A(u) = \frac{G(u) + G^{-1}(u)}{2}$$

and

$$G(u+v) = G(u)\ G(v).$$

If $u = (u_1,\ldots,u_m)$ then

$$G(u) = \exp\{u_1T_1+\ldots+u_mT_m\}$$

for some complex matrices $T_1,\ldots,T_m$ of order m. From the condition $A(u)\ A(v) = A(v)\ A(u)$ we obtain

$$T_j\, T_k = T_k\, T_j \;;\; j,\, k = 1,\ldots,m.$$

We let

$$\text{chu } T = \frac{\exp\{u_1T_1+\ldots+u_mT_m\} + \exp\{-u_1T_1-\ldots-u_mT_m\}}{2}$$

It is clear that $A(u) = \text{chu } T$ and $\text{chu } T \text{ shv } T = \text{shv } T \text{ chu } T$. From (10) we get

$$A \text{ chu } T = \text{chu } T\, A \tag{12}$$

and because of the reality of the matrices $A(u)$ we deduce the reality of matrices T_j^2, $j = 1,\ldots,m$. Further it is clear that

$$[A(u)]^T = A(u).$$

9. Let us show now that all solutions h of the equation

$$h(x+y) + h(x-y) = 2 \text{ chy } T\, h(x) \tag{14}$$

have the form

$$h(x) = Qx + \frac{1}{2}[\exp\{\sum_{s+1}^{m} x_iT_i\} - \exp\{-\sum_{s+1}^{m} x_iT_i\}]\, b_1 \tag{15}$$

for some $b_1 \varepsilon C^{m-s}$ and matrix Q of order s. Here s is the number of identity matrices T_j in the representation of $G(u)$.

In fact if $h(x) = (h^1(x), h^2(x))^T$ where vector $h^1(x)$ is of dimension s and $h^2(x)$ of dimension m-s then

$$h^1(u+v) + h^1(u-v) = Q\, h^1(u)$$

and

$$h^2(u+v) + h^2(u-v) = 2 \text{ ch } v\, T\, h^2(u).$$

These equations prove (15). Since $m > 1$ we can rewrite (15) in the form

$$h(u) = \text{shu } T\, b\;,\; b \,\varepsilon\, C^m \tag{16}$$

with new matrices $T_1,\ldots,T_m$ such that $T_i^2 = I$, $i = 1,\ldots,s$.

Because of the form G and the properties of the function h we see that the subgroup K is a trivial one.

10. At last we come to an equation

$$\varphi(x+y) - \varphi(x-y) = (A\,\mathrm{sh}x\ T\ b,\ \mathrm{sh}y\ T\ b)$$

It is not difficult to check up that

$$\varphi(x) = (A\ \mathrm{sh}\ \frac{x}{2}\ T\ b,\ \mathrm{sh}\ \frac{x}{2}\ T\ b)$$

with the matrix A satisfying (12) and (13) is a solution of (7). The theorem is proved.

The optimal "universal" equivariant estimator $\tilde{f}$ may be found from the relation

$$\sum_{1}^{n} \mathrm{sh}(x_j - \tilde{f}(x_1,\ldots,x_n))\ T\ b = 0$$

or

$$\mathrm{ch}\ \tilde{f}\ (x_1,\ldots,x_n)\ T \sum_{1}^{n} \mathrm{sh}\ x_j\ T\ b = \tag{17}$$
$$= \mathrm{sh}\ \tilde{f}\ (x_1,\ldots,x_n)\ T \sum_{1}^{n} \mathrm{ch}\ x_i\ T\ b.$$

It is known that the best equivariant estimator of a location parameter is in general inadmissible if the quality of statistical procedure is measured by a risk function corresponding to one loss function W. Because of "universal" character of estimators (17) it seems to be more natural to measure quality of $\tilde{f}$ by all loss functions under which $\tilde{f}$ is optimal. With such order in the set of all estimators $\tilde{f}$ turns to be absolutely admissible.

Let us note that completely symmetrical distributions in multivariate case include normal law and more generally all linear transforms of the products of one dimensional completely symmetrical distributions (2).

REFERENCES

[1] J. Aczel. [1966). Lectures on Functional Equations and Their Applications, Academic Press.
[2] A. M. Kagan, Yu. V. Linnik, C. R. Rao (1973). Characterization Problems in Mathematical Statistics. John Wiley and Sons, N. Y.
[3] L. Rejto (1972). Studia Sci. Math. Hung. 7, 331-336.
[4] A. L. Rukhin (1970). Soviet Math. Dokl. II, I, 89-92.

APPROXIMATION THEORY, MOMENT PROBLEMS AND DISTRIBUTION FUNCTIONS

M. S. Ramanujan

The University of Michigan, Ann Arbor, Michigan, U.S.A.

SUMMARY. Results concerning distribution functions, moment problems and approximation theory are discussed.

KEY WORDS. Approximation theory, moment problem, summability methods, negative binomial kernel, Poisson kernel.

1. INTRODUCTION. f(x) be a real valued function defined on [0,1]; the n-th Bernstein polynomial $B_n(f;x)$ associated with f is defined by

$$B_n(f;x) = \sum_{k=0}^{n} p_{nk}(x)\ f(\frac{k}{n}),\quad p_{nk}(x) = \binom{n}{k}\ (1-x)^{n-k}\ x^k;$$

Bernstein (1912) introduced these polynomials to provide a proof of the famous approximation theorem of Weierstrass; more exactly, if f is a continuous function on [0,1] then $B_n(f;x) \to f(x)$ uniformly on [0,1].

The above result on uniform approximation is our starting point. Various proofs of this result are available and we cite two particularly simple ones available in the standard texts of Feller (1966) and of Lorentz (1953).

In this exposition, we shall make no effort to go into problems like that of rate of convergence of $B_n(f;x)$ to f(x); instead we shall attempt to make a selected survey of how this

G. P. Patil et al. (eds.), Statistical Distributions in Scientific Work, Vol. 1, 185-192.

theorem of Bernstein can be looked upon as one generating some generalizations in the theory of approximation, how it is related to the so called moment problems on finite intervals, to classical summability theory and to several other related problems. The choice of these topics is dictated mostly by the author's personal tastes, experiences and limitations.

2. APPROXIMATION THEORY. Let us now get back to the Bernstein polynomials and the proof of the Weierstrass theorem; two major (but simple) steps in the proof are:

$$\text{(i)} \quad \sum_{k=0}^{n} p_{nk}(x) \equiv 1 \quad \text{and} \quad \text{(ii)} \quad \sum_{|\frac{k}{n} - x| > \delta} p_{nk}(x) \leq \frac{1}{4n\delta^2} .$$

We then notice that $p_{nk}(x)$ is the probability that an event e of probability x will occur exactly k times in n independent trials; also (ii) above is only a reformulation of the well-known Bernoulli theorem of large numbers. One further observation stemming from the author's early love viz summability theory is that the matrix

$$P = (p_{nk}),\ p_{nk} = \begin{cases} p_{nk}(x), & n \geq k \\ 0 \quad , & n < k \end{cases}$$

corresponds to the Euler method E(x); see Hardy (1949).

We shall now look at a few analogues and extensions of the Bernstein theorem. The first and perhaps the most natural replacement for $p_{nk}(x)$ is the kernel $m_{nk}(x) = \binom{k}{n} (1-x)^{n+1} x^{k-n}$, $n = 1,2,\ldots,$. $k = n, n+1,\ldots,$ and $M_n(f;x) = \sum_{k \geq n} m_{nk}(x) f(\frac{k-n}{k})$, $n = 1,2,\ldots,$. To the probabilist, obviously, $m_{nk}(x)$ corresponds to the Pascal distribution or the negative binomial probability, see Patil and Joshi (1968); to the summabilist this corresponds to the Hardy-Littlewood-Fekete-Taylor method T(x) in its sequence-to-sequence form [see Hardy (1949)]. Meyer-König and Zeller (1960) proved that for functions $f \in C[0,1]$, $M_n(f;x)$ converges uniformly to f(x) on [0,1]; see also Arato and Renyi (1957) for a probabilistic proof.

Using the kernel of the Poisson distribution, Szasz (1950) considered, for each complex valued function f(x), the function (= power series)

$$S_n(f;x) = e^{-nx} \sum_{k=0}^{\infty} \frac{(nx)^k}{k!} f(\tfrac{k}{n}) .$$

In summability theory this corresponds to the method of Borel's exponential means.

Cheney and Sharma (1964) start with the identity

$$1 = (1-x)^{n+1} \exp(\tfrac{tx}{1-x}) \sum_{k=0}^{\infty} L_k^{(\alpha)}(t)\, x^k, \quad \alpha > -1$$

where $L_k^{(\alpha)}$ denotes the Laguerre polynomial of degree k and define

$$P_n(f;x) = (1-x)^{n+1} \exp(\tfrac{tx}{1-x}) \sum_k L_k^{\alpha}(t)\, x^k f(\tfrac{k}{k+n})$$

where $t \leq 0$; they show that for $f \in C[0,1]$, if $a < 1$ and $\frac{t}{n} \to 0$ then $P_n(f;x)$ converges uniformly to $f(x)$ on $[0,a]$; for $t = 0$ this corresponds to the Meyer-König-Zeller theorem using negative binomial probability kernel. In this context see also Pethe and Jain (1972).

One of the main abstractions which incidentally is also a major tool in proving many of the uniform approximation theorems is the following result due to Korovkin (1960): If $\{T_n\}$ is a sequence of positive linear operators on $C[0,1]$ and if $T_n f \to f$ for $f = 1, x, x^2$ then $T_n f \to f$ for each $f \in C[0,1]$.

A generalization of the Bernstein theorem, due to Jakimovski and Ramanujan (1964) is the following: Let α be real, $f(x)$ be bounded in $\beta \leq x \leq 1$, where $\beta = \inf_{m,n} (\frac{n+\alpha}{m+\alpha})$. Let

$$B_m(\alpha,f,x) = \sum_{n=0}^{m} \binom{m+\alpha}{n-n} (1-x)^{m-n} x^{n+\alpha} f(\tfrac{n+\alpha}{m+\alpha}) .$$

Then for f continuous on $0 < a \leq x \leq b \leq 1$, $B_m(\alpha,f,x)$ converges uniformly to f. We shall point out later the use of this in certain moment problems.

Before quitting the topic of approximation we shall point out two generalizations of the Bernstein theorem to abstract set-ups. The first is in the theory of semigroups of operators; a major source of reference for this is the treatise of Hille and Phillips (1957); the result we mention is the following, due essentially to Kendall (1954): If for $\xi \geq 0$ $T(\xi)$ is a strongly continuous semigroup of operators on a Banach space X then $\lim_{n\to\infty} ||[(1-\xi)I + \xi T(\frac{1}{n})]^n x - T(\xi)x|| = 0$ for each $x \varepsilon X$ and each $\xi \varepsilon [0,1]$ uniformly in ξ. The reader is invited to look also into Feller (1966) for a nice brief motivation.

The second extends the Bernstein theorem to functions continuous on $I = [0,1]$ and taking values in a Frechet space E; for details see Ramanujan (1967).

3. SUMMABILITY METHODS AND MOMENT PROBLEMS. Let us start with an infinite matrix $A = (a_{nk})$ of reals or complexes. The matrix A is said to be convergence preserving or a K-matrix (K, for Kojima) if for each $x = (x_k) \varepsilon c$, the space of convergent sequences, the sequence $y = (y_n)$, $y_n = \sum_k a_{nk}x_k$ exists and $y \varepsilon c$; if, in addition $\lim x_n = \lim y_n$ then A is said to be limit preserving or a T-matrix (T, for Toeplitz). The exact conditions for A to be a K or T-matrix are well-known; see, for instance, Hardy (1949). The well-known Cesaro, Hölder, Euler methods are all limit preserving. Hausdorff (1921) introduced a general class of methods (H,μ_n) defined as follows: let $\{\mu_n\}$ be a sequence of reals; define $\Delta^k\mu_n = \Delta^{k-1}\mu_n - \Delta^{k-1}\mu_{n+1}$, $k = 0,.,...$ where $\Delta^0\mu_n = \mu_n$ and $\Delta^{-1}\mu_n = 0$; define the matrix $(H,\mu)_{n,k} = h_{nk} = \begin{cases} \binom{n}{k} \Delta^{n-k}\mu_k, & n \geq k \\ 0 & n < k \end{cases}$. The matrix (H,μ) is called the Hausdorff matrix corresponding to μ_n; $\mu_n = (n+1)^{-k}$, $\binom{n+k}{k}$ and p^n give respectively the Hölder method (H,k), the Cesaro method (C,k) and the Euler method (e,p). Given $\{\mu_n\}$ it is called a moment sequence (or, more accurately a Hausdorff moment sequence) if there exists on [0,1] a function g of normalized bounded variation so that for each $n = 0,.,...,$ $\mu_n = \int_0^1 t^n \, dg(t)$ holds. A beautiful result of Hausdorff's is that (H,μ) is a K-matrix iff $\{\mu_n\}$ is a moment sequence or (equivalently) is the difference of two completely monotone sequences.

Another important class of summability methods are the quasi-Hausdorff methods (H^*,μ_n); its matrix $H^* = (h^*_{nk})$ is given by

$$h^*_{nk} = \binom{k}{n} \Delta^{k-n} \mu_{n+1}, \ k \geq n \text{ and } = 0, \ k < n.$$

Thus it is very closely related to the transpose of the Hausdorff matrix; for $\mu_n = p^n$ one obtains the Taylor method. Ramanujan (1953, 1957) carries out a systematic investigation of these and proves in particular that (H^*, μ_{n+1}) is a K-matrix $\Leftrightarrow$ $\{\mu_n\}$ is a Hausdorff moment sequence.

The Hausdorff moment problem is the following: given $\{\mu_n\}$ find necessary and sufficient conditions on $\{\mu_n\}$ so that there exists a distribution function g of BV[0,1] such that the $\{\mu_n\}$ are the moments of g about the origin, i.e., $\mu_n = \int_0^1 t^n dg(t)$. Thus we now have two (equivalent) solutions to this problem, due to Hausdorff and Ramanujan as mentioned above.

At this stage we start looking at the above moment problem from a functional analytical viewpoint; such an approach generates further inquiries. If we look at C[0,1] as a Banach space, as is well-known, its topological dual can be (isometrically) identified with the space of normalized BV[0,1]; this is the Riesz theorem. Now the moment problem admits the following description: given a sequence $\{\mu_n\}$ of reals, find exact conditions so that there exists a linear, continuous functional L on C[0,1] such that at the "polynomial points" $1, t, t^2, \ldots, L(t^n) = \mu_n$, $n = 0,1,2,\ldots$. It is possible to give purely functional analytical proofs of the solutions indicated earlier utilizing (i) the uniform approximation of continuous functions via $B_n(f;x)$ or $M_n(f;x)$ i.e., looking at dense subspaces of C[0,1], (ii) constructing linear continuous functionals L on the dense subspaces such that $L(t^n) = \mu_n$ and (iii) extending these functionals to C[0,1] via Hahn-Banach theorems. An amplification of these can be found in Lorentz (1953). This approach has certain ease in the generalizations. Instead of C[0,1] one may look at some standard function spaces like $L_p[0,1]$ $p \geq 1$ or at some Lorentz spaces Λ or X(C), (1953); for various moment problems of this type, initiated by Hausdorff himself, see [(1921), (1953), (1964)]; in particular we point out that moment problems of the type $\nu_n = \int_0^1 t^{n+\alpha} f(t)dt$, $f \in L_p[0,1]$ are discussed in Jakimovski and Ramanujan (1964).

Let us now indicate a generalization of the moment problems in a different direction. In the problems discussed above we considered spaces $S[0,1]$ of functions $f: [0,1] \to \mathbb{R}$ and belonging to class S, and we looked at their duals, namely at maps $L: S[0,1] \to \mathbb{R}$. The generalization we aim at is to look at $S[I,E]$ of functions belonging to S, defined on $I = [0,1]$ and taking values in a Frechet space; $S[I,E]$ is topologized by using the seminorms of E and then we look at maps $L: S[1,E] \to F$ where F is another Frechet space, L is linear and continuous; the $\{\mu_n\}$ are no longer reals but elements of $L(E,F)$ and, roughly, we are asking for suitable conditions so that $\mu_n x = L(t^n x)$, for each $x \in E$ and each n. This involves the theory of vector measures and representations of continuous linear maps on $S[I,E] \to F$. Hausdorff moment problems for this set up are discussed by Goodrich (1966), Kurtz and Tucker (1965) and Ramanujan (1967) and their relation to operator valued Hausdorff summability matrices are pointed out by them. L_p-moment problems for operators are discussed by Leviatan and Ramanujan (1970).

4. APPLICATIONS OF THE MOMENT PROBLEM. The solution to the L_p-moment problem has a surprising application; suppose $f \in L_p(0,2\pi)$, $1 < p \leq 2$; then its sequence of Fourier coefficients $c_n = (2\pi)^{-1} \int_0^1 f(t)\, e^{-int} dt$, $n = 0, \pm 1, \pm 2, \ldots$ is in ℓ_q, $1/p + 1/q = 1$; however, ℓ_q is not exhausted by such Fourier coefficients. So, one may ask: given $c \in \ell_q$ find necessary and sufficient conditions that it is the sequence of Fourier coefficients of a suitable $f \in L_p$. The only interesting cases are for $q \geq 2$. This question has been answered by Rooney (1960) using Hausdorff's solution of the L_p-moment problem (derived using Bernstein polynomials) and by Ramanujan (1962) using his alternate solution of the L_p problem via negative binomial kernels.

Finally we look at an application of the Hausdorff moment problem to mean ergodic theorems. As is well-known ergodic theory, analytical theory of semigroups of operators, Markov processes and the theory of Brownian motions are all inter-related (for instance, see Yosida (1965)). One version of the mean ergodic theorem is the following: Let E be a Banach space and $v \in L(E,E)$. Suppose $\{||v^n||\}$ is bounded. Let $T_n x = \frac{(I+v+v^2+ \ldots + v^{n-1})x}{n}$ for x and n. If $\{T_n x\}$ is weakly relatively compact then $T_n x \to Px$, pointwise, where P is the projection onto the null space of (I-V). In the above theorem we can look upon $T_n x$ as the Cesaro transform of $\{x, vx, v^2x, \ldots\}$. Yosida (1938) relaxed the uniform boundedness condition on $\{||v^n||\}$ to the growth condition, $||v^n|| = o(n)$. One may also replace the Cesaro method above by a suitable Hausdorff method. In this direction, Kurtz and Tucker (1968) proved an ergodic theorem, retaining $||v^n|| = O(1)$. A two-fold generalization of the mean ergodic theorem is the following result which is a particular case of a more general result of Leviatan and Ramanujan (1971): Suppose (μ_n) is a (Hausdorff) moment sequence corresponding to b.v. function g on [0,1] and let G(t) denote the variation of g on [0,t]; suppose $\int_0^1 \frac{dg(t)}{\sqrt{t(1-t)}} < \infty$ and $||v^n|| \leq \gamma_n$, $\gamma_n = o(\sqrt{n})$. Then if $(T_n x)$ denotes the (H,μ_n)-transform of $\{v^n x\}$, the conclusion of the mean ergodic theorem holds under the hypothesis of weak relative compactness of $\{T_n x\}$. This reveals how the boundedness hypothesis on $\{||v^n||\}$ can be weakened to a condition related to the mass function g. The proof makes effective use of the moment representation of $\{\mu_n\}$.

REFERENCES.

[1] Arato, M. and Renyi, A. (1957). Acta Math. Acad Sci. Hung 8, 91-98.

[2] Bernstein, S. (1912). Commun. Soc. Math. Kharkow 13, 1-2.

[3] Cheney, E. W. and Sharma, A. (1964). Canad. J. Math. 16, 241-252.

[4] Chung, K. L. (1962). Math. Scand. 10, 153-162.

[5] Feller, W. (1966). Introduction to Probability Theory and its Applications, II. Wiley, New York.

[6] Goodrich, R. K. (1966). Ph.D. Thesis, Univ. of Utah.

[7] Hardy, G. H. (1949). Divergent Series, Oxford.

[8] Hausdorff, F. (1921). Math. Z. 9, 74-109, 280-299.

[9] Hille, E. and Phillips, R. S. (1957). Functional Analysis and Semigroups, A.M.S. Colloq. Publ. 31.

[10] Jakimovski, A. and Ramanujan, M.S. (1964). Math. Z. 84, 143-153.

[11] Kendall, D. G. (1954). Math. Scand. 2, 185-186.

[12] Korovkin, P. P. (1960). Linear Operators and Approximation Theory (Transl.), Hindustan Publ.

[13] Kurtz, L. and Tucker, D. H. (1965). Proc. Amer. Math. Soc. 16, 419-428.

[14] Kurtz, L. and Tucker, D. H. (1968). Pacific J. Math. 27, 539-545.

[15] Leviatan, D. and Ramanujan, M. S. (1970). Indiana U. Math. J. 20, 97-105.

[16] Leviatan, D. and Ramanujan, M. S. (1971). Studia Math. 39, 113-117.
[17] Lorentz, G. G. (1953). Bernstein Polynomials. Toronto.
[18] Meyer-König, W. and Zeller, K. (1960). Studia Math. 19, 89-94.
[19] Patil, G. P. and Joshi, S. W. (1968). A Dictionary and Bibliography of Discrete Distributions. Hafner, New York.
[20] Pethe, S. P. and Jain, G. C. (1972). Canad. Math. Bull. 15, 551-557.
[21] Ramanujan, M. S. (1953). J. Indian Math. Soc. 17, 47-53.
[22] Ramanujan, M. S. (1957). Quart. J. Math. (Oxford) 8, 197-213.
[23] Ramanujan, M. S. (1964). Archiv der Math. 15, 71-75.
[24] Ramanujan, M. S. (1962). Indian J. Math. 4, 87-91.
[25] Ramanujan, M. S. (1967). Rend. Circ. Mat. Palermo 16, 353-362.
[26] Rooney, P. G. (1960). Proc. Amer. Math. Soc. 11, 762-768.
[27] Szasz, O. (1950). J. Res. Nat. Bur. Standards 45, 239-345.
[28] Yosida, K. (1938). Proc. Imp. Acad. Tokyo 14, 292-294.
[29] Yosida, K. (1965). Functional Analysis, Springer.

KURTOSIS AND DEPARTURE FROM NORMALITY

C. C. Heyde

Department of Statistics, Australian National University, Canberra, Australia

SUMMARY. The coefficients of skewness and kurtosis are traditional measures of departure from normality which have been widely used, particularly in an empirical context. Theoretical disadvantages of the quantities have been often mentioned in the literature; these are here emphasized by an example of a family of non-symmetric distributions, all of whose odd order moments vanish, which have the same moments, the first four coinciding with those of the unit normal law. Nevertheless, if one restricts the class of distributions under consideration to a class L_2 of mixtures of normals, then the kurtosis appears as a distance in a metric space setting [Keilson and Steutel (1974)]. It is shown here that, in this metric space setting, the kurtosis can also be used in a bound on both the uniform metric for the distance between a distribution function and the unit normal distribution function and a non-uniform bound. A similar, and simpler, bound is also given in the case of more general mixtures.

KEY WORDS. Departure from normality, distance between distribution functions, bounds on uniform metric, mixtures of distributions.

1. INTRODUCTION. Let X be a random variable which possesses finite moments at least up to the fourth, $\mu_r = E(X-EX)^r$, r=1,2,3,4. Early statistical literature often included discussion of the

G. P. Patil et al. (eds.), Statistical Distributions in Scientific Work, Vol. 1, 193-201.

measures $\gamma_1=\mu_3/\mu_2^{3/2}$ and $\gamma_2=(\mu_4/\mu_2^2)-3$, which are called the skewness and kurtosis respectively [Kendall and Stuart (1958) pp. 85,86; 92,93]. Smallness of these quantities, which are zero if X is normally distributed, has frequently been used in the literature to justify assertions of approximate normality, especially in data analysis [e.g. Fisher (1958) pp. 52-3; Kendall and Stuart (1961) p. 461], but also sometimes in a theoretical context [e.g. Kendall and Stuart (1966) pp. 351-2, the material from which is basically attributable to Bienaymé [see Heyde and Seneta (1972)]].

Of course the shortcomings of such an approach are well known. It is well known that in general no distribution is determined by a finite set of moments [bounds on the distribution functions can be supplied, however, e.g. Wald (1939)]. Examples abound of cases where a full set of moments does not determine the distribution [e.g. Feller (1971) p. 227; Kendall and Stuart (1958) p. 109]. To illustrate the possibilities we shall construct a family of non-symmetric distributions, all with the same moments, whose odd order moments vanish and whose first four moments are the same as those of the unit normal law.

First note that the distribution with frequency function

$$f(x) = \frac{1}{48}\alpha^4 \exp(-\alpha|x|^{1/4})\{1-\varepsilon \operatorname{sgn} x \sin(\alpha|x|^{1/4})\} \qquad (-\infty<x<\infty) \tag{1}$$

$\alpha > 0$, $\varepsilon \neq 0$, $|\varepsilon| < 1$, is non-symmetric and has moments

$$\mu_{2n} = \frac{1}{6}\alpha^{-8n}(8n+3)!, \quad \mu_{2n+1} = 0 \quad (n = 0,1,2,\ldots).$$

[For a discussion of such distributions see Heyde (1963).] Now let X_1 be a random variable with distribution

$$P(X_1 = 1) = P(X_1 = -1) = \frac{1}{2}$$

and let X_2 be independent of X_1 and have its distribution given by (1) with α chosen so that $EX_2^2 = 1$. For a constant A, $0<A<1$, yet to be specified, set

$$X = A^{1/2} X_1 + (1-A)^{1/2} X_2.$$

Then, X has a non-symmetric distribution, has all its odd order moments zero, $EX^2 = 1$ and

$$EX^4 = A^2 + 6A(1-A) + (1-A)^2 EX_2^4 .$$

We can then make $EX^4 = 3$ by choosing

$$A = (\gamma_2(X_2)-2)^{-1}\left[\gamma_2(X_2) - \{2\gamma_2(X_2)\}^{1/2}\right]$$

where $\gamma_2(X_2) = EX_2^4 - 3$, the kurtosis of X_2.

Of course if we specify a class of distributions and confine attention to members of the class, it may become quite a different matter to specify a member of the class on the basis of a finite number of moments. A member of the Pearson family of distributions is, for example, specified by a knowledge of at most four moments.

Now, many of the limit theorems of probability and statistics describe the convergence in distribution of a sequence of random variables. These limit results can be rightly criticized in the absence of a suitable measure of departure from the limit distribution. Estimates of such measures of departure as the uniform metric are often difficult to obtain and sometimes may be of limited value by virtue of the breadth of the class of distributions under consideration.

Keilson and Steutel (1974) have adopted the interesting approach to this problem of restricting the space of distributions considered in such a way that simple measures of distance can be given. They found, for example, that for a certain broad class $\mathcal{D}$ of distributions, the kurtosis appears as a distance in a metric space setting. They focus particular attention on the space L_2 of mixtures of symmetric normal distributions. The random variable X with characteristic function $\phi_X(t)$ belongs to L_2 if and only if ϕ_X is of the form

$$\phi_X(t) = \int_0^\infty e^{-(1/2)t^2w^2}\, dF_W(w),$$

where F_W is the distribution function of some non-negative random variable. Equivalently, if W has the distribution function F_W and is independent of N, whose distribution is normal N(0,1), then

$$X \overset{d}{=} NW.$$

A discussion of this surprisingly broad and important class is given in Keilson and Steutel (1974) and reference mentioned therein. The class consists, in fact, of the real characteristic functions $\phi(t)$ such that $\phi(|t|^{1/2})$ is completely monotone on $(0,\infty)$. It contains, for example, the symmetric stable distributions and the Laplace and t distributions and their mixtures.

Let $X_1 \overset{d}{=} NW_1$ and $X_2 \overset{d}{=} NW_2$ be members of L_2 with distribution functions F_1 and F_2 respectively and write μ_{W_1} and μ_{W_2} for the probability measures corresponding to $W_1^{1/2}$ and $W_2^{1/2}$ respectively. It is shown in Section 5 of Keilson and Steutel (1974) that the subspace of L_2 with $EX^2 = 1$ and $EX^4 < \infty$, and distance function

$$\rho_2\,(F_1,F_2) = \int (w-1)^2 \mid \mu_{W_1}\,(dw) - \mu_{W_2}\,(dw) \mid,$$

is a metric space. Furthermore, if we confine attention to departure from normality, the distance becomes

$$\begin{aligned} \rho_2\,(F,\Phi) = \int (w-1)^2\, \mu_W\,(dw) &= E(W^2-1)^2 \\ &= E(W^4-1) \\ &= \frac{1}{3}\, EX^4 - 1 = \frac{1}{3}\,\gamma_2. \end{aligned} \tag{2}$$

The key to this result is easy to see. If $\gamma_2=0$, then $EW^4=1=(EW^2)^2$, which forces the distribution of W to be degenerate at unity.

The interesting result (2) affords a simple way to assess departure from normality (of course all members of L_2 are symmetric so skewness considerations do not arise). Unfortunately, a criterion such as $\gamma_2/3$, although quantitative, provides us with little concrete information. We shall go on to show, however, that we can upper bound the uniform metric using γ_2 to measure departure from normality.

2. BOUNDS FOR DEPARTURE FROM NORMALITY.

We shall begin by obtaining the following result.

Theorem 1. If X is a random variable belonging to L_2 with $EX^2=1$ and $EX^4<\infty$, then

$$\sup_x \mid P(X \leq x) - \Phi(x) \mid \leq C\gamma_2^{\frac{1}{5}}$$

for some universal constant C with $C \leq 3^{\frac{3}{5}}.5.\,(2\pi)^{-\frac{2}{5}}\,\pi^{-1}$.

This result should be compared with the relevant case of that of Zolotarev (1965) which it sharpens at the expense of restricting the class of distributions under consideration.

Proof. Let X have characteristic function f. Since X belongs to L_2, there exists a non-negative random variable W such that $f(t) = E\{\exp(-\frac{1}{2}t^2W^2)\}$, i.e. $X \overset{d}{=} NW$, where N has a normal N(0,1) distribution and is independent of W.

Expanding $\exp(-\frac{1}{2}t^2u)$ in a Taylor series about the point u=1, we have

$$e^{-\frac{1}{2}t^2u} = e^{-\frac{1}{2}t^2} + (u-1)\left(-\frac{1}{2}t^2 e^{-\frac{1}{2}t^2}\right) + \frac{1}{8}(u-1)^2 t^4 e^{-\frac{1}{2}t^2\xi_0} \tag{3}$$

where $\xi_0 = \xi_0(t,u)$ satisfies $0 \leq \xi_0 = 1+\theta(u-1)$ for some $\theta=\theta(t,u)$ with $|\theta| \leq 1$. Then, using (3) we have

$$\begin{aligned} \left|f(t) - e^{-\frac{1}{2}t^2}\right| &= \left|Ee^{-\frac{1}{2}t^2W^2} - e^{-\frac{1}{2}t^2}\right| \\ &= \left|-\frac{1}{2}t^2 e^{-\frac{1}{2}t^2} E(W^2-1) + \frac{1}{8}t^4 E(W^2-1)^2 \times e^{-\frac{1}{2}t^2\xi_0(t,W^2)}\right| \\ &= \frac{1}{8}t^4 E\left[(W^2-1)^2 e^{-\frac{1}{2}t^2\xi_0(t,W^2)}\right] \\ &\leq \frac{1}{8}t^4 E(W^2-1)^2 = \frac{1}{24}t^4\gamma_2, \end{aligned} \tag{4}$$

since $EX^2 = EW^2 = 1$ and

$$\gamma_2 = EX^4-3 = 3EW^4-3 = 3E(W^2-1)^2.$$

Now write $\Delta = \sup_x | P(X \leq x) - \Phi(x) |$. A well known bound due to Esseen [e.g. Feller (1966), p. 512] gives for any $T > 0$,

$$\Delta \leq \frac{1}{\pi}\int_{-T}^{T} \left|\frac{f(t) - e^{-\frac{1}{2}t^2}}{t}\right| dt + \frac{24}{T\pi\sqrt{2\pi}},$$

so that, using (4),

$$\Delta \leq \frac{1}{48\pi} \gamma_2 T^4 + \frac{24}{T\pi\sqrt{2\pi}} .$$

Next, choose $T^5 = A\gamma_2^{-1}$, where A is yet to be fixed. We have

$$\Delta \leq \gamma_2^{\frac{1}{5}} \inf_{A>0} \left\{ \frac{1}{48\pi} A^{\frac{4}{5}} + \frac{24}{\pi\sqrt{2\pi}} A^{-\frac{1}{5}} \right\},$$

which yields

$$\Delta \leq 3^{\frac{3}{5}}.5.(2\pi)^{-\frac{2}{5}}.\pi^{-1}\gamma_2^{\frac{1}{5}} .$$

This completes the proof.

The constant $3^{\frac{3}{5}}.5.(2\pi)^{-\frac{2}{5}}.\pi^{-1}$ may be improved somewhat using the ideas of Zolotarev (1967). A more important issue is, however, that of improving the exponent $\frac{1}{5}$ of γ_2. The present value is unfortunate since in the central limit context for sums S_n of independent and identically distributed random variables with zero mean and variance one which belong to L_2, the uniform metric $\sup_x |P(S_n \leq xn^{\frac{1}{2}}) - \Phi(x)|$ will be $O(n^{-1}) = O(\gamma_2(S_n))$. This follows from results due to Cramér [e.g. Gnedenko and Kolmogorov (1954), p. 220].

Keilson and Steutel (1974) discuss the kurtosis in a metric space setting in the context of the class $\mathcal{D}$ defined as follows. A random variable Z belongs to $\mathcal{D}$ if $Z_1 - Z_2$ belongs to L_2, where Z_1 and Z_2 are independent, each with the distribution of Z. Of course if Z belongs to $\mathcal{D}$, the result of Theorem 1 gives a bound in terms of $\gamma_2(Z)$ for the uniform metric which measures the departure of the distribution of $Z_1 - Z_2$ from normality.

The uniform metric is not particularly useful for the estimating of large deviation probabilities. It is, however, possible to use Theorem 1 to obtain a non-uniform bound on the difference between the distribution functions F and Φ. This is done in the next theorem.

Theorem 2. Suppose that X is a random variable belonging to L_2 with $EX^2 = 1$ and $EX^4 < \infty$. If $\gamma_2 < e^{5/2}$, then

$$(1 + x^4)|P(X \leq x) - \Phi(x)| \leq A(\log \gamma_2^{-1})^2 \gamma_2^{\frac{1}{5}} + \gamma_2 \quad (5)$$

where

$$A \leq (K\sqrt{\frac{2}{\pi}} + 5C) \frac{4}{25}$$

with

$$K = \sup_{a \geq 1} a^{-3} e^{(1/2)a^2} \int_a^\infty y^4 e^{-(1/2)y^2} dy \quad (6)$$

and C the constant in Theorem 1.

Proof. Again write $\Delta = \sup_x |P(X \leq x) - \Phi(x)|$, and define K as in (6). It has been shown by Kolodiazhniy [see Petrov (1972) p. 154] that for a > 0,

$$(1+x^4)|P(X \leq x) - \Phi(x)| \leq \int_{|y| \geq a} y^4 d\,\Phi(y) + \gamma_2 + 5a^4\Delta$$

$$\leq K\sqrt{\frac{2}{\pi}}\, a^4 e^{-(1/2)a^2} + \gamma_2 + 5a^4\Delta.$$

Thus, using Theorem 1,

$$(1+x^4)\ |P(X \leq x) - \Phi(x)| \leq K\sqrt{\frac{2}{\pi}}\, a^4 e^{-(1/2)a^2} + \gamma_2 + 5a^4\, C\gamma_2^{\frac{1}{5}}, \quad (7)$$

where $C \leq 3^{\frac{3}{5}}.5.\ (2\pi)^{-\frac{2}{5}}.\ \pi^{-1}$. Then, for $\gamma_2 < e^{5/2}$ we take

$a = \frac{2}{5}\left[\log \gamma_2^{-1}\right]^{1/2}$ in (7) and (5) follows. This completes the proof.

3. BOUNDS FOR MORE GENERAL MIXTURES. For a general class $\mathcal{K}$ of mixtures whose kernel distribution is that of a random variable Y we can obtain an analogue of Theorem 1 under minimal conditions. Let X belong to $\mathcal{K}$ with mixing distribution that of a random variable W, so that

$$X \stackrel{d}{=} YW \tag{8}$$

where Y and W are independent.

Theorem 3. Suppose that $E|Y| < \infty$ and that either

(i) the distribution function of Y is absolutely continuous with density bounded in absolute value by m, or

(ii) the distribution function of Y is of lattice type with step h.

If $E|X| < \infty$, and in the case of (ii) the distribution functions of X and Y have a common lattice of points of discontinuity, we have

$$\sup_x |P(X \leq x) - P(Y \leq x)| \leq CE|X-Y| = CE|Y|E|W-1|,$$

where $C \leq 8\pi^{-1}(3m)^{1/2}$ in the case of (i) and $C \leq \frac{1}{2}\pi$ in the case of (ii).

Proof. Using (8), we have

$$\begin{aligned} |E\, e^{itX} - E\, e^{itY}| &= |\, E\, e^{itYW} - E\, e^{itY}| \\ &= |\, E\{e^{itY}(e^{itY(W-1)} - 1)\}| \\ &\leq E|e^{itY(W-1)} - 1| \\ &< |t|E|Y|E|W-1|. \end{aligned}$$

The required result in the case of (i) follows an analysis similar to that given in the proof of Theorem 1. In the case of (ii) we use Theorem 2 of Tsaregradskii (1958) which gives

$$\sup_x |P(X \leq x) - P(Y \leq x)| \leq \frac{1}{4}\int_{-\pi}^{\pi} |E\, e^{itX} - E\, e^{itY}||t|^{-1}dt.$$

REFERENCES

[1] Feller, W. (1966). An Introduction to Probability Theory and Its Applications, Vol. II. Wiley, New York.

[2] Fisher, R. A. (1958). Statistical Methods for Research Workers, 13th ed. Oliver and Boyd, Edinburgh.

[3] Gnedenko, B. V. and Kolmogorov, A. N. (1954). Limit Distributions for Sums of Independent Random Variables. Addison-Wesley, Cambridge, Mass.
[4] Heyde, C. C. (1963). Quart. J. Math. Oxford 14, 97-105.
[5] Heyde, C. C. and Seneta, E. (1972). Biometrika 59, 680-683.
[6] Keilson, J. and Steutel, F. W. (1974). Ann. Probab. 2, 112-130.
[7] Kendall, M. G. and Stuart, A. (1958). The Advanced Theory of Statistics, Vol. 1. Griffin, London.
[8] Kendall, M. G. and Stuart, A. (1961). The Advanced Theory of Statistics, Vol. 2. Griffin, London.
[9] Kendall, M. G. and Stuart, A. (1966). The Advanced Theory of Statistics, Vol. 3. Griffin, London.
[10] Petrov, V. V. (1972). Sums of Independent Random Variables. Izdat. Mauka, Moscow. (In Russian)
[11] Tsaregradskii, I. P. (1958). Theor. Probab. Appl. 3, 434-438.
[12] Wald, A. (1939). Trans. Amer. Math. Soc. 46, 280-306.
[13] Zolotarev, V. M. (1965). Theor. Probab. Appl. 10, 472-479.
[14] Zolotarev, V. M. (1967). Z. Wahrscheinlichkeitstheorie verw. Geb. 8, 332-342.

CONVERGENCE OF SEQUENCES OF TRANSFORMATIONS OF DISTRIBUTION FUNCTIONS AND SOME MOMENT PROBLEMS

W. L. Harkness

Department of Statistics, The Pennsylvania State University, University Park, Pa., U.S.A.

SUMMARY. Sequences of iterated transformations of univariate distribution functions (d.f.'s) are defined and their basic structural properties are determined. Limiting d.f.'s are obtained; these d.f.'s satisfy functional-integral equations. Solutions of these equations are exhibited, which involve problems of indeterminate moment sequences. The results are generalized to the bivariate case.

KEY WORDS. Sequences of distribution functions, limit theorems, moment problems, functional-integral equations, growth rates of moment sequences.

1. INTRODUCTION. Let F be the distribution function (d.f.) of a non-negative random variable (r.v.) X. Assume that the moments $\mu_n = \int_0^\infty x^n \, dF(x)$ are finite for all n. For the fixed non-negative integer n, define the sequence $\{G_N\}$ of absolutely continuous d.f.'s as follows: for $x > 0$, put

$$G_1(x) = \int_0^x y^n \, [1-F(y)]dy/\int_0^\infty y^n[1-F(y)]dy$$

and recursively,

$$G_N(x) = \int_0^x y^n \, [1-G_{N-1}(y)]/\int_0^\infty y^n \, [1-G_{N-1}(y)]dy;$$

$G_N(x) = 0$ for $x \leq 0$.

G. P. Patil et al. (eds.), Statistical Distributions in Scientific Work, Vol. 1, 203-211.

In the particular case when $n = 0$, a number of problems related to properties of the sequence $\{G_N\}$ have been studied extensively by Harkness and Shantaram (1969) and van Beek and Braat (1973). A bivariate extension has been considered by Shantaram (1970). For a stationary renewal process with continuous d.f. F, one can define a r.v. Y (called the overshoot of the process) which represents the distance from a fixed point, independent of the process, until the next renewal. It is known [see van Beek and Braat (1973)] that the d.f. of Y is given by G_1 (when $n = 0$). The sequence $\{G_N\}$, defined recursively above, can thus be interpreted as a sequence of d.f.'s corresponding to stationary renewal processes.

2. MOMENTS, CHARACTERISTIC FUNCTIONS (c.f.'s). For the case $n = 0$, the moments and c.f.'s of G_N are given by Harkness and Shantaram (1969). The results are as follows:

(i) The c.f. ϕ_{G_1} of G_1 is given by

$$\phi_{G_1}(t) = (it\mu_1)^{-1}\,[\phi_F(t)-1],\text{ for } t \neq 0 \text{ with } \phi_{G_1}(0) = 1. \tag{2.1}$$

(ii) The moments $\mu_{k,1}$ of G_1 are given by

$$\mu_{k,1} = \mu_{k+1}/(k+1)\mu_1,\quad k = 1, 2, \ldots \tag{2.2}$$

while for arbitrary N, the moments $\mu_{k,N}$ of G_N satisfy the recurrence relation $\mu_{k,N} = \mu_{k+1,N-1}/(k+1)\mu_{1,N-1}$ and are given explicitly by

$$\mu_{k,N} = k!N!\mu_{N+k}/(N+k)!\mu_k,\quad k,N = 1, 2, \ldots . \tag{2.3}$$

Using the recurrence relation (valid for $t \neq 0$), $\phi_{G_N}(t) = [i\,t\,\mu_{1,N-1}]^{-1}\,[\phi_{G_{N-1}}(t) - 1]$, one can show that the c.f. ϕ_{G_N} of G_N is given explicitly by

$$\phi_{G_N}(t) = \frac{N!}{(it)^N\mu_N}\,[\phi_F(t) - \sum_{j=0}^{N-1}\mu_j\,\frac{(it)^j}{j!}],\quad t \neq 0, \tag{2.4}$$

with $\phi_{G_N}(0) = 1$.

For an arbitrary but fixed positive integer n, we find that the kth moment $\mu_{k,N}^{(n)}$ of G_N is given by

$$\mu^{(n)}_{k,N} = N!(n+1)^N \mu_{N(n+1)+k}/\mu_{N(n+1)} \prod_{j=1}^{N} [(n+1)j+k] \tag{2.5}$$

and the c.f. $\phi^{(n)}_{G_1}$ of G_1 has the form

$$\phi^{(n)}_{G_1}(t) = \frac{(n+1)!}{(it)^{n+1}\mu_{n+1}} [\sum_{k=0}^{n} (-1)^{n-k} \phi_F^{(k)}(t) \frac{t^k}{k!} + (-1)^{n+1} t^{n+1}] \tag{2.6}$$

where $\phi^{(k)}_F(t)$ is the kth derivative of ϕ_F, with $\phi^{(0)}_F(t) = \phi_F(t)$.

3. GROWTH RATES OF MOMENT SEQUENCES. Let $\{\mu_n\}$ be a moment sequence of a non-negative r.v. with d.f. F. Then Harkness and Shantaram (1969) have established the following results:

(i) $\{\mu_{n+1}/\mu_n\}^k$ is a monotonically non-decreasing sequence.

(ii) $\mu_{n+k}/\mu_n \geq (\mu_{n+1}/\mu_n)^k$ for $k, n \geq 1$.

(iii) If F is "concentrated" on $[a,b]$, where $0 \leq a < b < \infty$, with $b = \inf \{x | F(x) = 1\}$, then $\mu_n^{1/n} \to b$ and $\mu_{n+k}/\mu_n \to b^k$, $k = 1, 2, \ldots$.

(iv) If $F(x) < 1$ for all x, then $\mu_{n+1}/\mu_n \to +\infty$.

(v) If $\mu_{n+2}\mu_n/\mu^2_{n+1} \to L < \infty$, then for each $k > 0$, $0 \leq r \leq k$, $\mu_{n+k}\mu_n/\mu_{n+r}\mu_{n+k-r} \to L^{r(k-r)}$.

(vi) If F has an analytic c.f., then $\liminf_{n\to\infty} \mu_{n+2}\mu_n/\mu^2_{n+1} = 1$.

Theorem 1. Let F be a finite d.f. on $[a,b]$, where $b = \inf\{x | F(x) = 1\}$. Then for $x > 0$,

$$\lim_{N \to \infty} G_N(x/N) = G(x) = 1 - e^{-x/b} .$$

Theorem 2. Let $\{c_N\}$ be a sequence of positive real numbers such that $H_N(x) \equiv G_N(c_N x) \to G(x)$, with G a proper d.f. and assume that $\limsup_{N \to \infty} \frac{c_N}{c_{N-1}} = L$ (finite). Then

(i) The sequence $\{b_N\}$, where $b_N = \int_0^\infty [1 - H_N(u)]\, du$, is bounded and converges to $b = \int_0^\infty [1 - G(u)]\, du < \infty$.

(ii) $c_N/c_{N-1} \to L$ as $N \to \infty$.

(iii) G is continuous and concave on $[0,\infty)$ and $H_N'(x) \to G'(x)$ for $x > 0$.

(iv) $L \geq 1$ and equality holds if F has an analytic c.f.

(v) The kth moment of H_N converges for each k to the kth moment $\mu_k(G)$ of G (finite for all k) where

$$\mu_k(G) = k\, b^k\, L^{k(k-1)/2} = \nu_k, \text{ say.}$$

(vi) G satisfies the functional equation $G(x) = \frac{1}{b}\int_0^{Lx} [1 - G(u)]du$.

(vii) $c_N \sim \mu_{N+1}/b(N+1)\mu_N$.

(viii) If $L = 1$, $G(x) = 1 - e^{-x/b}$.

(ix) $L = 1$ if either (a) $\limsup_{N\to\infty} c_N < \infty$ or (b) $\mu_{N+1}/\mu_N = O(N)$.

(x) $\delta_N = \mu_N\mu_{N+2}/\mu_{N+1}^2 \to L$ as $N \to \infty$.

The question naturally arises as to when a sequence $\{c_N\}$ exists satisfying the hypotheses of Theorem 2. If $\delta_N = \mu_{N+2}\mu_N/\mu_{N+1}^2 \to 1$ as $\to \infty$, then $G_N(c_N x) \to G(x)$ with $c_N = \mu_{N+1}/(N+1)\mu_N$; in this case, $L = 1$. This raises another important problem, namely, which d.f.'s have the property that $\delta_N \to 1$ as $N \to \infty$? One large class of d.f.'s with this property is the class $\mathcal{F}^*$ of d.f.'s F having Increasing Hazard Rate (IHR); F has IHR if $\ell n\ 1 - F(x)$ is concave. In fact, if $F \in \mathcal{F}$, then $1 \leq \delta_N \leq 1 + (1+N)^{-1}$ [see Harkness and Shantaram (1969)]. The Weibull distribution $F_\beta(x) = 1 - e^{-x^\beta}$, for $x > 0$, $\beta > 0$ has IHR for $\beta > 1$, so that $\delta_N \to 1$. However, for $0 < \beta < 1$, $F_\beta \notin \mathcal{F}^*$ but nevertheless $\delta_N \to 1$, showing that $\mathcal{F}^*$ is a proper subset of the family $\mathcal{F}$ of all d.f.'s for which $\delta_N \to 1$. One might conjecture that $\mathcal{F} = \{F{:}F$ has an analytic c.f.$\}$. However, $\mu_N = (4N + 3)!/6$ is the Nth moment of a d.f. F which is not uniquely determined by $\{\mu_N\}$ and hence does not have an analytic c.f. [c.f., Widder (1941), p. 126]. It is easily seen that for this sequence, $\delta_N \to 1$ and $c_N \simeq 256N^3$.

4. THE FUNCTIONAL EQUATION $G(x) = \frac{1}{b}\int_0^{Lx} [1 - G(u)]du$, $L > 1$. It is not hard to show that the d.f. G of the r.v. $Z = UV$, where U has density $\theta^{-1} e^{-u/\theta}$ with $\theta = bL^{-1/2}$ and $V = e^{X\sqrt{\ell n\, L}}$ with $X \sim N(0,1)$ (so that V has a log-normal distribution), has k^{th} moment ν_k, when U and V are independent, and that G does satisfy the functional equation above.

Explicitly,

$$1 - G(z) = \int_0^\infty (2\pi \ell n\, L)^{-1/2}\ v\ e^{-(\ell n\, v)^2/2\ell n\, L - z/v\theta} dv.$$

Shantaram and Harkness (1972) have shown that if $G(x) = 1 - f(x)$, with $f(x) = \int_0^\infty e^{-tx} g(t)dt$, where g is such that $g(t) = Lt\, g(Lt)$ for all $t > 0$, then G is a solution of the equation. Let ϕ be an <u>arbitrary continuous function</u> on $[L^{-1}, 1]$ for t in this interval. Define $g(t) = \phi(t)$ for $L^{-1} \leq t \leq 1$ and extend g to $[0, \infty]$ by setting $g(t) = Lt\, g(Lt)$. In this way, one can construct (uncountably) many d.f.'s satisfying the equation with all having the same moments.

One such g is the following:

$$g(t) = (2\pi \ell n\, L)^{-1/2} t^{-1} e^{-(\ell n\, t\theta)^2/2\ell n\, L} \equiv \phi(t) \text{ with } \theta = 1^{-1/2};$$

this g yields the solution presented earlier.

5. CONVERGENCE OF G_N FOR POSITIVE INTEGRAL n. Let n be an arbitrary but fixed positive integer; G_N now depends on n but this dependence is suppressed in the notation. Recall that, for $N = 1,2, \ldots$, $G_N(x) = \int_0^x y^n[1 - G_{N-1}(y)]/\int_0^\infty y^n[1 - G_N{-1}(y)]dy$; $G_0(x) = F(x)$. The analogue of Theorem 1 is the following.

<u>Theorem 3</u>. Let F be a d.f. on $[a,b]$, where $b = \inf\{x: F(x) = 1\}$. Then for $x > 0$, $\lim_{N \to \infty} G_N(x/\gamma_N) = 1 - e^{-(\frac{x}{b})^{n+1}} \equiv H(x)$ where $\gamma_N = N^{1/n+1}$. H is the Weibull d.f.

<u>Proof</u>. Let W_N have d.f. G_N and $Z_N = \gamma_N W_N$; Z_N has d.f. H_N, where $H_N(x) \equiv G_N(x/\ _N)$. We show that the kth moment of Z_N

converges, as $N \to \infty$, to the kth moment of the Weibull distribution H. From (2.5), it follows that

$$E(Z_N^k) = \frac{N^{k/n+1} \; N!\,(n+1)^N \; \mu_{N(n+1)+k}}{\prod_{j=1}^{N} [(n+1)j+k]\mu_{N(n+1)}} = [N^a \prod_{j=1}^{n} (j/a+j)][\frac{\mu_{N(n+1)+k}}{\mu_{N(n+1)}}],$$

where $a = k\,\gamma_N$.

However, using (iii), Section 3, $\mu_{N(n+1)+k}/\mu_{N(n+1)} \to b^k$ as $N \to \infty$, while $N^a \prod_{j=1}^{N} (j/a+j) \to \Gamma(\frac{n+k+1}{n+1})$ [see equation 12.25, p. 66, Feller (1966)]. Thus, $E(Z_N^k) \to \Gamma(\frac{n+k+1}{n+1})\, b^k = \lambda_k$, say; λ_k is the kth moment of the Weibull distribution.

Now assume that for all $x > 0$, $F(x) < 1$. The extension of Theorem 2 for a positive integer n is straightforward, the proof paralleling that of Theorem 2 in every detail. Some modifications in the results (to reflect the new "G_N's") is necessary, however. The changes are as follows: replace $\limsup_{N \to \infty} c_N/c_{N-1} = 1$ by $\limsup c_N/c_{N-1} = L^* < \infty$ and (i) - (vii) by

(i') The sequence $\{b_N\}$, where $b_N = \int_0^\infty u^n[1 - H_N(u)]du$, is bounded and converges to $b = \int_0^\infty u^n[1 - G(u)]du < \infty$.

(ii') $c_N/c_{N-1} \to L^{n+1} = L^*$, where L is the same quantity appearing in Theorem 2 (corresponding to $n = 0$).

(iii') G is continuous on $[0, \infty]$ and $H_N'(x) \to G'(x)$ for $x > 0$.

(iv') $L = 1$ and equality holds if F has an analytic c.f.

(v') The kth moment of H_N converges for each k to the kth moment λ_k of G, where

$$\lambda_k = \Gamma(\frac{n+k+1}{n+1})\, L^{(n+1)k + k(k-1)/2}.$$

(vi') G satisfies the functional equation

$$G(x) = \int_0^{Lx} u^n[1 - G(u)]du/\int_0^\infty u^n[1 - G(u)]du.$$

(vii') $c_N \sim \mu_{N^*+1}/N^{1/n+1}\, \mu_{N^*}$ where $N^* = N(n+1)$.

(viii') If $L = 1$, $G(x) = 1 - e^{cx^{n+1}}$.

From the functional equation $G(x) = \int_0^{Lx} u^n[1 - G(u)]du / \int_0^{\infty} u^n[1 - G(u)]du$, it follows that the moments of G satisfy the recurrence relation $\lambda_{n+k+1} = (n+k+1)L^k . \lambda_{n+1} \lambda_k$; it is easily verified that, in fact, the moments λ_k as given by (v') do satisfy this relation.

6. BIVARIATE EXTENSION. Let $F(x,y)$ be the bivariate d.f. of non-negative r.v.'s X and Y. Shantaram (1970) has shown that

$$\mu_{m+1,n+1} = E(X^{m+1}Y^{n+1}) = (m+1)(n+1) \int_0^{\infty} \int_0^{\infty} x^m y^n \bar{F}(x,y)\, dxdy$$

where

$$\bar{F}(x,y) = 1 - F_1(x) - F_2(y) + F(x,y) = P\{X > x,\ Y > y\}$$

with F_1 and F_2 being the (marginal) d.f.'s of X and Y, respectively. Assume that $\mu_{m,n} < \infty$ for all positive integers m and n. Define, recursively, the sequence of absolutely continuous bivariate d.f.'s G_N by

$$G_N(x,y) = \int_0^x \int_0^y \bar{G}_{N-1}(u,v)\, dvdu / \int_0^{\infty} \int_0^{\infty} \bar{G}_{N-1}(u,v)\, dvdu.$$

Shantaram (1970) has established bivariate analogues of all of the results given in the first three sections. In particular, he has proven

(i) that the c.f. of G_1 is given by

$$\phi_{G_1}(t_1,t_2) = (i^2\mu_{1,1}t_1t_2)^{-1}[1 - \phi_1(t_1) - \phi_2(t_2) + \phi(t_1,t_2)]$$

where ϕ_1 and ϕ_2 are the (marginal) c.f.'s of X and Y, respectively.

(ii) that the cross-product moment $\mu_{ij,N}$ of G_N is equal to

$$\mu_{ij,N} = \frac{N!N!i!j!}{(N+i)!(N+j)!} \mu_{N+i,N+j}/\mu_{NN}.$$

(iii) several inequalities concerning growth rates of bivariate moments.

With regard to extensions of Theorems 1 and 2, in particular, he has proven the following.

Theorem 4. If $F(x,y)$ is a finite d.f. on the rectangle $[0,a] \times [0,b]$, i.e., $F(a,b) = 1$ but $F(x,y) < 1$ for $x < a$ or $y < b$, then

$$\lim_{N \to \infty} G_N(x/N, y/N) = G(x,y) = [1 - e^{-x/a}][1 - e^{-y/b}]$$

for $\min(x,y) > 0$ and zero elsewhere.

Theorem 5. Let $\{c_N\}$, $\{d_N\}$ be sequences of positive real numbers such that $H_N(x,y) \equiv G_N(c_N x, d_N y) \to G(x,y)$, with G a proper d.f. If $\limsup c_N/c_{N-1} = L_1 < \infty$ and $\limsup d_N/d_{N-1} = L_2 < \infty$, then

(i) the sequence $\{b_N\}$, where $b_N = \int_0^\infty \int_0^\infty \overline{H}_N(x,y)dxdy$, is bounded and converges to $b = \int_0^\infty \overline{G}(x,y)dxdy$.

(ii) $c_N/c_{N-1} \to L_1$ and $d_N/d_{N-1} \to L_2$, as $N \to \infty$.

(iii) G is continuous and the convergence $H_N(x,y)$ to G is uniform in x and y.

(iv) $L_1 L_2 \geq 1$ and equality holds if F has an analytic c.f.

(v) the (ij)th moment of H_N converges for each i and j to the (ij)th moment λ_{ij} of G.

(vi) G satisfies the functional equation

$$G(x,y) = \frac{1}{b} \int_0^{L_1 x} \int_0^{L_2 y} \overline{G}(u,v)dvdu \tag{6.1}$$

(vii) $c_N d_N \sim \mu_{N+1,N+1}/b(N+1)^2 \mu_{N,N}$.

Using (vi), it is easily shown that the moments $\lambda_{m,n}$ of G must satisfy the recursion relationship

$$\lambda_{m+1,n+1} = (m+1)(n+1)b\, L_1^m\, L_2^n\, \lambda_{m,n}. \tag{6.2}$$

A very interesting problem is that of determining the solutions of the equation (6.1). For the special case $L_1 = L_2 = 1$, Puri and Rubin (1974) have shown that the only absolutely continuous distributions satisfying (6.1) are the ones which are mixtures of exponential distributions with pdf $g(x,y)$ given by

$$g(x,y) = b^{-1} \int_0^\infty \int_0^\infty \exp[1-u_1 x - u_2 y]\, G(du_1,du_2) \qquad (6.3)$$

for $x,y > 0$, where the probability measure G is concentrated on the set $A = \{u_1 u_2 = b^{-1},\ u_i > 0,\ i = 1,2\}$. Clearly, a pdf given by (6.3) satisfies (6.1). By direct calculations, the moments of a distribution having a pdf of this form are given by

$$\lambda_{m,n} = m!n!b^{-1} \int_0^\infty \int_0^\infty u_1^{-m} u_2^{-n}\, G(du_1, du_2); \qquad (6.4)$$

it is easily verified that they satisfy the recurrence relation in (6.2). Finally, we note that if X and Y are independent r.v.'s having d.f.'s G_1 and G_2 satisfying the functional equations

$$G_i(Z) = b^{-1/2} \int_0^{L_i Z} [1 - G_i(u)]\, du$$

then, of course, $G(x,y) = G_1(x)G_2(y)$ is also a solution of (6.1). The moments of this G, given by

$$\lambda_{m,n} = m!n!\, b^{\frac{m+n}{2}} L_1^{m(m+1)/2} L_2^{n(n+1)/2},$$

also satisfy (6.2). However, it does not follow, a fortiori, that a d.f. having these moments is the product of its marginal d.f.'s G_1, G_2.

REFERENCES

[1] van Beek, P. and Braat, J. (1973). Stochastic Processes and Their Applications <u>1</u>, 1-10.

[2] Feller, W. (1966). <u>An Introduction to Probability Theory and Its Applications</u>. Vol. 1 (3rd ed.), Wiley: New York.

[3] Harkness, W. L. and Shantaram, R. (1969). Pac. J. Math. <u>31</u>, 403-415.

[4] Puri, P. S. and Rubin, H. (1974). Ann. Prob. <u>2</u>, 738-740.

[5] Shantaram, R. (1970). Pac. J. Math. <u>33</u>, 217-232.

[6] Shantaram, R. and Harkness, W. L. (1972). Ann. Math. Statist. <u>43</u>, 2067-2071.

[7] Steutel, F. W. (1971). <u>Preservation of Infinite Divisibility Under Mixing</u>. Mathematisch Centre, Amsterdam.

[8] Widder, D. V. (1941). <u>The Laplace Transform</u>. Princeton University Press, Princeton, New Jersey.

WEAK CONVERGENCE FOR EXPONENTIAL AND MONOTONE LIKELIHOOD RATIO FAMILIES AND THE CONVERGENCE OF CONFIDENCE LIMITS

Bernard Harris and Andrew P. Soms*

Mathematics Research Center, University of Wisconsin, Madison, Wisconsin, U.S.A./G. D. Searle & Co., Skokie, Illinois, U.S.A.

SUMMARY. Two convergence theorems for one-sided confidence limits for exponential and monotone likelihood ratio families are given. In addition, for some special exponential families, the class of discrete limit distributions is characterized and a uniqueness result obtained. Examples of the above results are given. Applications to hypothesis testing and confidence limits are indicated.

KEYWORDS. Weak convergence, confidence limits, limit distributions, exponential family, monotone likelihood ratio family.

1. INTRODUCTION. In this paper the weak convergence of certain sequences of distributions in discrete exponential families is studied. A procedure is given which often simplifies the determination of the limiting distribution. A number of examples are presented to illustrate the technique and also to show that such sequences arise naturally in statistical theory and practice.

Initially, this investigation was motivated by the following considerations. Let $F(x;\theta,n)$ be a given family of distribution functions, where $n = (n_1,n_2,\ldots,n_k)$ and each n_i is a

*Sponsored by the United States Army under Contract No. DA-31-124-ARO-D-462 and das Mathematisches Institut der Technischen Universität München.

G. P. Patil et al. (eds.), Statistical Distributions in Scientific Work, Vol. 1, 213-226.

positive integer. Let n and x be given. Further, let α, $0 < \alpha < 1$, be specified. The objective is to solve the equation $F(x;\theta,n) = \alpha$ in θ. Frequently, $F(x;\theta,n)$ is of such a nature that the given equation may be difficult to solve explicitly. One way of coping with this situation is to replace $F(x;\theta,n)$ by another family of distributions $F(x,\gamma)$, γ a specified function of n and θ; this replacement being justified by considerations of weak convergence. In this way an "approximate solution" to the above equation is obtained. To the best of our knowledge, the study of such approximations has not been previously treated from a general and mathematically precise point of view in the literature.

In section 2, we treat discrete limits of discrete exponential families. Continuous limits of discrete monotone likelihood ratio families are discussed in section 3. The fourth section is concerned with applications to confidence intervals and tests of hypotheses. Specific examples appropriate to the considerations of this paper may be found in R. J. Buehler (1957), B. Harris (1971), B. Harris and A. P. Soms (1974) and D. Hwang (1971).

2. DISCRETE LIMITS OF DISCRETE EXPONENTIAL FAMILIES AND CONVERGENCE OF ONE-SIDED CONFIDENCE LIMITS.

Let X_n be a set of non-negative integer valued random variables with probability density functions

$$P_{\theta_n}\{X_n = x\} = p_{n,\theta_n}(x) = e^{-C(n,\theta_n)} e^{\theta_n x} h(x,n), \qquad (2.1)$$

where $n = (n_1,n_2,\ldots,n_k)$, $\theta \in \Theta$, an interval (possibly infinite) on the real line. By $n \to \infty$, we will mean $n_i \to \infty$, $i = 1,2,\ldots,k$. The carrier set of $p_{n,\theta_n}(x)$ will be assumed to be a set of consecutive non-negative integers of cardinality > 1, the lower limit of which we designate by L_n. The upper limit will be denoted by U_n, where, when there is no upper limit, we define $U_n = \infty$. Hence $0 \leq L_n < U_n \leq \infty$.

We further assume that there exist real numbers $g_n > 0$, such that for

$$\ell_n(x+1) = h(x+1,n)/h(x,n)g_n, \quad L_n \leq x < U_n, \qquad (2.2)$$

we have

$$\lim_{n\to\infty} \ell_n(x+1) = \ell(x+1) > 0, \tag{2.3}$$

for all x with $0 \leq \bar{L} = \limsup_n L_n \leq x < \liminf_n U_n = \underline{U} \leq \infty$.
We now obtain the following theorem.

Theorem 2.1. A necessary and sufficient condition for $p_{n,\theta_n}(x)$ to have a non-degenerate discrete limiting distribution $p(x)$ is that

(a) there exists a least non-negative integer L such that $p_{n,\theta_n}(L) \to p(L) > 0$,

(b) $$\lim_{n\to\infty} g_n e^{\theta_n} = \lambda, \ 0 < \lambda < \infty,$$

and

(c) $$\sum_{i=0}^{U-L} \lambda^i \left(\prod_{j=L+1}^{L+i} \ell(j)\right) p(L) = 1,$$

where U is the upper limit of the carrier set of p(x), or ∞ when there is no upper limit.

If $p_{n,\theta_n}(x) \to p(x)$, then $\bar{L} = L$ and $\underline{U} = U$.

For $p_{n,\theta_n}(x)$ to have a degenerate limiting distribution, it is necessary that (a) is satisfied and that

(d) $$\lim_{n\to\infty} g_n e^{\theta_n} = 0 \text{ or } \infty.$$

Proof. Necessity: Since $p_{n,\theta_n}(x) \to p(x)$, x = 0,1,... and p(x) is a probability distribution, it follows that $p(x) \geq 0$, x=0,1,... and $\sum_{x=0}^{\infty} p(x) = 1$. Hence there clearly is a least non-negative integer L with $p(L) > 0$. Further, the hypothesis that

$$p_{n,\theta_n}(x) \to p(x) ,$$

a probability distribution insures $L \geq \overline{L}$ and hence $\overline{L} < \infty$. To see this, note that there is an N such that for $\min_{1 \leq i \leq k} n_i \geq N$, there are infinitely many L_n with $L_n = \overline{L}$ and no L_n such that $L_n > \overline{L}$. If $\overline{L} > 0$, then there are subsequences $p_{n_\nu, \theta_{n_\nu}}(x)$, $0 \leq x < \overline{L}$, such that $p_{n_\nu, \theta_{n_\nu}}(x) = 0$ for all n_ν, and since $\lim_{n\to\infty} p_{n,\theta_n}(x)$ exists for all $x = 0,1,\ldots$; for such x, we have $p(x) = 0$. Similarly, if $\underline{U} < \infty$, then $p(x) = 0$ for $x > \underline{U}$.

For $\underline{U} < \infty$ and $i > 0$ such that $\overline{L} < L + i \leq \underline{U} < \infty$, we have

$$p(L+i) = \lim_{n\to\infty} \left\{ p_{n,\theta_n}(L)\, e^{i\theta_n} \prod_{j=1}^{i} \frac{h(L+j,n)}{h(L+j-1,n)} \right\}$$

$$= \lim_{n\to\infty} \left\{ p_{n,\theta_n}(L)\, (g_n e^{\theta_n})^i \prod_{j=1}^{i} \frac{h(L+j,n)}{h(L+j-1,n)g_n} \right\} \qquad (2.4)$$

Then, if $p(x)$ is non-degenerate, we must have $p(x_0) > 0$ for some $x_0 > L$. Thus, from (2.2), (2.3) and (2.4), we have for $x_0 = L + i_0$ and $p(L + i_0) > 0$,

$$p(L + i_0) = p(L) \prod_{j=1}^{i_0} \ell(L + j) \lim_{n\to\infty} (g_n e^{\theta_n})^{i_0}$$

and thus $\lim_{n\to\infty} g_n e^{\theta_n}$ exists and is positive. We denote the limit by λ. Then it is obvious that $p(L), p(L + 1),\ldots,p(\underline{U})$ are all positive. Hence $\underline{U} = U$. When $\underline{U} = \infty$, the above argument estab- $\lim_{n\to\infty} g_n e^{\theta_n}$ exists and is positive and that $p(L+i) > 0$ for all $i \geq 0$. Similarly, if $L > \overline{L}$, then

$$p(L - 1) = p(L)/\lambda\, \ell(L) > 0,$$

a contradiction. Thus, $L = \overline{L}$. Finally, (c) is a trivial consequence of (2.2), (2.3), and (2.4), and (b), since for $0 \leq i \leq U - L$,

$$p(L + i) = p(L)\, \lambda^i \prod_{j=1}^{i} \ell(L + j),$$

and the sum over i is unity since p(x) is a probability distribution.

Now assume that p(x) is degenerate, then clearly $p(L) = 1$ and $p(x) = 0$ for all $x \neq L$. Hence, for $L < \underline{U}$,

$$p(L+1) = p(L)\, \ell(L+1) \lim_{n\to\infty} (g_n e^{\theta_n}) = 0,$$

and hence $\lim_{n\to\infty} (g_n e^{\theta_n}) = 0$. Similarly for $L > \overline{L}$,

$$p(L-1) = p(L)\, \{\ell(L) \lim_{n\to\infty} (g_n e^{\theta_n})\}^{-1} = 0$$

and $\lim_{n\to\infty} (g_n e^{\theta_n}) = \infty$. Further, we have actually shown that $\lim_{n\to\infty} (g_n e^{\theta_n}) = 0$ implies $L = \overline{L}$ and $\lim_{n\to\infty} (g_n e^{\theta_n}) = \infty$ implies $L = \underline{U}$, since for $\overline{L} + 1 \leq L \leq \underline{U} - 1$, we have deduced that $\lim_{n\to\infty} g_n e^{\theta_n}$ is both 0 and ∞, a contradiction.

Sufficiency: If $p_{n,\theta_n}(L) \to p(L)$

$$p_{n,\theta_n}(L+i) = p_{n,\theta_n}(L)\, (g_n e^{\theta_n})^i \prod_{j=1}^{i} \frac{h(L+j,n)}{h(L+j-1,n)g(n)},$$

from (2.3) and hypothesis (b), we have that

$$\lim_{n\to\infty} p_{n,\theta_n}(L+i) = p(L)\, \lambda^i\, \ell(L+1)\, \ell(L+2) \ldots \ell(L+i),$$

for $L < L + i \leq \underline{U}$. Hypothesis (c) insures that this limit is a probability distribution on $L, L + 1, \ldots, \underline{U}$. Further, it follows as before that $L = \underline{L}$ and that p(x) is the limit of $p_{n,\theta_n}(x)$ and is non-degenerate.

Remarks. It is easily seen that when $p(x)$ is non-degenerate,

$$\lambda = p(L+1)/p(L)\ \ell(L+1) . \tag{2.5}$$

Further $p(x)$ is also a distribution of exponential type.

Example 1. Let $B(M_n,p_n)$ denote the binomial distributions with parameters M_n, p_n, $n = 1,2,\ldots$. Then $h(x+1,n)/h(x,n) = (M_n-x)/(x+1)$, $x = 0,1,\ldots,M_n-1$. If $M_n \to \infty$ as $n \to \infty$, we can take $g_n = M_n$ and then $\ell(x+1) = 1/(x+1)$, $x = 0,1,2,\ldots$. Consequently, $g_n e^{\theta_n} = M_n p_n/(1-p_n) \to \lambda > 0$, if and only if $M_n p_n \to \lambda$, and thus when $M_n \to \infty$, the only possible non-degenerate discrete limiting distribution is the Poisson distribution. If M_n tends to a finite limit $M > 0$ as n tends to infinity, then we can take $g_n = 1$ and $\ell(x+1) = (M-x)/(x+1)$, $x = 0,1,\ldots,M-1$. Then $g_n e^{\theta_n} = M_n p_n/(1-p_n) \to \lambda > 0$ if and only if $\lim_{n\to\infty} p_n = p$, $0 < p < 1$. Thus, in this case the limit distribution is the binomial distribution with parameters M,p.

Degenerate limits arise when $M_n \to \infty$ and $M_n p_n \to 0$ (then the limiting distribution is degenerate at zero), when $M_n \to M < \infty$ and $p_n \to 0$ (degenerate at zero) or $p_n \to 1$ (degenerate at M).

Note that while the example is in fact very elementary, nevertheless the totality of possible discrete limiting distributions when M_n has a limit ($+\infty$ or $< \infty$) is very quickly enumerated by using the preceding theorem.

In the succeeding examples, we restrict attention to non-trivial limiting distributions.

Example 2. We now consider the negative binomial distribution defined by

$$p_{r_n,\theta_n}(x) = \binom{r_n+x-1}{x} p_n^{r_n}(1-p_n)^x ,$$

$$r_n > 0,\ 0 < p_n < 1,\ x = 0,1,2,\ldots . \tag{2.6}$$

Then $h(x+1,n)/h(x,n) = (r_n+x)/(x+1)$, $x = 0,1,2,\ldots$. Thus for $r_n \to \infty$ as $n \to \infty$, it suffices to take $g(n) = r_n$; then

$\ell(x+1) = 1/(x+1)$, $x = 0,1,2,\ldots$, and $g_n e^{\theta_n} = r_n(1-p_n)$, so that

$g_n e^{\theta_n} \to \lambda > 0$ if and only if $r_n(1-p_n) \to \lambda$. Hence, from Theorem 2.1, the only non-degenerate discrete limiting distribution is the Poisson with parameter λ.

Example 3. We now consider the following family of distributions.

$$p_{n,\gamma_n}(x) = \gamma_n^x \prod_{j=1}^{k} \binom{n_j}{x+u_j} \Big/ \sum_y \gamma_n^y \prod_{j=1}^{k} \binom{n_j}{y+u_j}, \quad k \geq 2 \tag{2.7}$$

where $u_1 = 0$, for $j = 2,3,\ldots,k$; u_j are specified integers with $-n_1 \leq u_j \leq n_j$; $\max_{1\leq j\leq k} (-u_j) \leq x \leq \min_{1\leq j\leq k} (n_j-u_j)$, and the sum in the denominator is over the range of x. The parameter γ_n is a non-negative real number. Here $h(x+1),n)/h(x,n) =$

$\prod_{j=1}^{k} \{n_j-x-u_j)/(x+u_j+1)\}$, $\max_{1\leq j\leq k} (-u_j) \leq x < \min_{1\leq j\leq k} (n_j-u_j)$. Therefore

we can take $g_n = \prod_{j=1}^{k} n_j$ and $\ell(x+1) = \prod_{j=1}^{k} (1 + x + u_j)^{-1}$. We

assume $n \to \infty$, and require that $g_n e^{\theta_n} = (\Pi n_j)\gamma_n \to \lambda > 0$; then we have

$$p(L) = \lambda^L / \prod_{j=1}^{k} (L+u_j)!h(u_2,u_3,\ldots,u_k;\lambda),$$

where $L = \max_{1\_j\_k} (-u_j)$ and $h(u_2,\ldots,u_k;\lambda = \sum_{y=L}^{\infty} \lambda^y / \prod_{j=1}^{k} (y+u_j)!$.

The distribution thus obtained is the "generalized incomplete modified Bessel distribution" (GIMB) with

$$p(x) = \lambda^x / \prod_{j=1}^{k} (x+u_j)!h(u_2,u_3,\ldots,u_k;\lambda),$$

$L \leq x < \infty$ and has been extensively studied in B. Harris and A. P. Soms (1974). This provides an example with $L \neq 0$ and also in this case the resulting limit distribution is not infinitely divisible, even though the carrier set is unbounded.

Example 4. Here we let n, γ_n index the family of distributions

$$p_{n,\gamma_n}(x) = a_{xn}\gamma_n^x / \sum_{y=0}^{\infty} a_{yn}\gamma_n^y, \quad x = 0,1,2,\ldots,$$

where

$$a_{xn} = \prod_{i=1}^{k} \binom{n_j+u_j+x-1}{u_j+x}, \quad u_1 = 0,$$

and $u_2,\ldots,u_k$ are specified non-negative integers, The asymptotic distribution for $k = 2$ was previously obtained by D. Hwang (1971). Here we obtain Hwang's result by applying Theorem 2.1.

We have that $h(x+1,n)/h(x,n) = \prod_{j=1}^{k} (n_j+u_j+x)/(u_j+x)$, so that we can take $g(n) = \prod_{j=1}^{k} n_j$ and hence $\ell(x+1) = \prod_{j=1}^{k} (u_j+x+1)^{-1}$. Then we require that $\gamma_n \pi(n_j) \to \lambda > 0$. It follows readily that $p(0) = 1/\prod_{j=1}^{k} u_j!\, h(u_2,\ldots,u_k;\lambda)$ and that the limiting distributions is the GIMB distribution (with $L = 0$) and is the unique non-degenerate discrete limiting distribution.

The next theorem provides an application of Theorem 2.1 to convergence of one-sided confidence limits.

Theorem 2.2. Let α and x be fixed where $0 < \alpha < 1$ and x is a non-negative integer. For each n, let $p_{n,\theta_n}(x)$ be a family of distributions indexed by $\theta \in \Theta$ and of the form (2.1). We assume that the equations $\sum_{t \leq x} p_n(t,\theta) = F(x;\theta,n) = \alpha$ has a solution $\overline{\theta}_n$ in θ for every n. Further, we assume that for all sequences $\{g_n e^{\theta_n}\}$, with $\lim_{n\to\infty} g_n e^{\theta_n} = \lambda > 0$, $p_{n,\theta_n}(x)$ has the non-degenerate discrete limit distribution $p_\lambda(x)$ of the form given in Theorem 2.1. Then

$$\lim_{n\to\infty} g_n e^{\bar{\theta}_n} = \bar{\lambda}, \tag{2.8}$$

where $\bar{\lambda}$ is the solution of $\sum_{t\leq x} p_\lambda(x) = \alpha$.

Proof. Since $F(x;\bar{\theta}_n,n) = \alpha < 1$, we have $x < U_n$, and

$$1 \geq p_{n,\bar{\theta}_n}(x+1) = [g(n)e^{\bar{\theta}_n} h(x+1,n)/h(x,n)g(n)]p_{n,\bar{\theta}_n}(x).$$

Consider the set $\{g_n e^{\bar{\theta}_n}\}$ and assume that there is a subsequence with $\lim_{n\to\infty} g_{n_\nu} e^{\bar{\theta}_{n_\nu}} = \infty$. Then there is an N such that for all $n_i > N$, $p_{n_\nu \bar{\theta}_{n_\nu}}(x-i) \to 0$ for all $i \geq 0$ by virtue of (2.3). Hence this contradicts

$$\sum_{t\leq x} p_{n_\nu \bar{\theta}_{n_\nu}}(t) = \alpha.$$

Thus, the set $g_n e^{\bar{\theta}_n}$ is bounded. Let $\bar{\lambda}$ be an accumulation point. Then there is a subsequence for which $\sum_{t\leq x} p_{\bar{\lambda}}(t) = F_{\bar{\lambda}}(x) = \alpha$. Since $p_\lambda(x)$ is an exponential family of distributions, it has strict monotone likelihood ratio and the solution of $F_\lambda(x) = \alpha$, if there is a solution, must be unique. Hence there cannot be two distinct subsequences converging to different limits.

Remark. Thus for the Poisson limit to the binomial distribution $n\bar{p} \to \bar{\lambda}$. For $x = 0$, this was explicitly demonstrated by Buehler (1957).

3. CONTINUOUS LIMITS OF MONOTONE LIKELIHOOD RATIO FAMILIES AND CONVERGENCE OF ONE-SIDED CONFIDENCE LIMITS. Let $F(x,\theta) = \sum_{t\leq x} p(t,\theta)$ be a discrete probability distribution on the

non-negative integers. We assume that $F(x,\theta)$ has a strictly monotone likelihood ratio in θ. We further assume that $F(x,\theta) = \alpha$ has a solution in θ for all x and all α, $0 < \alpha < 1$. It is further assumed that there exist functions $\mu(\theta)$, $\sigma(\theta)$ such that

$$L((X - \mu(\theta))/\sigma(\theta)) \to L(Y),$$

as $\theta \to \infty$, where $P\{Y \leq y\} = G(y)$ is a continuous probability distribution, $-\infty < y < \infty$. Further, there is a unique x_α such that $G(x_\alpha) = \alpha$ and that for all integers $x \geq 0$ and all α, $0 < \alpha < 1$, the equation

$$q_x(\theta) = (x - \mu(\theta))/\sigma(\theta) = x_\alpha$$

has a unique solution in θ, and thus $q^{-1}(x_\alpha)$ exists. This solution is a monotonically increasing function of x and tends to ∞ as x tends to ∞. In addition, for given x_α, we require that for each $\varepsilon > 0$, there exists $x_0(\varepsilon)$ and $\delta(\varepsilon)$, $\lim_{\varepsilon\to 0} \delta(\varepsilon) = 0$, such that $x > x_0(\varepsilon)$ implies that

$$\frac{q_x^{-1}(x_\alpha + \varepsilon) - q_x^{-1}(x_\alpha - \varepsilon)}{q_x^{-1}(x_\alpha - \varepsilon)} < \delta(\varepsilon) .$$

Then we have

Theorem 3.1. Let $\bar{\theta}_x$ be the unique solution of $\alpha = F(x,\theta)$ in θ and let θ_x^* be the unique solution of $q_x(\theta) = x_\alpha$, where $\alpha = G(x_\alpha)$, then

$$\lim_{x\to\infty} \theta_x^*/\bar{\theta}_x = 1.$$

Proof. Since $F(x,\theta)$ has a strict monotone likelihood ratio, we have $\lim_{x\to\infty} \bar{\theta}_x = \infty$. Further since $G(x)$ is a continuous distribution, the convergence in distribution to $G(x)$ is uniform in x; thus

$$\alpha = P_{\bar{\theta}_x}\left\{\frac{(X-\mu(\bar{\theta}_x)}{\sigma(\bar{\theta}_x)} \leq \frac{x-\mu(\bar{\theta}(x))}{\sigma(\bar{\theta}_x)}\right\} \to G\left(\frac{x-\mu(\bar{\theta}_x)}{\sigma(\bar{\theta}_x)}\right)$$

uniformly in $(x-\mu(\bar{\theta}_x))/\sigma(\bar{\theta}_x)$; equivalently, given $\varepsilon > 0$, there exists an x_0 such that for $x > x_0$

$$\left| G(x_\alpha) - G\left(\frac{x-\mu(\bar{\theta}_x)}{\sigma(\bar{\theta}_x)}\right) \right| < \varepsilon.$$

Thus there exists an x_1 such that for $x > x_1$, we have

$$\left| x_\alpha - \frac{x-\mu(\bar{\theta}_x)}{\sigma(\bar{\theta}_x)} \right| < \varepsilon,$$

or

$$|x_\alpha - q_x(\bar{\theta}_x)| < \varepsilon,$$

and

$$|x_\alpha - q_x(\theta_x^*)| < \varepsilon.$$

Hence

$$\frac{|\bar{\theta}_x - \theta_x^*|}{\bar{\theta}_x} \leq \frac{q_x^{-1}(x_\alpha+\varepsilon) - q_x^{-1}(x_\alpha-\varepsilon)}{q_x^{-1}(x_\alpha-\varepsilon)},$$

establishing the conclusion.

The conditions of this theorem and the assumptions enumerated at the start of this section are satisfied by the Poisson distribution with $\mu(\theta) = \theta$, $\sigma(\theta) = \sqrt{\theta}$ and the GIMB distributions with $\mu(\theta) = \theta^{1/k}$ and $\sigma(\theta) = \theta^{1/2}/\sqrt{k}$. Two cases not covered by this theorem are the normal limits to the binomial and negative binomial distributions.

4. APPLICATIONS TO HYPOTHESIS TESTING AND CONFIDENCE INTERVALS. This section provides a brief treatment of how the methods of section 2 may be applied. When the conditions of Theorems 2.1 and 2.2 are satisfied, to test $\theta \leq \theta_0$, we use

$\lambda \leq g(n) e^{\theta_0} = \lambda_0$ and for the corresponding $1 - \alpha$ confidence intervals on e^{θ}, we use $\bar{\lambda}/g(n)$, where $\bar{\lambda}$ is a $1 - \alpha$ upper confidence limit on λ. From Theorem 2.2, $g(n) e^{\bar{\theta}_n} \rightarrow \bar{\lambda}$ for each x; hence, if $g(n) e^{\theta_n}$ converges monotonically to λ, the true coverage probabilites converge to the nominal coverage probabilities, namely $1-\alpha$. From Theorem 2.1, it follows that for any sequence of hypotheses H_n: $e^{\theta} \leq \frac{c}{g(n)}$ and any sequence (n,θ_n) such that $g(n) e^{\theta_n} \rightarrow \lambda_1$, the true power of the test at (n,θ_n) converges to the power of the test $\lambda \leq c$ at λ_1. One may also regard (2.11) as the conditional distribution of $U_1 = X$, given $U_i = X_i - X_1 = u_i$, $2 \leq i \leq k$, where X_i are independently exponentially distributed by $P\{X_i = x\} = e^{-C(n_i,\theta_i)\,\theta_i x} h(x,n_i)$, and assume the convergence of section two. Then we may, as above, construct approximate tests of the hypothesis $\prod_{i=1}^{k} \theta_i \leq c$ and obtain approximate upper confidence limits on $\prod_{i=1}^{k} \theta_i$, with coverage and power converging as above.

Asymptotically optimal properties of the tests and confidence interval may be treated, as in Harris and Soms (1974), based on the limiting exponential type distribution.

We conclude with a discussion of limiting properties of the hypergeometric distribution.

Example 5. The hypergeometric distribution is given by

$$p_n(x) = P(X = x) = \binom{r}{x}\binom{N-r}{M-x}/\binom{N}{M} ,$$

where N,r,M are positive integers, and $\max(0,M+r-N \leq x \leq \min(r,M)$. Under the assumption $N \rightarrow \infty$, $r \rightarrow \infty$, $M \rightarrow \infty$ we have for $L_n \leq x < U_n$, $n = (N,r,M)$,

$$\frac{p_n(x+1)}{p_n(x)} = \frac{(r-x)(M-x)}{(x+1)(N-r-M+x+1)} = \frac{Mr(1-\frac{x}{r})(1-\frac{x}{M})}{N(x+1)(1+\frac{x+1-r-M}{N})} .$$

Then we can take $g_n = N/Mr$ and $\ell(x+1) = 1/(x+1)$; with these identifications, it can be shown by methods similar to those of section 2, that if $\frac{r+M}{N} \to 0$ as $n \to \infty$, the unique limiting discrete distribution is the Poisson distribution with $\lambda = \frac{Mr}{N}$, whenever $\frac{Mr}{N} \to \lambda > 0$. Let $\bar{r}$ be the unique value of r such that

$$\sum_{i=\max(0,M+r-N)}^{x} p_n(i) \leq \alpha, \quad \sum_{i=\max(0,M+r-N)}^{x+1} p_n(i) > \alpha, \; 0 < \alpha < 1,$$

where x is fixed; then $M\bar{r}/N \to \bar{\lambda}$, where $p_{\bar{\lambda}}(x) = \sum_{i=0}^{x} e^{-\bar{\lambda}} \bar{\lambda}^i/i! = \alpha$. To see this observe that $M\bar{r}/N$ is bounded and therefore $M-N+\bar{r} = N(\frac{M+\bar{r}}{N} - 1) < 0$, for n sufficiently large. Denoting $p_n(x)$ with $r = \bar{r}$ by $p_{n,\bar{r}}(x)$ we can assume that

$$\sum_{i=0}^{x} p_{n,\bar{r}}(i) \leq \alpha \text{ and } \sum_{i=0}^{x+1} p_{n,\bar{r}}(i) > \alpha.$$

Further, we must have $\bar{r} \to \infty$, since otherwise $p_{\bar{r}}(x)$ is asymptotically degenerate and also, $\bar{r}/N \to 0$. We note further that $p_{n,\bar{r}+1}(i) - p_{n,\bar{r}}(i) \to 0$, since

$$|p_{n,\bar{r}+1}(i) - p_{n,\bar{r}}(i)| = \frac{1}{\binom{N}{M}} \left| \binom{\bar{r}}{i}\binom{N-\bar{r}-1}{M-i}\left(\frac{\bar{r}+1}{\bar{r}+1-i} - \frac{N-\bar{r}}{N-\bar{r}+i-M}\right) \right|$$

$$= \frac{N-M}{N} \frac{\binom{\bar{r}}{i}\binom{N-\bar{r}-1}{M-i}}{\binom{N-1}{M}} \left| \frac{\bar{r}+1}{\bar{r}+1-i} - \frac{N-\bar{r}}{N-\bar{r}+i-M} \right|$$

$$\leq \left| \frac{Ni-M\bar{r}+i-M}{(\bar{r}+1-i)(N-\bar{r}+i-M)} \right|$$

$$\leq \left| \frac{1}{(\bar{r}+1-i)(1-\frac{\bar{r}}{N}-\frac{M}{N}+\frac{i}{N})} \right| \left(\frac{\bar{r}\,M+M+Ni+i}{N}\right),$$

and tends to zero as $n \to \infty$. The convergence is clearly uniform for $0 \leq i \leq x$ and hence $\sum_{i=0}^{x} p_{n,\bar{r}}(i) \to \alpha$ and it follows that $\frac{M\bar{r}}{N} \to \bar{\lambda}$. The basic difficulty in this example is occasioned by the fact that r is discrete.

REFERENCES

[1] Buehler, R. J. (1957). J. Amer. Statist. Assoc. 52, 482-493.

[2] Harris, B. (1971). J. Amer. Statist. Assoc. 66, 609-617.

[3] Harris, B. and Soms, A. P. (1974). J. Amer. Statist. Assoc. 69, 259-263.

[4] Harris, B. and Soms, A. P. (1974). Tables of the two factor and three factor generalized incomplete modified Bessel distributions. In Selected Tables in Mathematical Statistics, H. L. Harter and D. B. Owen (eds.). Vol. 3. To appear.

[5] Hwang, D. (1971). Interval estimation of functions of Bernoulli parameters with reliability and biomedical applications. Technical Report #152, University of Minnesota School of Statistics.

ON EFFICIENCY AND EXPONENTIAL FAMILIES IN STOCHASTIC PROCESS ESTIMATION

C. C. Heyde and P. D. Feigin

Department of Statistics, Australian National University, Canberra, Australia

SUMMARY. A general definition of efficiency for stochastic process estimation is proposed and some of its ramifications are explored. Of particular importance in the definition is the form of the derivative of the logarithm of the likelihood. The question of the simplest possible form for this leads on to a discussion of extensions of the concepts of sufficiency and exponential families, the latter in a Markov process context. The paper concludes with several illustrative examples.

KEY WORDS. Stochastic process estimation; efficiency; sufficiency; exponential family; maximum likelihood; Martingale limit theory; branching process; first order autoregression; power series distributions.

1. A GENERAL DEFINITION OF EFFICIENCY.

Suppose that we have a sample $X_1, X_2, \ldots, X_n$ of consecutive observations from some stochastic process whose distribution depends on a single parameter θ, $\theta \in \Theta \subset R$. Let $L_n(\theta)$ be the likelihood function associated with $X_1, \ldots, X_n$ and suppose that $L_n(\theta)$ is differentiable with respect to θ and $E(d \log L_n(\theta)/d\theta)^2 < \infty$ for each n. Suppose in addition that if $P_n(X_1, \ldots, X_n)(= L_n(\theta))$ is the joint probability (density) function of $X_1, \ldots, X_n$, then $\sum_{x_n} P_n(x_1, \ldots, x_n)$ $(\int P_n(x_1, \ldots, x_n)\, dx_n)$ can be differentiated twice with respect to θ under the summation (integration) sign. Then, writing F_k for

G. P. Patil et al. (eds.), Statistical Distributions in Scientific Work, Vol. 1, 227-240.

the σ-field generated by $X_1,\ldots,X_k$, $k \geq 1$, taking F_o as the trivial σ-field, and $L_o = 1$, we have

$$E\left(\frac{d \log L_n(\theta)}{d\theta} \,|F_{n-1}\right) = \frac{d \log L_{n-1}(\theta)}{d\theta} \quad \text{a.s., } n \geq 1. \tag{1}$$

The condition (1) is precisely that $\{U_n = (d \log L_n(\theta)/d\theta) = \sum_1^n u_i, F_n, n \geq 1\}$ is a zero mean martingale.

Next, we set

$$I_n(\theta) = \sum_{k=1}^{n} E\left[\left(\frac{d \log L_k(\theta)}{d\theta} - \frac{d \log L_{k-1}(\theta)}{d\theta}\right)^2 |F_{k-1}\right]. \tag{2}$$

This is a form of conditional information which reduces to the standard Fisher information in the case where the X_i's are independent. We can think of

$$E\left[\left(\frac{d \log L_k(\theta)}{d\theta} - \frac{d \log L_{k-1}(\theta)}{d\theta}\right)^2 |F_{k-1}\right]$$

$$= E\left[\left(\frac{d \log L_k(\theta)}{d\theta}\right)^2 |F_{k-1}\right] - \left(\frac{d \log L_{k-1}(\theta)}{d\theta}\right)^2$$

$$= E(u_k^2|F_{k-1})$$

as the information contained in $X_1,\ldots,X_k$ which is not contained in $X_1,\ldots,X_{k-1}$ for given $X_1,\ldots,X_{k-1}$. We also have, writing $v_k = (du_k/d\theta)$, that under the conditions imposed above,

$$E(u_k^2|F_{k-1}) = -E(v_k|F_{k-1}). \tag{3}$$

Definition. We shall say that an estimator T_n of θ is asymptotically efficient if

$$I_n^{\frac{1}{2}}(\theta)\left[T_n - \theta - \beta(\theta)\, I_n^{-1}(\theta)\, \frac{d \log L_n(\theta)}{d\theta}\right] \to 0 \tag{4}$$

in probability for some β which does not involve the observations.

In the standard case of independent and identically distributed observations, this definition reduces to that of Rao (1965) pp. 285, 286. Furthermore, the motivation behind the general definition is the same as for the independence case. Under the conditions we have imposed, $\{(d \log L_n(\theta)/d\theta) = \sum_1^n u_i\}$ is a martingale and a law of large numbers for martingales gives consistency of T_n for θ if $I_n(\theta) \to \infty$ a.s. as $n \to \infty$. A central limit result for martingales gives that

$$I_n^{-\frac{1}{2}}(\theta) \frac{d \log L_n(\theta)}{d\theta} = \left(\sum_1^n u_i\right) \Big/ \left[\sum_1^n E(u_i^2 | F_{i-1})\right]^{\frac{1}{2}} \tag{5}$$

converges in distribution to $N(0,1)$ under certain regularity conditions which we shall mention later. This result, together with (4), will assure that $I_n^{\frac{1}{2}}(\theta)(T_n - \theta) \to N(0, \beta^2(\theta))$ in distribution. Interestingly, the random norming $I_n^{\frac{1}{2}}(\theta)$ in (4) and (5) gives rise to much more general central limit results than are provided by the constant norming

$$\left[E\, I_n(\theta)\right]^{\frac{1}{2}} = \left[E\left(\sum_1^n u_i\right)^2\right]^{\frac{1}{2}} = \left[\sum_1^n E u_i^2\right]^{\frac{1}{2}} .$$

To see that consistency of T_n for θ follows from (4) under the condition $I_n(\theta) \to \infty$ a.s. as $n \to \infty$ we shall show that

$$\left[I_n(\theta)\right]^{-1} \frac{d \log L_n(\theta)}{d\theta} = \left[\sum_1^n E(u_i^2 | F_{i-1})\right]^{-1} \sum_1^n u_i$$

converges a.s. to zero. This follows, for example, from Proposition IV-6-2 of Neveu (1970) provided $I_n(\theta) \to \infty$ a.s. and

$$\sum_1^\infty [I_n(\theta)]^{-2} E(u_n^2 | F_{n-1}) < \infty \text{ a.s.}$$

This last condition is automatically satisfied, however, since for $\{a_j\}$ a sequence of positive constants and writing $b_n = \sum_1^n a_j$,

$$\sum_1^\infty \left(\sum_1^\infty a_j\right)^{-2} a_n = \sum_1^\infty b_n^{-2} (b_n - b_{n-1}) \qquad (b_o = 0)$$

$$= \sum_1^\infty b_n (b_n^{-2} - b_{n+1}^{-2})$$

$$= \sum_1^\infty b_n (b_n^{-1} - b_{n+1}^{-1})(b_n^{-1} + b_{n+1}^{-1})$$

$$\leq 2 \sum_{1}^{\infty} (b_n^{-1} - b_{n+1}^{-1}) \leq 2b_1^{-1} < \infty .$$

The question of asymptotic normality in (5) cannot at this stage be provided with such a general answer. Various sufficient conditions have been given which ensure that

$$\left[\sum_{1}^{n} E(u_i^2 | F_{i-1})\right]^{-\frac{1}{2}} \sum_{1}^{n} u_i \to N(0,1)$$

in distribution. One such set of conditions is that, writing $s_n^2 = \sum_{1}^{n} Eu_i^2 = EI_n(\theta)$,

$$s_n^{-2} I_n(\theta) \overset{p}{\to} \eta, \tag{6}$$

"p" denoting "in probability", where $\eta > 0$ a.s., while

$$\text{(i)} \quad s_n^{-2} \sum_{1}^{n} E(u_i^2 \, I(|u_i| \geq \varepsilon \, s_n | F_{i-1}) \overset{p}{\to} 0 \tag{7}$$

as $n \to \infty$, $\forall \, \varepsilon > 0$, or

(ii) $(s_{n-r}^2 / s_n^2) \to c^{-r}$ as $n \to \infty$ for some $c > 1$ and for fixed $j \leq r$,

$$E[\exp (it \, u_{n-j} s_n^{-1} s_{n-r-1} \, I_{n-r-1}^{-\frac{1}{2}}(\theta)) | F_{n-j-1})] \overset{p}{\to} \exp \{-\tfrac{1}{2} t^2 c^{-j} (1-c^{-1})\} \tag{8}$$

[Hall and Heyde (1974)].

An example where (6) and (7) are required is given by the case where $\{u_i\}$ is (strictly) stationary but not ergodic. In the context of a stationary process $\{X_i\}$ it is useful to redefine the basic quantities slightly differently by replacing at each stage the σ-fields F_k by σ-fields $\mathcal{G}_k$ with $\mathcal{G}_k$ generated by $X_k, X_{k-1}, \ldots$. Then, stationarity of the process $\{X_i\}$ gives stationarity of the process $\{u_i\}$ and, of course, the martingale property is not disturbed. It is useful to avoid the assumption of ergodicity since such an assumption is uncheckable on the basis of a single realization of the process $\{X_i\}$.

Two examples where the use of (6) and (8) is appropriate are given in Section 3.

We should note that if the condition (4) is strengthened to convergence in the mean of order 2, then equivalent forms are

$$\operatorname{corr}\left\{I_n^{\frac{1}{2}}(\theta)(T_n-\theta),\ I_n^{-\frac{1}{2}}(\theta)\frac{d \log L_n(\theta)}{d\theta}\right\} \to 1,$$

or alternatively,

$$E\left\{(T_n-\theta)\frac{d \log L_n(\theta)}{d\theta}\right\} \Big/ \left[E\{I_n(\theta)(T_n-\theta)^2\}\ E\left\{I_n^{-1}(\theta)\left(\frac{d \log L_n(\theta)}{d\theta}\right)^2\right\}\right]^{\frac{1}{2}} \to 1.$$

The latter effectively contains the Cramér-Rao lower bound formulation of asymptotic efficiency.

Further discussion of the general properties of asymptotically efficient estimators will be given elsewhere.

To check asymptotic efficiency via the definition (4), the vital piece of information is a suitably tractable expression for $(d \log L_n(\theta)/d\theta)$. This will of course be in its simplest form when

$$\frac{d \log L_n(\theta)}{d\theta} = J_n(\theta)(\hat{\theta}_n-\theta), \quad \theta \in \Theta, \tag{9}$$

and we shall focus attention on this functional form, recalling that in the case of independent X_i's such results are obtained if, and only if, the probability (density) functions belong to the exponential family.

Suppose now that (9) holds. With a view to (4) we need to know when $J_n(\theta) = I_n(\theta)$, which would give asymptotic efficiency of the maximum likelihood estimator $\hat{\theta}_n$. In addition, when (9) holds with $J_n(\theta) = I_n(\theta)$ we have that $\hat{\theta}_n$ is strongly consistent for θ provided $I_n(\theta) \to \infty$ a.s. and also $I_n^{\frac{1}{2}}(\theta)(\hat{\theta}_n-\theta) \to N(0,1)$ in distribution under certain additional conditions cited above. In fact, when (9) holds we have $J_n(\theta) = I_n(\theta)$ if and only if $J_n(\theta)$ is F_{n-1} measurable and $(d/d\theta)\ J_{n-1}(\theta)\ J_n^{-1}(\theta) = 0$, $\forall n > 1$. Furthermore, the condition $J_n(\theta) = I_n(\theta)$, $\forall n \geq 1$, is equivalent to the condition that $J_n(\theta) = \phi(\theta)\ H_n\ (X_1,\ldots,X_{n-1})$ for some functions ϕ and H_n where ϕ does not involve the X_i's and H_n does not involve θ.

To establish these last results we first suppose that $J_n(\theta) = I_n(\theta)$, $\forall n \geq 1$. Then, under the regularity conditions imposed above,

$$E(u_n \hat{\theta}_n | F_{n-1}) = \frac{d}{d\theta} E(\hat{\theta}_n | F_{n-1})$$

$$= \frac{d}{d\theta} E(I_n^{-1}(\theta) U_n + \theta | F_{n-1})$$

$$= \frac{d}{d\theta} [I_n^{-1}(\theta) U_{n-1} + \theta]$$

$$= (\hat{\theta}_{n-1} - \theta) \frac{d}{d\theta} [I_{n-1}(\theta)\, I_n^{-1}(\theta)] + 1 - I_{n-1}(\theta)\, I_n^{-1}(\theta),$$

while

$$E(u_n \hat{\theta}_n | F_{n-1}) = E[u_n (I_n^{-1}(\theta)\, U_n + \theta) | F_{n-1}]$$

$$= I_n^{-1}(\theta)\, E(u_n^2 | F_{n-1}) = 1 - I_{n-1}(\theta)\, I_n^{-1}(\theta),$$

so that $(d/d\theta)\, I_{n-1}(\theta)\, I_n^{-1}(\theta) = 0$. On the other hand, if $J_n(\theta)$ is F_{n-1} measurable and $(d/d\theta)\, J_{n-1}(\theta)\, J_n^{-1}(\theta) = 0$, the arguments just presented give

$$E(u_n \hat{\theta}_n | F_{n-1}) = 1 - J_{n-1}(\theta)\, J_n^{-1}(\theta)$$

and

$$E(u_n \hat{\theta}_n | F_{n-1}) = J_n^{-1}(\theta)\, E(u_n^2 | F_{n-1})$$

which yield $J_n(\theta) = I_n(\theta)$. These results ensure that if $J_n(\theta) = \phi(\theta) H_n(X_1, \ldots, X_{n-1})$, then $I_n(\theta) = J_n(\theta)$. On the other hand, if $I_n(\theta) = J_n(\theta)$, $\forall n \geq 1$, we have that $(d/d\theta) J_{n-1}(\theta) J_n^{-1}(\theta) = 0$, $\forall n > 1$, and hence

$$\frac{d}{d\theta} \log J_n(\theta) = \frac{d}{d\theta} \log J_{n-1}(\theta) = \ldots = \frac{d}{d\theta} \log J_1(\theta)$$

$$= \frac{d}{d\theta} \log I_1(\theta) = c(\theta),$$

say. The general solution of this system is the form $J_n(\theta) = \phi(\theta) H_n(X_1, \ldots, X_{n-1})$, $J_n(\theta)$ being F_{n-1} measurable.

At this point, it seems appropriate to discuss an extension to stochastic processes of the concept of sufficiency. For independent observations, sufficiency can be characterized by the factorization theorem for the joint probability (density) function of a sample of size n. That is, a statistic T_n is sufficient for θ if and only if

$$f(X_1,\ldots,X_n|\theta) = g(T_n,\theta)\; h_n(X_1,\ldots,X_n) \tag{10}$$

where h_n does not involve θ. The form (10) remains appropriate in the stochastic process situation and the following calculation shows that $(H_n(X_1,\ldots,X_{n-1}),\ \hat{\theta}_n)$ is minimal - sufficient for θ when (9) holds with $J_n(\theta) = I_n(\theta)$. In fact,

$$\frac{d \log L_n(\theta)}{d\theta} = I_n(\theta)(\hat{\theta}_n-\theta) = \phi(\theta)H_n(X_1,\ldots,X_{n-1})(\hat{\theta}_n-\theta)$$

implies

$$\log L_n(\theta) = H_n(X_1,\ldots,X_{n-1})(\hat{\theta}_n\Phi(\theta) - \psi(\theta)) + K_n(X_1,\ldots,X_n) \tag{11}$$

where

$$\frac{d}{d\theta}\Phi(\theta) = \phi(\theta),\ \frac{d}{d\theta}\psi(\theta) = \theta\,\phi\,(\theta),$$

and the function K_n does not involve θ. Clearly (11) corresponds to the form (10) with $T_n = (H_n(X_1,\ldots,X_{n-1}),\ \hat{\theta}_n)$. This last statistic is minimal - sufficient in the usual sense that any other sufficient statistic must be a function of H_n and θ_n.

2. THE MARKOV CASE: CONDITIONAL EXPONENTIAL FAMILIES.

In the remainder of this paper we shall confine our attention to the case where the stochastic process $\{X_i\}$ under consideration is a Markov process. There is a substantial literature on the use of the likelihood function in estimation for such processes; see for example the books of Billingsley (1961) and Roussas (1972), but all this work is constrained by assumptions of stationarity and ergodicity so that the same constant normings as in the classical random sampling (independence) case can be used. We shall avoid such assumptions and concentrate on families which give rise to the form (9) with $J_n(\theta) = I_n(\theta),\ \theta \in \Theta$.

We shall deal with the time-homogeneous Markov process whose conditional probability (density) function of X_n given X_{n-1} is $f(X_n|X_{n-1},\ \theta)$. Then,

$$L_n(\theta) = \prod_{i=1}^{n} f(X_i | X_{i-1}, \theta),$$

so that

$$\frac{d \log L_n(\theta)}{d\theta} = \sum_{i=1}^{n} \frac{d}{d\theta} \log f(X_i | X_{i-1}, \theta) = \sum_{i=1}^{n} u_i$$

and

$$u_i = \frac{d}{d\theta} \log f(X_i | X_{i-1}, \theta).$$

Suppose that

$$\frac{d \log L_n(\theta)}{d\theta} = I_n(\theta)(\hat{\theta}_n - \theta), \quad \forall n \geq 1. \tag{12}$$

Then, taking $n = 1$, we have

$$\frac{d}{d\theta} \log f(X_1 | X_o, \theta) = I_1(\theta)(\hat{\theta}_1 - \theta) = \phi(\theta) H(X_o)(\hat{\theta}_1 - \theta).$$

Clearly there must be a single root $\tilde{\theta} = m(x,y)$ of the equation $(d/d\theta)\ f\ (x|y,\theta) = 0$ and then

$$\frac{d}{d\theta} \log f(x|y,\theta) = \phi(\theta) H(y)[m(x,y) - \theta]. \tag{13}$$

This equation defines what we shall call a conditional exponential family for the problem under consideration.

Conversely, if (13) is satisfied we have

$$\frac{d \log L_n(\theta)}{d\theta} = \phi(\theta) \sum_{i=1}^{n} H(X_{i-1})[m(X_i, X_{i-1}) - \theta] \tag{14}$$

and

$$u_i = \phi(\theta)\ H(X_{i-1})[m(X_i, X_{i-1}) - \theta] = \frac{d}{d\theta} \log f(X_i | X_{i-1}, \theta)$$

so that, almost surely,

$$\begin{aligned} 0 = E(u_i | F_{i-1}) &= E\{\phi(\theta) H(X_{i-1})[m(X_i, X_{i-1}) - \theta] | F_{i-1}\} \\ &= \phi(\theta) H(X_{i-1})[E(m(X_i, X_{i-1}) | F_{i-1}) - \theta] \end{aligned}$$

and

$$E(m(X_i, X_{i-1}) | F_{i-1}) = \theta$$

Also, from (3),

$$E(u_i^2 | F_{i-1}) = - E\left(\frac{d^2}{d\theta^2} \log f(X_i | X_{i-1}, \theta) | F_{i-1}\right)$$

$$= - E[\phi'(\theta) H(X_{i-1})(m(X_i, X_{i-1}) - \theta) | F_{i-1}]$$

$$+ \phi(\theta) H(X_{i-1})$$

$$= \phi(\theta) H(X_{i-1}) \text{ a.s.,} \tag{15}$$

so that (14) can be rewritten as

$$\frac{d \log L_n(\theta)}{d\theta} = I_n(\theta)(\hat{\theta}_n - \theta) \tag{16}$$

where

$$\hat{\theta}_n = \left[\sum_{i=1}^{n} H(X_{i-1})\right]^{-1} \sum_{i=1}^{n} H(X_{i-1}) m(X_i, X_{i-1}). \tag{17}$$

The conditional exponential form is thus necessary and sufficient for the derivative of the logarithm of the likelihood to be expressible in the form (12). Notice also that $\hat{\theta}_n$ is strongly consistent for θ provided $\sum_{i=1}^{\infty} H(X_{i-1})$ diverges a.s.

3. EXAMPLES. In this section we shall give a discussion of the conditional exponential families defined by (13) for two estimation problems. These examples have been deliberately chosen as ones in which the norming using random variables $I_n(\theta)$ cannot conveniently be replaced by a constant norming using $EI_n(\theta)$. The full force of the ideas of Section 1 is thus illustrated.

The first problem we shall consider is that of the estimation of the mean θ of the offspring distribution of a supercritical Galton - Watson branching process on the basis of a sample $\{X_0, X_1, \ldots, X_n\}$ of consecutive generation sizes. Here we have $1 < \theta = E(X_1 | X_o = 1) < \infty$ and we shall suppose that $\sigma^2 = \text{var}(X_1 | X_o = 1) < \infty$.

In this case we have [from work of Harris (1948)]

$$\hat{\theta}_n = (Y_n - X_o) Y_{n-1}^{-1}$$

where $Y_n = \sum_{k=o}^{n} X_k$, so that $\hat{\theta}_1 = X_1 X_o^{-1}$ and hence $m(x,y) = xy^{-1}$ for integers $x \geq 0$ and $y \geq 1$. The form (13) then gives

$$\frac{d}{d\theta} \log f(x|y,\theta) = \phi(\theta)\ H(y)y^{-1}(x-\theta y).$$

Now

$$u_i = \frac{d}{d\theta} \log f(X_i|X_{i-1},\theta) = \phi(\theta)\ H(X_{i-1})X_{i-1}^{-1}(X_i-\theta X_{i-1})$$

so that

$$E(u_i^2|F_{i-1}) = \phi^2(\theta)\ H^2(X_{i-1})\ X_{i-1}^{-2}\ E[(X_i-\theta X_{i-1})^2|F_{i-1}]$$

$$= \phi^2(\theta)\ \sigma^2 H^2(X_{i-1})X_{i-1}^{-1}$$

$$= \phi(\theta)\ H(X_{i-1})$$

using (15). Thus,

$$H(X_{i-1}) = X_{i-1}\ \sigma^{-2}\ [\phi(\theta)]^{-1}$$

and

$$\frac{d}{d\theta} \log f(x|y,\theta) = \sigma^{-2}\ (x-\theta y). \tag{18}$$

Taking $y = 1$ and writing $p_j = P(X_1 = j|X_o = 1) = f(j|1,\theta)$, we then have

$$\frac{dp_j}{d\theta} = \sigma^{-2}\ (j-\theta)\ p_j. \tag{19}$$

It is easily checked that the family of power series distributions is characterized by the property (19). These are the distributions for which

$$p_j = a_j\ \lambda^j\{f(\lambda)\}^{-1}, \quad j = 0,1,2,\ldots;\ \lambda > 0,$$

where $a_j \geq 0$ and $f(\lambda) = \sum_{j=o}^{\infty} a_j\lambda^j$. For details and references concerning the class of power series distributions see Ord (1972), Chapter 6 and Johnson and Kotz (1969), Chapter 2. We readily find that

$$\theta = \lambda f'(\lambda)\{f(\lambda)\}^{-1}, \quad \sigma^2 = \{(d/d\theta) \log \lambda\}^{-1}$$

and also that (18) is satisfied since

$$\frac{d}{d\theta} f(k_1 | k_o, \theta) = \frac{d}{d\theta} \sum_{\substack{j_1=o \\ j_1 + \dots + j_{k_o} = k_1}}^{\infty} \dots \sum_{j_{k_o}=o}^{\infty} p_{j_1} p_{j_2} \dots p_{j_{k_o}}$$

$$= \sum_{\substack{j_1=o \\ j_1 + \dots + j_{k_o} = k_1}}^{\infty} \dots \sum_{j_{k_o}=0}^{\infty} \left[\sum_{r=1}^{k_o} \frac{d \log p_{j_r}}{d\theta} \right] p_{j_1} \dots p_{j_{k_o}}$$

$$= \sigma^{-2} (k_1 - \theta k_o) f(k_1 | k_o, \theta)$$

upon using the condition (19).

The power series distributions are just the one-parameter exponential family and they form the conditional exponential family in the branching process estimation problem. A detailed discussion of this estimation problem has been given in Heyde (1974). This includes discussion of the strong consistency of $\hat{\theta}_n$ and asymptotic normality of $I_n^{\frac{1}{2}}(\theta)(\hat{\theta}_n - \theta)$ (which can be established via (6) and (8)).

The next problem we shall consider is that of estimating the parameter θ in a first order autoregression

$$X_i = \theta X_{i-1} + \varepsilon_i$$

where the ε_i's are independent and identically distributed random variables with mean zero and variance σ^2 and ε_i is independent of X_{i-1}. We wish to estimate θ on the basis of a sample $X_o, X_1, \dots, X_n$. The process is not subjected to a stability condition of the kind $|\theta| < 1$.

Let $g(x)$ denote the density (probability) function of ε_i. We have

$$f(X_i | X_{i-1}, \theta) = g(X_i - \theta X_{i-1})$$

so that

$$L_n(\theta) = \prod_{i=1}^{n} g(X_i - \theta X_{i-1})$$

and

$$u_i = \frac{d}{d\theta} \log f(X_i | X_{i-1}, \theta) = - X_{i-1} \frac{g'(X_i - \theta X_{i-1})}{g(X_i - \theta X_{i-1})} .$$

Also,

$$v_i = \frac{du_i}{d\theta} = X_{i-1}^2 \frac{g(X_i - \theta X_{i-1}) g''(X_i - \theta X_{i-1}) - (g'(X_i - \theta X_{i-1}))^2}{g^2(X_i - \theta X_{i-1})} ,$$

so that, using (3),

$$\begin{aligned} E(u_i^2 | F_{i-1}) &= - E(v_i | F_{i-1}) \\ &= - X_{i-1}^2 \, E \left\{ \frac{g(\varepsilon_1) g''(\varepsilon_1) - (g'(\varepsilon_1))^2}{g^2(\varepsilon_1)} \right\} \\ &= c \, X_{i-1}^2 , \end{aligned}$$

where

$$c = E \left\{ \frac{(g'(\varepsilon_1))^2 - g(\varepsilon_1) g''(\varepsilon_1)}{g^2(\varepsilon_1)} \right\}$$

which must be positive since $E(u_i^2 | F_{i-1}) > 0$ a.s.

If (13) holds we have from (15) that

$$\phi(\theta) \, H(X_{i-1}) = c \, X_{i-1}^2 .$$

Furthermore, since

$$\frac{d}{d\theta} f(x | y, \theta) = - y \frac{g'(x - \theta y)}{g(x - \theta y)} ,$$

we have that

$$x - m(x,y) \, y = m$$

where m is the (unique from (13)) mode of the distribution of ε. Then, (13) gives

$$- y \frac{g'(x - \theta y)}{g(x - \theta y)} = c \, y^2 \left(\frac{x-m}{y} - \theta\right),$$

which yields

$$\frac{d}{dz} g(z) = - c \, (z-m) \, g(z)$$

and hence

$$g(z) = Be^{\frac{1}{2}c(z-m)^2}, \quad -\infty < z < \infty.$$

The distribution of ε is thus normal. Since ε has mean zero and variance σ^2 we must have $m = 0$, $c = \sigma^{-2}$, $B = (2\pi\sigma^2)^{-\frac{1}{2}}$. The normal distribution is therefore the only member of the conditional exponential family for this problem.

A detailed discussion of this estimation problem has been given by Anderson (1959). This includes a discussion of consistency and the asymptotic behavior of

$$I_n^{\frac{1}{2}}(\theta)(\hat{\theta}_n-\theta) = \sigma^{-1}\left(\sum_{i=1}^{n} X_{i-1}^2\right)^{\frac{1}{2}} \left[\sum_{i=1}^{n} X_i X_{i-1} \left(\sum_{i=1}^{n} X_{i-1}^2\right)^{-1} - \theta\right].$$

It is interesting to note that the expression $\sum_{i=1}^{n} X_i X_{i-1} \Big/ \sum_{i=1}^{n} X_{i-1}^2$ continues to be a consistent estimator of θ whatever the distribution of ε and that

$$\left(\sum_{i=1}^{n} X_{i-1}^2\right)^{\frac{1}{2}} \left[\sum_{i=1}^{n} X_i X_{i-1} \left(\sum_{i=1}^{n} X_{i-1}^2\right)^{-1} - \theta\right]$$

converges in distribution to a proper limit law whatever ε. This limit law is normal if $|\theta| < 1$ but if $|\theta| > 1$ it is normal if and only if ε is normally distributed. Of course, if ε is not normally distributed, the estimator $\left(\sum_{i=1}^{n} X_{i-1}^2\right)^{-1} \sum_{i=1}^{n} X_i X_{i-1}$ will not in general bear any special relation to the maximum likelihood estimator.

REFERENCES

[1] Anderson, T. W. (1959). Ann. Math. Statist. 30, 676-687.

[2] Billingsley, P. (1961). Statistical Inference for Markov Processes. University of Chicago Press, Chicago.

[3] Hall, P. G. and Heyde, C. C. (1974). Manuscript under preparation.

[4] Harris, T. E. (1948). Ann. Math. Statist. 19, 474-494.

[5] Heyde, C. C. (1974). Remarks on efficiency in estimation for branching processes. To appear in Biometrika.

[6] Johnson, N. L. and Kotz, S. (1969). Discrete Distributions. Houghton Mifflin, Boston.

[7] Neveu, J. (1970). Calcul des Probabilites. 2nd ed. Masson et Cie, Paris.
[8] Ord, J. K. (1972). Families of Frequency Distributions. Griffin, London.
[9] Rao, C. R. (1965). Linear Statistical Inference and Its Applications. Wiley, New York.
[10] Roussas, G. G. (1972). Contiguity of Probability Measures. Cambridge University Press, Cambridge.

A LAGRANGIAN GAMMA DISTRIBUTION

D. L. Nelson and P. C. Consul

Boeing Computer Services, Inc., Seattle, Washington, U.S.A. and Division of Statistics, The University of Calgary, Alberta, Canada

SUMMARY. A new generalization of the gamma distribution is developed. This generalization, called the Lagrangian gamma, is the distribution of the time between occurrences of a generalized Poisson process. Moments of the distribution are derived, plots of the density are reproduced, and methods for parameter estimation are discussed.

KEY WORDS. Generalized Poisson process, Lagrangian gamma distribution.

1. INTRODUCTION. There exist a number of generalizations of the gamma distribution, perhaps most notable among these being the one by Stacy (1962). The present paper develops a different generalization of the gamma, one that has its basis in the discrete Lagrange distributions developed recently by Consul and Shenton (1972). One of these, the generalized or Lagrangian Poisson (LP) distribution, has been presented in detail by Consul and Jain (1973a, 1973b) in the form

$$p(k;\lambda_1,\lambda_2) = \lambda_1(\lambda_1+k\lambda_2)^{k-1}\, e^{-(\lambda_1+k\lambda_2)}/k! \qquad (1.1)$$

$$(\lambda_1>0,\ |\lambda_2|<1,\ \lambda_1+k\lambda_2>0,\ k=0,1,2,\ldots)$$

$$= 0 \text{ otherwise}$$

This distribution has a variance that is less than, equal to, or greater than the mean as λ_2 is negative, zero, or positive,

G. P. Patil et al. (eds.), Statistical Distributions in Scientific Work, Vol. 1, 241-246.

respectively. Consul and Jain (1973a) have shown that the LP can be used to describe data that follow binomial or negative binomial, as well as Poisson, distributions, by employing suitable parametric values.

In the next section we develop the continuous distribution of the waiting time between occurrences of a Lagrangian Poisson process. Since the time between occurrences of a classical Poisson process follows a gamma distribution, we call this new distribution the Lagrangian gamma (LG) distribution. It is shown that the classical gamma is a special case of the LG, and other useful properties of the LG are demonstrated.

2. DERIVATION OF THE LAGRANGIAN GAMMA DISTRIBUTION. Consider an LP process with density function given by (1.1). Define a and θ so that

$$\lambda_1 = at, \quad \lambda_2 = a\theta t,$$

where $t \geq 0$ is the time between occurrences of the LP process, and $a>0$. With this notation, the cumulative distribution of the random variable T, which is the time until the rth occurrence of the GP process, is given by

$$F(t;\, a,\, \theta,\, r) = P[\, T \leq t \,]$$

$$= 1 - \sum_{k=0}^{r-1} \frac{at[(1+k\theta)at]^{k-1}}{k!} e^{-(1+k\theta)at}.$$

Differentiation yields the density function of T:

$$f(t;a,\theta,r) = \sum_{k=0}^{r-1} \frac{a[(1+k\theta)at]^{k-1}}{k!} e^{-(1+k\theta)at}[(1+k\theta)at - k], \quad (2.1)$$

where $a>0$; $r=1,2,\ldots$; $\theta>-1/r$; and $t \geq 0$. Expression (2.1) is the one that we have adopted for the Lagrangian gamma density.

We now show that $\int f(t;a,\theta,r) = 1$.

$$\int_0^\infty f(t;a,\theta,r)dt = \sum_{k=0}^{r-1} \frac{a}{k!} \Big[(1+k\theta)^k a^k \int_0^\infty t^k e^{-(1+k\theta)at} dt$$

$$- k(1+k\theta)^{k-1} a^{k-1} \int_0^\infty t^{k-1} e^{-(1+k\theta)at} dt \Big]$$

$$= \sum_{k=0}^{r-1} \frac{a}{k!} \frac{k!}{a(1+k\theta)} - \sum_{k=1}^{r-1} \frac{a}{k!} \frac{k!}{a(1+k\theta)} = 1.$$

Irrespective of the value of θ, the special case $r = 1$ yields the classical exponential distribution $f(t;a,\theta,1) = ae^{-at}$. The special case $\theta = 0$ yields the following:

$$f(t;a,0,r) = \sum_{k=0}^{r-1} \frac{a(at)^{k-1}}{k!} e^{-at}(at-k)$$

$$= \frac{a(at)^{r-1}}{(r-1)!} e^{-at},$$

which is the classical gamma distribution for integer r. The LG distribution should, then, give a closer approximation to the true distribution of many phenomena that can be described by the usual gamma distribution.

3. MOMENTS OF THE DISTRIBUTION. Straightforward computation gives:

$$E[T] = \frac{1}{a} \sum_{k=0}^{r-1} \frac{1}{(1+k\theta)^2}. \tag{3.1}$$

This, then, is the mean time to the rth occurrence of a generalized Poisson process. Similarly,

$$E[T^2] = \frac{2}{a} E[T] + \frac{2(1-\theta)}{a^2} \sum_{k=0}^{r-1} \frac{k}{(1+k\theta)^3} \tag{3.2}$$

Higher moments can be derived in a straightforward manner, using the integral definition of the gamma function.

An alternative method of computing moments is through the moment generating function of the LG, which is obtained as

$$M(u) = E[e^{uT}]$$

$$= (1 - u/a)^{-1} + \frac{u}{a} \sum_{k=1}^{r-1} \frac{(1+k\theta)^{k-1}}{(1+k\theta-u/a)^{k+1}},$$

where the second term is understood to vanish for $r = 1$. It is easy to verify that $M(u) = (1 - u/a)^{-r}$, the moment generating function of the classical gamma distribution, when $\theta = 0$.

Note from equation (3.1) that the mean of the LG distribution increases as θ decreased and vice versa.

4. GRAPHICAL REPRESENTATION. The behavior of the Lagrangian gamma density for various values of r, a, and θ was investigated by computing tables for the following parametric values: $r = 2(1)10$, $a = 1(1)10$, $-1/r < \theta \leq 14/r$, and $t = 0(.1)14.6$. Figures 1 and 2 show graphs drawn from these tables for typical values of the parameters.

Figure 1 demonstrates that for even relatively small changes in θ, the LG density curve changes drastically, becoming more peaked as θ increases, for fixed a. As θ approaches $-1/r$, the density becomes very broad and flat. The fact that the mean of the distribution, as well as the mode, decreases as θ increases, is clearly recognizable. The curve for $\theta = 0$ is that of the classical gamma function.

Figure 2 shows densities for fixed θ and several values of a. As a increases for fixed θ, the mean, mode, and variance all decrease and the density becomes more peaked.

5. PARAMETER ESTIMATION. The estimation of a and θ in the LG distribution is made difficult by the summations necessary to compute the mean and mean square in equations (3.1) and (3.2). This will be especially true when the parameter r is large. When this is the case, however, it is possible to obtain parameter estimates using other functions of the random variable, namely $E[T^{-1}]$ and $E[T^{-2}]$.

Using equation (2.1) and assuming $r \geq 2$, we obtain

$$E[T^{-1}] = a \log(1+\theta) + \frac{a}{r-1} . \tag{5.1}$$

For $r \geq 3$,

$$E[T^{-2}] = a \int_0^\infty \left[\frac{1-e^{-a\theta t}}{t^2} + a(1+\theta)e^{-a\theta t} - a(1+2\theta)e^{-2a\theta t} \right] e^{-at} dt$$

$$+ \frac{a^3(1+2\theta)^2}{2} \int_0^\infty e^{-(1+2\theta)at} dt$$

$$+ \sum_{k=3}^{r-1} \int_0^\infty \frac{a^3(at)^{k-3}(1+k\theta)^{k-1}}{k!} [at(1+k\theta)-k]\, e^{-(1+k\theta)at} dt$$

Expanding the exponentials within the brackets in the first integral into a power series yields, upon simplification,

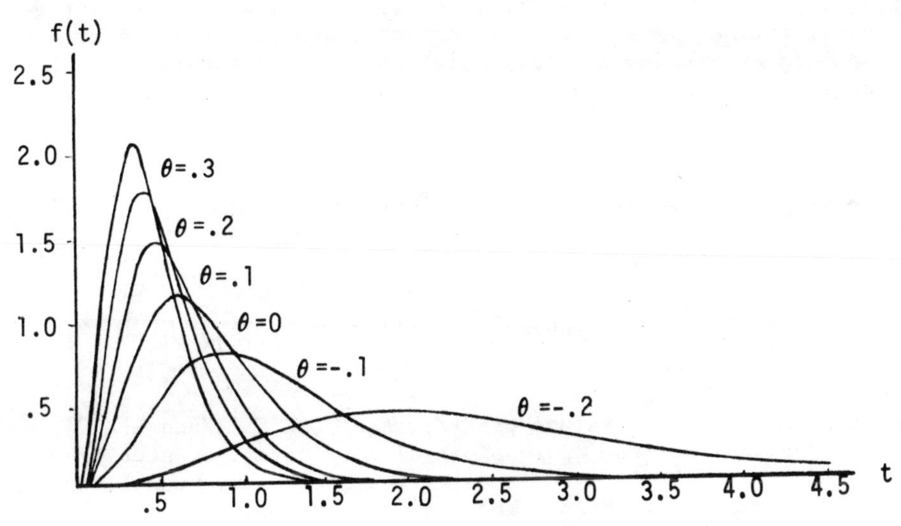

FIGURE 1. LAGRANGIAN GAMMA DENSITY FOR $r=4$, $a=5$, and $\theta=-.2, -.1, 0, .1, .2, .3$.

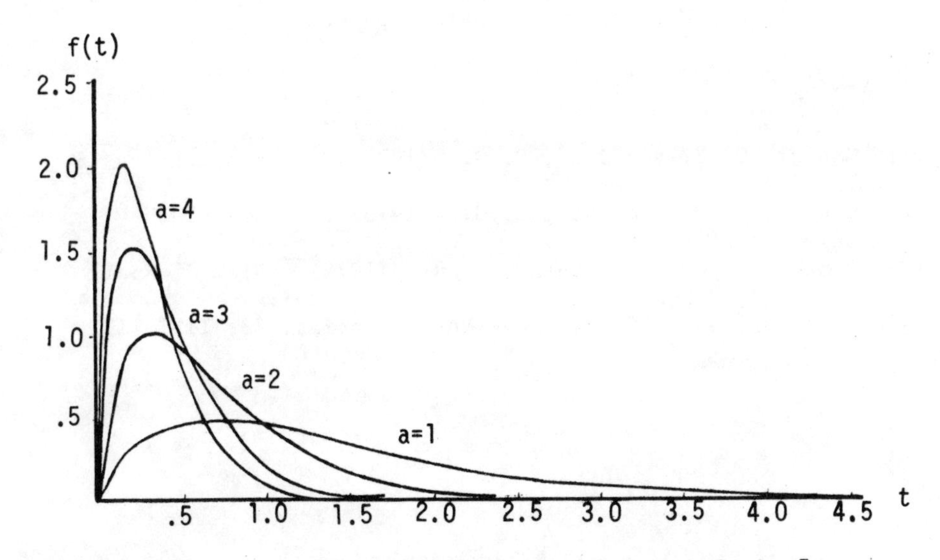

FIGURE 2. LAGRANGIAN GAMMA DENSITY FOR $r=2$, $\theta=.5$ and $a= 1, 2, 3, 4$.

$$E[T^{-2}] = a^2(1+2\theta) \quad \log \frac{1+2\theta}{1+\theta} + \frac{1}{r-2} - aE[T^{-1}]. \qquad (5.2)$$

Thus, for $r \geq 3$, equations (5.1) and (5.2) provide an alternative means of estimating a and θ when expectations are replaced by corresponding sample moments. Solution of these equations will generally require some machine algorithm, for example, a Newton or quasi-Newton algorithm.

6. ADDITIONAL COMMENTS. The transformation $x = at$ in equation (2.1) gives a standard form of the Lagrangian gamma distribution in two parameters only:

$$f(x;\theta,r) = \sum_{k=0}^{r-1} \frac{[(1+k\theta)x]^{k-1}}{k!} e^{-(1+k\theta)x}[(1+k\theta)x-k],$$

and the corresponding values of $E[X]$, $E[X^2]$, $E[X^{-1}]$ and $E[X^{-2}]$ can be derived from the expressions already given, by setting $a=1$. Thus a is merely a scale parameter in the LG, much like σ in the normal distribution. The standard form of the LG can be used for the tabulation of probability points.

The LG distribution should find wide application in waiting-line models and queuing theory, where the arrival process can be described by a generalized Poisson (including binomial and negative binomial) process. A numerical example and further properties of the LG will be published in a separate paper.

REFERENCES.

[1] Consul, P. C. and Jain, C. G. (1973a). Technometrics 15, 791-799.
[2] Consul, P. C. and Jain, C. G. (1973b). Biometrische Zeitschrift 15, 495-500.
[3] Consul, P. C. and Shenton, L. R. (1972). SIAM J. Appl. Math. 23, 239-248.
[4] Stacy, E. W. (1962). Ann. Math. Statist. 33, 1187-1192.

MULTIVARIATE DISTRIBUTIONS AT A CROSS ROAD*

Samuel Kotz

Department of Mathematics, Temple University,
Philadelphia, Pennsylvania, U.S.A.

SUMMARY. A selective survey of developments in the area of continuous multivariate statistical distributions during the twentieth century is presented with emphasis on non-normal models. Discrete multivariate distributions are not discussed. An extension of a useful bivariate family is indicated.

KEY WORDS. Multivariate distributions, survey, expansions, numerical methods, Farlie-Morgenstern family.

1. INTRODUCTION. From the inception of statistics as a self-contained science up to the present day, statistical distribution theory has been the cornerstone and a basic tool of statistical reasoning and applications. Recent advances in relatively "unorthodox" branches of statistics such as non-parametric methods, Bayesian (and subjective) inference, and computerization of statistical methodology have only increased the importance of statistical distribution theory as a basic operational device for applications. Most non-parametric methods derive their ultimate justification from asymptotic distributional theory; Bayesian methods have enhanced the application of distributional assumptions in statistical reasoning by placing additional emphasis on distributions on the parameter space; computerization of statistical techniques has made it feasible to study more efficiently

---

*Supported by the Air Force Office of Scientific Research under Grant AFOSR-74-2701.

G. P. Patil et al. (eds.), Statistical Distributions in Scientific Work, Vol. 1, 247-270.

many properties of distributions as well as to verify the degree of applicability of various distributions in practice.

However, in spite of developments in the field of computers, progress in the area of multivariate statistical models has been substantially less prominent than in other branches of mathematical statistics.

The remarks below attempt to indicate some of the reasons for this unhappy state of affairs. It is the purpose of this survey to assist in overcoming the existing hesitation and reluctance to tackle the problem in the field of non-normal multivariate models, by stressing some historically and practically significant contributions to this field conducted early in this century and proposing some generalizations to the existing models.

The following points should be kept in mind when one approaches a study or application of multivariate statistical models:

Uncertainty about the homogeneity of a sample may serve as an objection to the use of a single functional representation of a joint distribution and suggests the use of a mixture of distributions. Even if a sample is homogeneous, the "classical" translation systems of distributions and readily available transformations may be sufficient for various inferential purposes [Fisk (1973)].

The underlying assumption and the reference point (tacit or explicit) of any statistical analysis is the notion of the mathematical model. In practice, no model suggested for data will describe the data exactly; often we may make unduly restrictive assumptions about the parametric structure or we may ignore the knowledge latent in the data which cannot be expressed quantitatively [Finney (1974)]. (The widely accepted convention of models based on the normal distribution is undergoing a slow but consistent reevaluation nowadays. This process is less evident in the area of multivariate distributions mainly due to the excessive popularity among theoretical and practicing statisticians of the techniques of multivariate analysis which are rooted in the multinormal convention.)

One of the central problems of modern multivariate distribution theory is, no doubt, the proper and meaningful assessment of dependence between the components. Overemphasis on this aspect of the problem, however, may lead to situations where the increased sophistication of the theories is self-defeating. The increase of the sample size often yields additional dependence patterns which require incorporation of additional parameters in the model, and so on. This important problem has not as yet received sufficient attention in the literature devoted to multivariate models.

2. HISTORICAL SKETCH OF SOME NEGLECTED RESULTS. The study of multivariate distributions can be considered to have started with the work of Francis Galton (1885), "Family likeness in stature," with an appendix by J. D. H. Dickson, Proc. Roy. Soc. 40, 42-73, who introduced and developed the ideas of correlation and regression into the "bivariate normal surface" investigated earlier by such writers as P. S. Laplace, G. A. A. Flana, C. F. Gauss, and A. Bravais. Although subsequent workers developed the bivariate normal distribution theory to a high degree (it is still one of the few bivariate distributions available in any detail), it was soon realized that this frequency law was by no means universal. Various attempts were then made to describe analytically the "skew" (or non-normal) variation, whose presence in homogeneous material was observed to be at least as common as that of normality.

The investigation in these "skew bivariate frequency surfaces", as the subject came to be known, took on many diverse approaches. It will be seen subsequently that certain methods of research have more or less been exhausted without yielding much useful results, while many alternatives have not as yet been developed to any great detail. We shall briefly sketch some highlights of the investigations carried out mostly in the first third of the twentieth century. These results have been neglected--perhaps unjustly--by many modern authors.

Further details on the early history of this subject are given by Lancaster (1972).

(i) The Double Hypergeometric Series of K. Pearson (1923)

The earliest investigations of K. Pearson appear to have been prompted by certain observations exhibiting distinctly non-normal characteristics such as, for example, the correlation of cards in a certain game. By analogy with the bivariate normal surface, where a change of variables corresponding to a rotation of axis in the plane can separate two correlated variables into independent ones, he attempted to find a similar relation applicable to non-normal cases. He soon came to the conclusion, however, that for skew correlation surfaces these "axes of independent probability" do not in general coincide with the principal axes of the (equiprobability) contour system.

Pearson's next step was an attempt to extend the principle underlying his system of skew curves, i.e. to determine a family of surfaces from two general differential equations derived in connection with a certain double hypergeometric series. These equations, as given by Rhodes (1925) were of the general form:

$$\frac{1}{f}\frac{df}{dx} = \frac{\text{cubic in } x,y}{\text{Quartic in } x,y} \qquad \frac{1}{f}\frac{df}{dy} = \frac{\text{Another cubic in } x,y}{\text{Same Quartic in } x,y} \tag{1}$$

The above equations, however, were found to be non-integrable without imposing certain limitations on the constants occurring on the right sides. On the other hand, several special solutions as found by L. N. G. Filon, L. Isserlis and E. C. Rhodes [Pearson (1924)] all indicated that the respective correlation coefficients were dependent on the variation of these constants and besides, did not fit the theoretical frequencies (as computed from the original double hypergeometric series) to any reasonable degree. These surfaces, as Pearson pointed out again, were thus of little practical importance.

(ii) The Method of Skew Correlation and Non-Linear Regression

The failure of the approach described in (i) prompted the development of a more general method which involved no assumption as to the form of the frequency distribution. Such a method was suggested by a result due to Yule (1897) in a paper on multiple correlation, where it was shown that if the regression is linear, irrespective of the type of frequency, the multiple regression "plane" as reached by the method of least squares was identical in form with that derived from a multiple normal surface. The method of approaching the problem of correlation from the form of the regression curves was extended by Pearson (1905) to the case of non-linear regressions.

This method of non-linear regression, as studied by Pearson (1921) and later extended by Neyman (1926) also had some serious drawbacks. The main defect appeared to be that since no assumptions were made concerning the underlying frequency distributions, the probability of an individual observation falling within certain limits as measured from the regression curves could not be determined. For this (and a variety of other reasons) the non-linear regression method was not considered a generally satisfactory solution to the problem of skew correlations.

(iii) Other Methods

Among the many other methods considered by various authors, the most attention seems to have been concentrated on correlation surfaces with marginal distributions belonging to the Gram-Charlier type. One of these surfaces with Type A margins is of the form:

$$f(x,y) = \phi(x,y) + \sum_{(p+q)\geq 3}^{\infty} (-1)^{p+q} \frac{A_{pq}}{p!q!} \frac{\partial^{p+q}}{\partial_x^p \partial_y^q} \phi(x,y) \qquad (2)$$

$$\phi(x,y,\rho) = \frac{1}{2\pi\sqrt{1-\rho^2}} \exp\left\{-1/2 \frac{(x^2 - 2\rho xy + y^2)}{1-\rho^2}\right\} \qquad (3)$$

is the "normal surface".

The above surface, for example, had been considered from many diverse viewpoints, such as the approximation of special forms of F(x,y) [Charlier (1914), Jørgansen (1916), Pearson (1925), and Rhodes (1923)], determination of the differential coefficients [Pearson, (1925)], and approximation by considering the partial moment curves [Wicksell (1917a)], or curves of equal probability [Wicksell (1917b)]. Much other work has also been done on related topics, notably those of Edgeworth (1904), Wicksell (1917c), Steffenson (1922) and Van der Stok (1908) to name a few. More details are given in K. V. Mardia's monograph [(1970), Chapter 3].

Much of the earlier work described in the preceding section was handicapped by the lack of suitable data on which a test of goodness of fit could be made and comparative results could be derived. This drawback does not exist at present; this fact should encourage practicing statisticians to test the applicability of the classical methods described (or mentioned) in this section when the opportunity arises.

3. MORE RECENT CONTRIBUTIONS (1947 - 1960).

More recent works tend to specify on certain problems (or applications) rather than the purely theoretical approach used earlier. The period is also characterized by the introduction of a variety of specific bivariate distributions and first applications of numerical methods on computers.

(i) Generalization of Univariate Methods and Distributions

One possible approach to the problem of skew correlation surfaces is to seek some form of extending familiar univariate methods and distributions to higher dimensions. This approach has been investigated in considerable detail by Van Uven (1947, 1948), Sagrista (1952), and Risser (1950), among others. As Sagrista pointed out, a direct extension of Pearson's original first order equation into a system of partial differential equations is impractical, since the equations are difficult to solve and (in most cases) the solutions are too general in nature. One is thus limited to consider the notable special cases, a few of which will be outlined below.

Case (a). The first case studies by Sagrista (1952) and Van Uven (1947, 1948) considered the general problem from the point of view of the normal curve. It is known that the normal bivariate distribution

$$f = k \exp\left[-\frac{\alpha x_1^2 + 2\beta x_1 x_2 + \gamma x_2^2}{2}\right], \text{ (k is a normalizing constant)} \tag{4}$$

gives rise to the pair of equations

$$\frac{\partial f}{\partial x_1} = -f(\alpha x_1 + \beta x_2) \qquad \frac{\partial f}{\partial x_2} = -f(\beta x_1 + \gamma x_2) \tag{5}$$

and conversely. The bivariate normal surface is the unique solution of the system of equations (5). The foregoing seems to suggest that some useful bivariate surfaces may be derived by considering solutions of appropriate generalizations of (5).

One obvious generalization is to put the equations into the form

$$\frac{\partial f}{\partial x_i} = -f\Big(g_i(x_1,x_2)\Big), \qquad i = 1,2, \tag{6}$$

where $g_i(x_1,x_2)$, are homogeneous polynomials in x_1 and x_2 of order $n > 1$. This result, however, is not very useful since it only leads to a family of exponential distributions with homogeneous polynomial exponents of order $n + 1$. A somewhat more practical alternative is to consider instead the equations

$$\frac{\partial f}{\partial x_1} = -f^m(\alpha x_1 + \beta x_2), \quad m \neq 1 \ , \ \frac{\partial f}{\partial x_2} = -f^m(\beta x_1 + \gamma x_2) \tag{7}$$

which, on putting

$$m = 1 - \frac{1}{n} \tag{8}$$

and solving, give rise to the family of surfaces

$$f = k\left(1 - \frac{x_1^2}{a_1^2} - \frac{x_2^2}{a_2^2} + \frac{2rx_1x_2}{a_1a_2}\right)^n \tag{9}$$

$= K(1 + Q)^n$ say, where $Q = Q(x_1, x_2)$ is the quadratic form in (9) and a_i, k, K, r, n are constants.

The family of functions (9) appears to have been studied most extensively.

To begin with, if functions (9) were to quality as proper density functions for all arbitrary real n, it is necessary that $1 + Q \geq 0$. If, furthermore, it is assumed that Q is a positive definite quadratic form, then the condition of non-negativeness of a density function implies that the family (9) is only defined inside the elliptical regions $Q < 1$, centered at the origin, where $r^2 < 1$.

Secondly, in order that the integral defining the normalizing condition converge, it is necessary that $n + 1 > 0$.

By the symmetry of expression (9) in x_1 and x_2, it is obvious that all moments of odd order vanish. The even ordered moments all exist (for $n + 1 > 0$) and are computed in the usual manner to be

$$\mu_{f,g} = \frac{a^f b^g}{\pi(1 - r^2)^{\frac{f+g}{2}}} \frac{\Gamma(n + 2)}{\Gamma(\frac{f+g}{2} + n + 2)} \cdot \qquad (10)$$

$$\cdot \sum_{\nu=0}^{g/2} \binom{g}{2\nu} r^{g-\nu} (1 - r^2)^{\nu} \Gamma(\nu + 1/2) \Gamma\left(\frac{f + g - 2\nu + 1}{2}\right) .$$

The first few moments are:

$$\mu_{2,0} = \sigma_1^2 = \frac{a^2}{2(1 - r^2)(n + 2)}, \quad \mu_{0,2} = \sigma_2^2 = \frac{b^2}{2(1 - r^2)(n + 2)},$$

$$\mu_{1,1} = \frac{abr}{2(1 - r^2)(n + 2)}, \quad \mu_{4,0} = \frac{3a^4}{4(1 - r^2)(n + 2)(n + 3)}, \qquad (11)$$

when the correlation coefficient is r. From (11) one can verify the relationship

$$\frac{\mu_{f,1}}{\mu_{f+1,0}} = \frac{\mu_{1,1}}{\mu_{2,0}} = \frac{br}{a} \quad \text{for all } f=1,2,\ldots . \qquad (12)$$

Some other quantities of interest are listed below.

Marginal density functions:

$$f_1(s) = kb\sqrt{\pi}\ \frac{\Gamma(n+1)}{\Gamma(n+3/2)} \left(1 - \frac{1-r^2}{a^2}\, x^2\right)^{n+1/2};$$

$$-\frac{a}{\sqrt{1-r^2}} < x < \frac{a}{\sqrt{1-r^2}},$$

$$f_2(y) = ka\sqrt{\pi}\ \frac{\Gamma(n+1)}{\Gamma(n+3/2)} \left(1 - \frac{1-r^2}{b^2}\, y^2\right)^{n+1/2}; \tag{13}$$

$$-\frac{b}{\sqrt{1-r^2}} < y < \frac{b}{\sqrt{1-r^2}},$$

marginal moments = 0 for odd numbers, skewness = 0 ,

kurtosis $\dfrac{\mu_{4,0}}{\mu_{2,0}^2} - 3 = \dfrac{-3}{n+3} \qquad (<0)$;

regression function of y on x: $y = \dfrac{br}{a}\, x$

$$\text{of } x \text{ on } y\text{:} \quad x = \frac{ar}{b}\, y \tag{14}$$

conditional variance $\sigma^2(y|x) = \dfrac{b^2}{2n+3}\ \left(1 - \dfrac{1-r^2}{a^2}\, x^2\right)$.

Note that the regression functions are linear as a consequence of the relation between the moments (14). [For a generalization of density (9), using multidimensional matrices, see Goodman-Kotz (1973).]

Case (b). Another possible generalization of the system of equation (5) is by adding a suitable term to the right hand side:

$$\frac{\partial f}{\partial x} = f\left(-ax - by + n\,\frac{2ax + 2by}{h^2+ax^2+2bxy+cy^2}\right) \tag{15}$$

$$\frac{\partial f}{\partial y} = f\left(-ax - cy + n\,\frac{2bx + 2cy}{h^2 + ax^2 + 2bxy + cy^2}\right).$$

This method of generalization obviously includes the bivariate normal surface as a special case for n = 0. The quantities a, b, c, h, n are constants to be determined later.

The solution of the equations (15) gives rise to the family of surfaces

$$f = k\, e^{-\frac{ax^2 + 2bxy + cy^2}{2}} \cdot (h^2 + ax^2 + 2bxy + cy^2)^n \qquad (16)$$

$$-\infty < x < \infty \qquad -\infty < y < \infty$$

and, as before, the first task is to seek conditions on the constants so that (16) can become a family of proper density functions. These conditions are found to be

$$ac - b^2 > 0 \qquad a > 0 \qquad c > 0 \qquad (17)$$

(which is equivalent to taking $ax^2+2bxy+cy^2$ as a positive quadratic form).

Due to the symmetry of function (16) in x and y, the mean and all higher odd moments are again zero. The lower ordered moments are computed to be: $\mu_{2,0} = c\,Q$, $\mu_{0,2} = a\,Q$, $\mu_{1,1} = -b\,Q$, where

$$Q = \left[\frac{k\,\pi\, h^{2n+2}}{\sqrt{ac - b^2}} + \frac{2n + 2 - h^2}{2}\right]\left(\frac{1}{ac - b^2}\right)^{-1}, \qquad (18)$$

leading to the coefficient of correlation

$$r = \frac{\mu_{1,1}}{\sqrt{\mu_{2,0}\,\mu_{0,2}}} = \frac{-b}{\sqrt{ac}}, \qquad (19)$$

a quantity which is independent of h and k. The marginal density function for x is:

$$f_1(x) = \frac{2k}{\sqrt{c}}\, e^{-\frac{x^2 d}{2}} (h^2 + x^2 d)^{n+1/2}\, \phi(x,n) \qquad (20)$$

and the regression functions and conditional variances are of the form:

$$m_2(y) = -\frac{b}{a}y\,; \qquad \sigma^2(y|x) = \frac{h^2 + x^2 d}{c}\left[\frac{\phi(x,\, n + 1)}{\phi(x,n)} - 1\right]$$

$$\text{where} \quad d = \frac{ac - b^2}{c} \qquad (21)$$

and $$\phi(x,n) = \int_0^\infty e^{-\frac{h^2 + x^2 d}{2}} (1 + t^2)^n \, dt \tag{22}$$

is Laplace's Integral. It is noted that in all the above cases (15)-(20) the various quantities reduce to the bi-normal case for $n = 0$.

The special case $h = 0$ is of particular interest, since it results in a considerable simplification of all the quantities (15-20) considered above.

As Sagrista (1952) has pointed out, the method employed in the two preceding cases is a general systematic method for obtaining other probability surfaces. A similar procedure has been used by Risser (1950) to obtain generalization of Laplace's and Pearson's univariate systems.

Case (c) Extensions of Some Pearsonian Curves. In the two cases (a) and (b) described in the preceding section, the essential idea is to seek some practical generalization of established univariate procedures leading to families of distributions. An obvious alternative, then, is to seek generalizations of the univariate distributions themselves. This alternative approach has been studied by a very considerable number of authors and is discussed in Basu-Block (1974). For purposes of illustration of some less known results, the investigations of Cooper (1963) and McGraw and Wagner (1967) will be briefly discussed.

(i) The Multivariate Pearson Type II and Type VII Distribution. The familiar univariate normal distribution for a random variable X with mean μ and variance σ^2 has the density function

$$f(x) = \frac{1}{\sigma\sqrt{2\pi}} \exp\left[-\frac{(x-\mu)^2}{2\sigma^2}\right] \tag{23}$$

Hence an n-dimensional random variable $X = (X_1, X_2, \ldots, X_n)$ with mean vector $\underset{\sim}{\mu} = (\mu_1, \mu_2, \ldots, \mu_n)$ and covariance matrix $\Sigma = \{\sigma_{ij}\}$ has the n-multinormal distribution if the density is of the form

$$f(\underset{\sim}{x}) = \frac{1}{(2\pi)^{n/2}|\Sigma|} \exp\left[-1/2 \cdot (\underset{\sim}{x}-\underset{\sim}{\mu})^T \Sigma^{-1} (\underset{\sim}{x}-\underset{\sim}{\mu})\right] \tag{24}$$

where $|\Sigma| = \det \Sigma$ and T denotes transposition. The method of generalization from (23) to (24) is obvious and can be readily adapted for other suitable distributions. For example, the Pearson Type II distribution with parameter m and defined symmetrically about the mean μ has the density:

$$p(x) = \frac{w}{B(1/2,\ m+1)} \left[1 - w^2\ (x-\mu)^2\right]^m \qquad \frac{-1}{w} \leq x - \mu \leq \frac{1}{w} \tag{25}$$

where $B(1/2,\ m+1) = \Gamma(1/2)\ \Gamma(m+1)/\Gamma(m+3/2)$ is the beta function and $m \geq 0$. A corresponding expression for an n dimensional variable X with mean vector $\underset{\sim}{\mu}$ and covariance matrix Σ is seen to be

$$p(\underset{\sim}{x}) = \begin{cases} \dfrac{\Gamma(m+\frac{n}{2}+1)}{\Gamma(m+1)\,(\pi)^{m/2}}\ |W|^{1/2} \cdot \left[1 - (\underset{\sim}{x}-\underset{\sim}{\mu})^T\ W(\underset{\sim}{x}-\underset{\sim}{\mu})\right]^m & \text{over region G} \\ 0 & \text{elsewhere } (m \geq 0) \end{cases} \tag{26}$$

where region G is the interior of the hyper-ellipsoid

$$(\underset{\sim}{x}-\underset{\sim}{\mu})^T\ W\ (\underset{\sim}{x}-\underset{\sim}{\mu}) = 1 \tag{27}$$

and W (referred to as a scaling matrix) is a scalar multiple of the inverse covariance matrix:

$$W = \frac{1}{(2m + n + 2)}\ \Sigma^{-1} \tag{28}$$

The distribution (26) is clearly a multivariate analogue of (25) and extends the bivariate form (9). It should be noted, though, that the general U-shaped distribution defined for (25) and $-1 < m < 0$ is not included as a special case of (26).

It is well known that all moments of the distribution (25) exist and that the univariate Type II distribution can assume many different forms by changing the continuous parameter m (e.g. rectangular for $m = 0$, elliptic for $m = 1/2$, parabolic for $m = 1$, normal for $m \to \infty$, etc.). These properties are also valid for the multivariate analogue, inasmuch as all 1-dimensional marginal functions of (26) are of the form (25). More specifically, it can be shown that all marginal densities of the Type II distribution (26) remain of the same type, except that the parameter m increases by 1/2 for each dimension change. For example, a distribution having parameter $m = m'$ in n-space would have a marginal distribution in a k-space having parameter

$$m = m' + 1/2\ (n - k). \tag{29}$$

In going to a k-dimensional subspace within the original subspace, the corresponding submatrices of the covariance matrix and of the inverse scaling matrix are also preserved.

Finally, it is worth mentioning that for spaces of high dimensions, the univariate marginals for (26) approach the normal distribution. This property is generally not true for extensions of other Pearsonian distributions, e.g. the Type VII discussed below.

(ii) The Pearson Type VII Distribution. In accordance with previous examples, the univariate Pearson Type VII distribution with density

$$p(x) = \frac{1}{B(1/2,\ m-1/2)} \left[1 + w^2\ (w - \mu)^2\right]^{-m} \quad 2m > 1,\ -\infty < x < \infty,$$

has as one possible multivariate extension the n-dimensional distribution with density

$$p(\underset{\sim}{x}) = \frac{\Gamma(m)}{\pi^{\frac{n}{2}}\Gamma(m - \frac{n}{2})} |W|^{1/2} \left[1 + (\underset{\sim}{x} - \underset{\sim}{\mu})^T\ W(\underset{\sim}{x} - \underset{\sim}{\mu})\right]^{-m} \quad (2m > n) \qquad (30)$$

for all $-\infty < x < \infty$. (The quantities X, $\underset{\sim}{\mu}$ and W are as defined previously for the Type II distribution.) The condition $2m > n$ is necessary to ensure that $p(\underset{\sim}{x})$ be integrable and in fact, a density function. If $2m > n+q$, the qth moment exists. In particular, if $2m > n+2$, the covariance matrix Σ exists and W is again a scalar multiple of Σ^{-1}, i.e.

$$W = \frac{1}{(2m - n - 2)}\ \Sigma^{-1} \ . \qquad (31)$$

It is to be noted that even if Σ does not exist, W still represents a "scaling matrix". That is to say, the general significance of W is that it is a matrix which if applied as a transformation on the space, transforms a distribution with ellipsoidal constant probability contours into a spherically symmetric distribution.

Marginal functions of the Type VII distribution are again functions of the same form, although the parameter of a k-dimensional marginal of an n-dimensional distributions, with $m = m'$ originally, is now

$$m = m' - 1/2(n - k) \ . \qquad (32)$$

In other words, the parameter m decreases by 1/2 for each decrease in dimension. The relationships between the marginal covariance matrices are still valid [provided (31) is used in place of (28)] as for the Type II case although, in this instance, the marginal functions do not approach normality for spaces of high dimensions.

4. THE CANONICAL FORM OF BIVARIATE DISTRIBUTIONS. The canonical method is essentially an extension of the procedure used by Fisher (1940), Hirshfeld (1935) and Hotelling (1936), among others, to derive certain results in connection with contingency tables and discrete distributions. This method presently appears to provide the most powerful unified attack on the structure of bivariate distributions. By considering the expansion of appropriate bivariate distributions in terms of their eigenfunctions, suitably defined, the method arrives at a fairly general way of constructing bivariate functions with prescribed marginal functions. The bivariate functions so constructed also possesses the maximal correlation between the two original random variables.

The following two definitions will be required below: (i) Canonical Variables: Given a set of (1-dimensional) marginal distributions functions $G(x)$, $H(y)$, the corresponding canonical variables or functions $\xi^{(i)}(x)$, $\eta^{(i)}(y)$, $i = 1,2,\ldots$, are two sets of orthonormal functions defined on G and H in a recursive manner such that the correlation between corresponding members of the two sets of maximal. Unity may be considered as a member of zero order of each set of variables. Symbolically, the orthogonal and normalizing conditions are:

$$\int \xi^{(i)}(x)\, dG(x) = \int \eta^{(i)}(y)\, dH(y) = 0 \; ; \quad i = 1, 2 \ldots \tag{33}$$

$$\int \left[\xi^{(i)}(x)\right]^2 dG(x) = \int \left[\eta^{(i)}(y)\right]^2 dH(y) = 1 \; ; \quad i = 1, 2 \ldots$$

$$\int \xi^{(i)}(x)\, \xi^{(j)}(x)\, dG(x) = \int \eta^{(i)}(y)\, \eta^{(j)}(y)\, dH(y) = 0, \; i \neq j,$$

and the maximization conditions are that

$$\rho_i = \text{corr}\,(\xi^{(i)}(x), \eta^{(i)}(y)) = \iint \xi^{(i)}(x)\, \eta^{(i)}(y)\, dF(x,y) \tag{34}$$

should be maximal for each i, given the preceding canonical variables and for a given bivariate distribution function $F(x,y)$. The ρ_i are defined to be the corresponding canonical correlations and can by convention be taken always to be positive [Lancaster (1963)].

(ii) ϕ^2-boundedness: If $F(x,y)$ is a bivariate distribution function with marginal distributions $G(x)$ and $H(y)$, then the quantity ϕ^2 is defined by the relationship:

$$\phi^2 = \int \left\{ dF(x,y) \Big/ \left[dG(x)\, dH(y)\right] \right\}^2 dG(x)\, dH(y) - 1, \tag{35}$$

where the integral is to be taken as zero if the point (x,y) does not correspond to points of increase of both G(x) and H(y).

Using these definitions, a fundamental result due to Lancaster (1958, 1963) can be stated as follows: If a bivariate distribution function F(x,y) is ϕ^2 bounded (i.e. (35) is bounded for the F, G, H in question) then complete sets of orthonormal functions $\{x^{(i)}\}$ and $\{y^{(i)}\}$ can be defined on the marginal distributions such that

$$dF(x,y) = \left\{1 + \sum_{r=1}^{\infty} \rho_r \, x^{(r)} \, y^{(r)}\right\} dG(x)\,dH(y) \qquad \text{a.e.} \tag{36}$$

and $\phi^2 = \sum_{r=1}^{\infty} \rho_r^{\,2}$.

Conversely, a bivariate distribution is completely characterized a.e. by its marginal distributions and the matrix of correlation of any pair of complete sets of orthonormal functions on the marginal distributions. From the above relationship it can also be shown that any bivariate surface that can be put in the canonical form (36) necessarily has simple linear regression functions:

$$x^{(i)} = \rho_i \, y^{(i)} \qquad\qquad y^{(i)} = \rho_i \, x^{(i)} \;, \tag{37}$$

the regressions of $x^{(i)}$ on $y^{(j)}$ and $y^{(i)}$ on $x^{(j)}$ being zero for $i \neq j$.

A few simple and illuminating examples have been given by Lancaster (1958) to illustrate the use of the formula (36).

One important special case of this formula arises when the marginal functions are normal and the canonical functions are Hermite polynomials, viz.

$$f(x,y) = \frac{1}{2\pi} e^{-\frac{x^2+y^2}{2}} \left[1 + \sum_{k=1}^{\infty} c_k \, H_k(x) \, H_k(y)\right] \quad \begin{matrix} -\infty < x < \infty \\ -\infty < y < \infty \,, \end{matrix} \tag{38}$$

where the series (38) converges <u>in the mean</u> if

$$\sum_{k=1}^{\infty} c_k^{\,2} < \infty \,. \tag{39}$$

The Hermite polynomials H_k, k=1,2,... are normalized so that they are orthonormal with respect to the normal density as the weight function

$$\frac{1}{\sqrt{2\pi}} \int_{-\infty}^{\infty} H_k(x)\, H_\ell(x)\, e^{-\frac{x^2}{2}}\, dx = \begin{cases} 0 & k \neq \ell \\ 1 & k = \ell \end{cases} \tag{40}$$

or, equivalently

$$H_k(x) = \frac{(-1)^k}{\sqrt{k!}} e^{\frac{x^2}{2}} \frac{d^k}{dx^k} e^{-\frac{x^2}{2}} = \frac{1}{\sqrt{k!}} \left[x^k - \frac{k(k-1)}{1.\ 2} x^{k-2} + \ldots\right] \tag{41}$$

In order to generate a family of bivariate density functions from (38), it is necessary to choose the coefficients c_k such that the series for f(x,y) is non negative, then c_k would be the correlation coefficient between $H_k(x)$ and $H_k(y)$; hence only expansions of the type (38) for which $|c_k| \leq 1$, k = 1, 2... are of interest.

It has been shown by Sarmanov (1966) that for both these conditions to hold, it is necessary and sufficient that the c_k's be the moments of some probability distribution concentrated within the interval [-1, 1]. Thus, for example, the trivial case $c_1 = 1$, $c_2 = c_3 = c_4 = \ldots = 0$ satisfies the above requirements.

Of more interest, perhaps, is the case where the c_k's assume particularly simple forms, say $c_k = R^k$ or zero for some $-1 < R < 1$. The series is then of the form:

$$f(x,y) = \frac{1}{2\pi} e^{-\frac{x^2+y^2}{2}} \left[1 + \sum_{j=1}^{\infty} R^{k_j} H_{k_j}(x)\, H_{k_j}(y)\right] \tag{42}$$

where $0 < k_1 < k_2 < \ldots.$ are integers. Only two choices of k_j for which the sum (42) is positive for all x and y are known:

(i) $k_j = j$, j = 1, 2 ... This yields the known expansion of the normal correlation surface.

(ii) $k_j = 2j$, j = 1, 2 ... In this case a linear combination of two bivariate normals with R and -R respectively is obtained, viz:

$$f(x,y;R) = \frac{1}{2}\phi(x,y;R) + \frac{1}{2}\phi(x,y;-R) , \tag{43}$$

where ϕ is the standard bivariate density given in (3).

The importance of the Hermite polynomial expansion (38) and the above examples is that they form an important subset of the "Fréchet class of distributions", or the class of bivariate non-trivial distributions that can be constructed from a given pair of marginal functions (see the next section). In the present case the expansion (38) encompasses those bivariate distributions with normal marginals. Although some choices of c_k other than those of (42) have to be made in order to derive other appropriate bivariate functions, the general polynomial form is nevertheless very useful in predicting the properties pertaining to the entire subclass of functions. It can be demonstrated, for example, that in a bivariate distribution, the occurrence of linear correlation does not necessarily imply the coincidence of the usual correlation coefficient with the maximal correlation coefficient [cf. Sarmanov (1958)].

5. A GENERALIZATION OF THE FARLIE-MORGENSTERN BIVARIATE FAMILY. Another method of constructing bivariate and multivariate distributions is to follow and extend the suggestion of Fréchet (1951) (already mentioned in the previous section) who was also concerned with bivariate distributions with specified marginals $F_{X_1}(\cdot)$, $F_{X_2}(\cdot)$. His ideas were further elaborated and modified by Morgenstern (1956) and Farlie (1960) and we shall refer to the general family of distributions of the form:

$$F_{X_1,X_2}(x_1,x_2) = F_{X_1}(x_1)\, F_{X_2}(x_2)\, [1+\alpha\{1-F_{X_1}(x_1)\}\{1-F_{X_2}(x_2)\} \tag{44}$$

with $|\alpha| < 1$ as the Farlie-Morgenstern family of bivariate distributions (FM distributions for short). The bivariate exponential case with $F_{X_i}(x_i) = 1 - e^{-x_i}$, $x_i \geq 0$, $i = 1,2$, was extensively studied by Gumbel (1960). We shall now try to generalize this family and its particular cases to an m-dimensional case.

A natural extension to the trivariate case is: (for brevity we put $F_{X_i} = F_i$)

$$F_{\underset{\sim}{X}}(\underset{\sim}{x}) = F_1(x_1)\, F_2(x_2)\, F_3(x_3)\{[1+\alpha_{12}\{1-F_1(x_1)\}\{1-F_2(x_2)\} +$$

$$\alpha_{13}\{1-F_1(x_1)\}\{1-F_3(x_3)\} +$$

$$\alpha_{23}\{1-F_2(x_2)\}\{1-F_3(x_3)\} +$$

$$\alpha_{123}\{1-F_1(x_1)\}\{1-F_2(x_2)\}\{1-F_3(x_3)\} . \qquad (45)$$

To keep the analogy with the bivariate case we don't include the terms of the form $\alpha_i(1-F_i(x_i))$ here. The corresponding m-variate extension would be:

$$F_{\underset{\sim}{X}}(\underset{\sim}{x}) = \left\{\prod_{j=1}^{m} F_j(x_j)\right\} \left[1+ \sum_{j_1<j_2} \alpha_{j_1 j_2} G_{j_1}(x_1) G_{j_2}(x_2) + \right.$$

$$\sum\sum\sum_{j_1<j_2<j_3} \alpha_{j_1 j_2 j_3} G_{j_1}(x_1) G_{j_2}(x_2) G_{j_3}(x_3) + \ldots \qquad (46)$$

$$\left. + \alpha_{12\ldots m} \prod_{j=1}^{m} G_j(x_j)\right] \quad \text{with} \quad G(\cdot) = 1-F(\cdot).$$

It is easy to verify that this definition satisfies the four basic necessary and sufficient conditions on an m-variate function to be a multivariate c.d.f. Indeed

1) $F_{\underset{\sim}{X}}(\cdot\ \cdot, -\infty, \cdot\ \cdot\ \cdot) = 0$

2) $F_{\underset{\sim}{X}}(+\infty, +\infty, \ldots, +\infty) = 1$

3) If one of the $F_r(x_r) = 1$ then

$$F_{\underset{\sim}{X}}(x_1,x_2,\ldots,x_{r-1},x_r,x_{r+1},\ldots,x_m) = F(x_1,x_2,\ldots,x_{r-1}, x_{r+1},\ldots,x_m)$$

4) If all $F_r(x_r) \equiv 1$, $r = 1,\ldots,m$, then $F_{\underset{\sim}{X}}(\underset{\sim}{x}) = 1$.

These extensions have the properties that

(i) the marginal joint distribution of any two variates is FM;
(ii) the marginal univariate distributions of X_j are $F_j(x_j)$.

Next we observe that the distribution can be equivalently defined in the terms of the m-dimensional "survival" function

$G_{\underset{\sim}{X}}(\underset{\sim}{x}) = 1 - F_{\underset{\sim}{X}}(\underset{\sim}{x})$. Elementary algebraic manipulations show that in the three-dimensional case we have the obvious notation

$$G_{123} = G_1G_2G_3\{1 + \sum_{j_1<j_2}\sum \alpha_{j_1j_2}(1-G_{j_1})(1-G_{j_2}) - \alpha_{123}(1-G_1)(1-G_2)(1-G_3)\} . \quad (47)$$

Similar calculations in the four dimensional case yield the following result (in the obvious notation):

$$G_{1234} = G_1G_2G_3G_4\{1+\sum_{j_1<j_2}\sum \alpha_{j_1j_2}(1-G_{j_1})(1-G_{j_2}) - \sum_{j_1<j_2<j_3}\sum\sum \alpha_{j_1j_2j_3}\prod_{i=1}^{3}(1-G_{j_i}) + \alpha_{1234}\prod_{i=1}^{4}(1-G_i)\} , \quad (48)$$

which shows that the symmetry is preserved up to the sign of the third term in the braces.

An application of the induction method immediately yields the conclusion that the symmetry relation

$$\begin{cases} F_{12} = F_1F_2\ \{1 + \alpha(1-F_1)(1-F_2)\} \\ G_{12} = G_1G_2\ \{1 + \alpha(1-G_1)(1-F_2)\} . \end{cases}$$

is "almost" extendable to the general m-dimensional case, with the exception that the signs of α's will alternate, the first α always being positive. Clearly the restrictions on the values of α's in the trivariate case are $|\alpha_{ij}| < 1$, since the joint distribution of X_i, X_j is FM with $\alpha = \alpha_{ij}$. (This however is only a necessary condition.)

In the general case the restrictions on the α's can be derived by using the condition $\frac{\partial F_{\underset{\approx}{X}}(\underset{\sim}{x})}{\partial x_i} \geq 0$ for all i and taking into account that $F_{\underset{\approx}{X}}() \to 1$ as all x_i's $\to \infty$ and $F \to 0$ as any of the x_i's $\to -\infty$. Necessary and sufficient conditions are obtained utilizing the linearity of the derivative in F_i.

The array (conditional) distributions for the FM family are most conveniently defined by the "survival" function of the type:

$$G_1(X_1 | \bigcap_{j=2}^{m} X_j > x_j) = \Pr[X_1 > x_1 | \bigcap_{i=2}^{m} X_i > x_i)] \tag{49}$$

Indeed in the bivariate case we have

$$G_1(X_1\ X_2 > x_2) = G_1(x_1)[1 + \alpha_{12} F_1(x_1) F_2(x_2)] \tag{50}$$

and in the general case:

$$G_1(X_1 | \bigcap_{j=2}^{m} X_2 > x_2) = G_1(x_1)[1 + \sum\sum_{j<j'} \alpha_{jj'} F_j(x_j) F_{j'}(x_{j'}) \tag{50'}$$

$$- \sum\sum\sum_{j<j'<j''} \alpha_{jj'j''} F_j(x_j) F_{j'}(x_{j'}) F_{j''}(x_{j''}) + \ldots]/D,$$

where D is a function of x_i, $i=2,\ldots,m$ and does not depend on x_1.

In particular if the marginals are exponential satisfying $F_i(x_i) = 1-e^{-x_i/\lambda_i}$ $(x_i \geq 0; \lambda_i > 0)$ then

$$G_1(X_1 | \bigcap_{j=2}^{m} X_j > x_j) = e^{-x_1/\lambda_1}[1 + \sum\sum_{j<j'} \alpha_{jj'}(1-e^{-x_j/\lambda_j})(1-e^{-x_{j'}/\lambda_{j'}})$$

$$- \sum\sum\sum_{j<j'<j''} \alpha_{jj'j''}(1-e^{-x_j/\lambda_j})(1-e^{-x_{j'}/\lambda_{j'}})(1-e^{-x_{j''}/\lambda_{j''}})$$

$$+ \ldots + (-1)^m \alpha_{1,2\ldots m} \prod_{j=1}^{m} (1-e^{-x_j/\lambda_j})]/D, \tag{51}$$

which can be represented in the form

$$G_1(X_1 | \bigcap_{i=2}^{m} X_i > x_i) = e^{-x_1/\lambda_1}[A(x_2,\ldots,x_m) + e^{-x_1/\lambda_1} B(x_2,\ldots,x_m)]. \tag{52}$$

where $A(x_2,\ldots,x_m)$ and $B(x_2,\ldots,x_m)$ do not depend on x_1 and are functions of sum of products of exponential functions and constants.

As a function of x_1 this survival distribution is thus

$$A\,e^{-x_1/\lambda_1} + B\,e^{-2x_1/\lambda_1} .$$

It is easy to verify that in view of the linearity of the expression in the brackets of (46) in each of F_i, finite and infinite mixtures (with respect to α's) of distributions belonging to the FM family of any dimensions are also FM distribution with the vector $\underset{\sim}{\alpha}_0$ corresponding to the mixture satisfying

$$\underset{\sim}{\alpha}_0 = E(\underset{\sim}{\alpha}) \qquad (53)$$

where $\underset{\sim}{\alpha}$ is the vector of the α's of various orders corresponding to component distributions in the mixture. Thus the FM family is closed under this kind of mixing. In particular if $E[\underset{\sim}{\alpha}] = \underset{\sim}{0}$, then the mixture is in fact a distribution if independent variables (although the components in general do not possess this property).

For the sake of brevity we shall omit the discussion of special multivariate models which arise for particular choices of distributions F_i in this general model and their interrelation with other well-known families of distributions. We intend to discuss these and related topics in a separate paper. (An interested reader can construct some illuminating examples without much difficulty by adapting the detailed arguments given in Sections 3 and 4.)

6. NUMERICAL METHODS. Due to a general lack of good (or even adequate) analytical approximations to the standard multivariate sampling distributions, it is quite often necessary to resort to numerical approximating techniques. Numerous such procedures for the univariate case are known and have received extensive treatment elsewhere. Unfortunately, this is not true in multivariate statistical analysis, where even the well-known Monte Carlo techniques are sometimes inadequate due to a limitation to a fixed dimensionality and sample size. There is thus a need for techniques which produce approximations of a general form and which can be used in a variety of situations. The pioneering work of Chambers (1967) is an important step forward in this direction and no doubt requires additional scrutinization. Chambers "revives" and readjusts two methods--this method of Edgeworth expansions and the method of perturbations for the case of multivariate distributions. The perturbation method yields approximations to the quantities rather than to the distributions of quantities arising in multivariate statistics.

The discussion of the Edgeworth expansion for the standardized mean of a univariate sample is an extensive topic and is treated in various classical books [e.g. Elderton and Johnson (1969)]. For our present purpose it is only necessary to recall that an Edgeworth expansion to order r essentially involves first approximating the characteristic function of a random variable to order r in $n^{-1/2}$. An inverse Fourier transform is then applied to the approximation and the result is used as an approximation to the probability density of the random variable. This method was extended by Chambers (1967) to give an algorithm for the systematic approximation of multivariate distributions. Where the density in question is available in closed form (as for example in the central case of Wishart distribution), the Edgeworth expansions are still useful for, say, the distributions of functions of some related quantities (e.g. the covariance matrix).

Chambers' investigations in the area of perturbation approximations were limited to applications to principal component analysis, canonical correlation analysis and discriminant analysis; their applicability to distributions involving eigenvectors and matrices remains a problem awaiting solution since the current theory does not lend itself to higher order approximations.

7. A BRIEF SKETCH OF SOME RECENT TRENDS. Two basic non-normal multivariate continuous distributions continue to receive special attention:

a) Exponential distributions which have been thoroughly investigated in the last 15 years starting with the work of Gumbel (1960) and undergoing intensive development following the contributions of Marshall and Olkin (1967). Reliability considerations and concepts dominate this area of research.

b) Gamma distributions which originated as early as the 1930's [see for example, S. D. Wicksell's work On Correlated Surfaces of Type III, in Biometrika, 1933]. This was continued in 1941 by Cheriyan and Kibble (Sankhyā) and received prominent attention in the sixties in view of the apparent infinite divisibility (not demonstrated yet in the most general case) and other probabilistic properties of this distribution. Its practical applicability was motivated mainly by cloud seeding experiments since the rainfall for various periods has for a long time been (conventionally) assumed (or demonstrated!) to be gamma distributed. We are now witnessing a merger between these two models indicating a healthy development and an appropriate generalization of the theory.

Two other "active" main topics associated with multivariate distributions are multivariate hazard rates and the closely related notions of dependence and association. Both draw their inspiration from the various multivariate exponential distributions which serve as border lines or starting points for a definition of these concepts.

Among general families of multivariate distributions, the m-dimensional (linear) exponential-type family developed by Bildikar and Patil in 1968 has received special attention in recent years in view of its appealing structural properties--both mathematically and intuitively [see e.g. Roux (1971)]. Cumulant generating functions are a useful tool in investigations connected with these distributions. A few families of multivariate discrete distributions have also appeared in the literature. The sum-symmetric power series distributions due to Patil (1967) and the unified class of multivariate hypergeometric distributions due to Janardan and Patil (1972) may be mentioned in this connection. Like multivariate continuous distributions, multivariate discrete distributions also provide a challenge in modeling and analysis. For additional references, see Patil (1965) and Patil and Joshi (1968, 1974).

The topic of general series representation of bivariate distributions (there are hardly any results for the multivariate case) which facilitates the study of the structure of general bivariate distributions, covered and emphasized in a previous section of this survey, continues to attract a number of capable investigators.

Relevant references and a more detailed discussion of the topics sketched in this section are given in Johnson and Kotz (1972), in particular, Chapters 34, 40 and 41.

REFERENCES

[1] Basu, A. P. and Block. H. W. (1974). In Statistical Distributions in Scientific Work, Vol. III, Characterizations and Applications, Patil, Kotz and Ord (eds.). Reidel, Dordrecht and Boston, pp. 399-421.
[2] Bildikar, S. and Patil, G. P. (1968). Ann. Math. Statist. 39, 1316-1326.
[3] Chambers, J. M. (1967). Biometrika 54, 367-383.
[4] Charlier, C. V. L. (1914). Arkiv für Math., Astr. och Fysik. 9, 1-18.
[5] Cooper, P. W. (1963). Biometrika 50, 439-448.
[6] Eagleson, G. K. (1964). Ann. Math. Statist. 35, 1208-1215.
[7] Edgeworth, F. Y. (1904). Trans. Camb. Phil. Soc. 20, 116-119.
[8] Elderton, W. P. and Johnson, N. L. (1969). Systems of Frequency Curves. Cambridge University Press, London.

[9] Farlie, D. J. G. (1960). Biometrika 47, 307-323.
[10] Finney, D. J. (1974). J. Roy. Statist. Soc. Ser A 137, 12-14.
[11] Fisher, R. A. (1940). Ann. Eugenics 10, 422-429.
[12] Fisk, P. R. (1973). J. Roy. Statist. Soc. Ser A 136, 622.
[13] Fréchet, M. (1951). Ann. Univ. Lyon, Sec A, Ser 3, 14, 53-77.
[14] Goodman, I. R. and Kotz, S. (1973). J. Multiv. Anal. 3.
[15] Gumbel, E. J. (1960). J. Amer. Statist. Assoc. 55, 698-707.
[16] Hirschfeld, H. O. (1935). Proc. Camb. Phil. Soc. 31, 520-524.
[17] Hotelling, H. (1936). Biometrika 28, 321-377.
[18] Janardan, K. G. and Patil, G. P. (1972). Sankhyā Ser A, 1-14.
[19] Johnson, N. L. and Kotz, S. (1972). Continuous Multivariate Distributions. J. Wiley, New York.
[20] Jørgensen, N. R. (1916). Undersøgelser over Frequensflader og Korrelation. Busck, Copenhagen.
[21] Lancaster, H. O. (1972). Math. Chronicle 2, 1-16.
[22] Lancaster, H. O. (1963). Ann. Math. Statist. 34, 532-538.
[23] Lancaster, H. O. (1958). Ann. Math. Statist. 29, 719-736.
[24] Mardia, K. V. (1970). Families of Bivariate Distributions. Hafner Publishing Co., Darien, Connecticut.
[25] Marshall, A. W. and Olkin, I. (1967). J. Amer. Statist. Assoc. 62, 30-44.
[26] McGraw, D. K. and Wagner, J. F. (1967). J.A.S.A. 62, 589-600.
[27] Morgenstern, D. (1956). Mitteilingsblatt für Math. Statist. 8, 234-235.
[28] Neyman, J. (1926). Biometrika 18, 257-262.
[29] Patil, G. P. and Joshi, S. W. (1974). Sum-symmetric power series distributions and minimum variance unbiased estimation. Theory Prob. Appl. 19.
[30] Patil, G. P. and Joshi, S. W. (1968). Dictionary and Bibliography of Discrete Distributions. Oliver and Boyd, Edinburgh and Hafner, New York.
[31] Patil, G. P. (1967). Sankhyā Ser B 30, 355-366.
[32] Patil, G. P. (1965). Sankhyā Ser A 27, 259-270.
[33] Pearson, K. (1925). Biometrika 17, 268-313.
[34] Pearson, K. (1924). Biometrika 16, 172-182.
[35] Pearson, K. (1923). Biometrika 15, 222-230.
[36] Pearson, K. (1921). Biometrika 13, 296-300.
[37] Pearson, K. (1905). Drapers' Company Research Memoirs, Biometric Series II, 1-54.
[38] Rhodes, E. C. (1925). Biometrika 17, 314-324.
[39] Rhodes, E. C. (1923). Biometrika 14, 355-377.
[40] Risser, R. (1950). Bull. des Actuair, Franc. No. 191, 141-232.
[41] Roux, J. J. J. (1971). S. Afr. Statist. J. 5, 27-36.

[42] Sagrista, S. N. (1952). Trabajos Estadistica 3, 273-314. (Spanish).
[43] Sarmanov, O. V. and Bratoeva, Z. N. (1967). Theory Prob. Appl. 12, 520-531. (In Russian).
[44] Sarmanov, O. V. (1966). Dokl. Akad. Nauk. 168, 596-599.
[45] Sarmanov, O. V. (1958). Dokl. Akad. Nauk. 120, 715-718. (In Russian).
[46] Steffensen, L. (1922). Skand. Aktuar. Tidskr., 106-132.
[47] Van der Stok, J. P. (1908). Proc. Roy. Acad. Sci. Amsterdam 10, 799-817.
[48] Van Uven, M. J. (1947). Proc. Roy. Acad. Sci. Amsterdam I 50, 1063-1070; II 50, 1252-1264.
[49] Van Uven, M. J. (1948). Proc. Roy. Acad. Sci. Amsterdam III 51, 41-52; IV 51, 191-196.
[50] Wicksell, S. D. (1917a). Kungl. Svensk. Vet. Akad. Handl. 58, 1-48.
[51] Wicksell, S. D. (1917b). Skand. Aktuar. Tidskr. 2-3, 1-19.
[52] Wicksell, S. D. (1917c). Arkiv. für Math., Astr. och Fysik. 12.
[53] Yule, G. U. (1897). Proc. Roy. Soc. 60, 477-489.

DEPENDENCE CONCEPTS AND PROBABILITY INEQUALITIES

Kumar Jogdeo*

Mathematics Department, University of Illinois,
Urbana, Illinois, U.S.A.

SUMMARY. Concepts of positive dependence are reviewed and their applications are indicated. The role of these concepts to produce an ordering in the families of bivariate distributions is discussed. Recent results related to the concept of "association" are shown to have applications in deriving certain inequalities for the multivariate normal and related distributions. Finally a simple example illustrating somewhat unintuitive behavior of a pair of bivariate normal distributions is given.

KEY WORDS. Positive dependence, probability inequalities, ordered families of bivariate distributions, associated random variables, multivariate unimodality, contaminated independence model.

1. INTRODUCTION. The concept of stochastic independence is defined by equality relations and absence of it implies dependence. This negative way of describing dependence hardly provides a clue about the nature of dependence. To have any meaningful results it is necessary to have some more specific conditions. Of particular importance are conditions of "positive dependence". Intuitively one could say that random variables are positively dependent if they "hang together" in some sense. This intuitive notion is made precise by the following definitions for the bivariate distributions.

---

*Research and travel for the institute partially supported by NSF grant.

G. P. Patil et al. (eds.), Statistical Distributions in Scientific Work, Vol. 1, 271-279.

Let (X,Y) be a pair of real random variables with finite second moments. The pair (X,Y) or its joint distribution $F_{X,Y}$ is said to satisfy B_1 if Cov $(X,Y) \geq 0$. (1.1)

As is well known, the correlation coefficient measures linear dependence and B_1 trivially implies the following two inequalities. Let f_1, g_1 be a pair of nondecreasing linear functions defined on R then $\text{Cov}[f_1(X), g_1(Y)] \geqq 0$. (1.2)

More generally, if f_2 and g_2 are nondecreasing and linear on R^2 (that is, of the form $ax + by + c$ with $a \geq 0$, $b \geq 0$) then $\text{Cov}[f_2(X,Y), g_2(X,Y)] \geqq 0$. (1.3)

Removal of linearity restriction on f_i and g_i in (1.2) and (1.3) (which will be assumed henceforth) yield two distinct notions of positive dependence. Lehmann (1966) pointed out that (1.2) is equivalent to "positive quadrant dependence", $P[X \geqq a, Y \geqq b] \geqq P[X \geqq a]P[Y \geqq b]$, for every a,b. (1.2')

Note that (1.2') is a special case of (1.2) so that for condition (1.2) to hold, it suffices to look at the indicator functions which are nondecreasing. In fact, the following formula due to Hoeffding (1940) makes the equivalence of (1.2) and (1.2') almost trivial. If U,V are two real random variables with finite covariance then $\text{Cov}[U,V] = \int\int_{-\infty} \{F_{U,V}(u,v) - F_U(u)F_V(v)\}\, du\, dv$.

Esary, Proschan and Walkup (1967) studied condition (1.3). With their terminology a pair (X,Y) is called "associated" if $F_{X,Y}$ satisfies (1.3). Alternatively, in this article, the distributions satisfying (1.2) and (1.3) will be said to belong to B_2 and B_3 respectively.

Another origin of (1.2) and (1.3) can be traced to an inequality due to Tchebyshev [see Lehmann (1966)] which states that if Z is a real random variable, f_1 and g_1 two nondecreasing measurable functions then $\text{Cov}[f_1(Z), g_1(Z)] \geqq 0$. It is clear that conditions (1.2) and (1.3) are natural generalizations of this.

The fourth condition, stronger than (1.1), (1.2) and (1.3), is that of regression dependence, $x_2 > x_1 \implies P[Y \geqq y | X = x_2] \geqq P[Y \geqq y | X = x_1]$, for every y. (1.4)

A pair (X,Y) or its distribution $F_{X,Y}$ satisfying (1.4) will be said to belong to B_4.

It is easy to see that $B_4 \subset B_3 \subset B_2 \subset B_1$.

The purpose of this article is to give a brief survey of recent results culminating from these concepts. Most of the articles cited in the present paper contain several applications and we will restrict to some new or typical ones. Positive quadrant dependence is considered in Section 1 and an application to bivariate t-distribution is given. In Section 2 ordered families are considered and basic results of Yanagimoto and Okamoto (1969) are reinterpreted in a framework which streamlines the development of the results in the present paper as well as makes these concepts transparent. Section 3 considers multivariate generalizations of these results. The concept of association seems to be suitable for this purpose. Family of multivariate normal distributions has been endowed with a parametrization, (the covariance matrix) which determines the dependence completely. This prompts the study of probability assigned to certain regions and its change with the change in correlation coefficients. Sidák (1967, 1968, 1971) has several interesting results in this area. In Section 4, these results are discussed and an example about circular region is stated.

2. BIVARIATE DEPENDENCE INEQUALITIES. One of the basic tools to prove various probability inequalities was derived by Lehmann (1966) by utilizing condition (1.2). A pair of measurable functions defined on $R^n \to R$ is said to be "concordant" if each is monotone in all arguments and the directions of monotonicity for each of the arguments is same for the two functions.

Theorem 2.1 (Lehmann). Let (X_i, Y_i) be n independent pairs of real random variables such that each pair is in B_2. Let f,g be a pair of concordant functions defined on $R^n \to R$. Then the pair $f(X_1,\ldots,X_n)$ and $g(Y_1,\ldots,Y_n)$ is in B_2.

This theorem has been shown to have numerous applications. As an illustration, we present a simpler proof of a result of Halperin (1967) and at the same time generalize it. Let (X_i, Y_i) be n independent pairs as before and suppose $(|X_i|, |Y_i|)$ are in B_2. This can be shown to be true for a bivariate normal distribution with zero means or, more generally, when

$$Y_i = \beta_i X_i + Z_i, \quad i = 1,\ldots,n, \tag{2.1}$$

where X_i and Z_i are independent and the distribution of Z_i is symmetric (about 0) and unimodal. From Theorem 2.1, it is easy to derive

<u>Corollary 2.1</u>. If (X_i, Y_i), $i = 1,\ldots n$ are independent pairs and satisfy (2.1), then the pair $|X_1| / \sum_2^n X_i^2$, $|Y_1| / \sum_2^n Y_i^2$ is in B_2.

The above property shows that if t_1 and t_2 are student statistics constructed from X and Y observations taken from a random sample form a bivariate normal population, then $P[|t_1| \leq C_1, |t_2| \leq C_2] \geq P[|t_1| \leq C_1]\, P[|t_2| \leq C_2]$. This inequality is useful to construct "conservative confidence intervals".

Recently Yanagimoto and Sibuya (1972) extended the notion of positive dependence which have applications in tests of symmetry.

3. ORDERED FAMILIES OF BIVARIATE DISTRIBUTIONS. When two distributions satisfy a condition of positive dependence it is natural to ask whether one exhibits this dependence in a stronger form than the other. Bivariate normal family possesses an ordering produced by correlation coefficient. In general, although a complete ordering is impossible, a partial ordering can be given. It is desirable that this ordering when specialized to the normal family, should be consistent with its natural ordering. Yanagimoto and Okamoto (1969) considered partial orderings in B_2 and B_4. [Partial ordering in B_3 corresponding to condition (1.3) can be shown to be inconsistent with the natural ordering in the normal family. This will be discussed in a separate communication.] The basic motivation of Yanagimoto and Okamoto was to apply this ordering to study monotonicity properties of power functions of certain nonparametric tests for testing independence.

Suppose F and G are two bivariate distributions having the same marginals. Then G is said to have larger quadrant dependence than F if

$$G(x,y) \geq F(x,y), \quad \text{for every } (x,y) \tag{3.1}$$

Note that neither G nor F have to be in B_2. However, under (3.1) if $F \in B_2$ then $G \in B_2$. As one would expect, if (X_i, Y_i) are n independent pairs and f and g are concordant functions then the joint distribution of $f(X_1,\ldots,X_n)$, $g(X_1,\ldots,X_n)$ has larger quadrant dependence if these pairs are chosen from populations having larger quadrant dependence.

Similar ordering in B_4 is somewhat involved. In the following we consider a new approach, hopefully more intuitive than the one given by Yanagimoto and Okamoto (1969). A distribution F is said to be "stochastically larger" than G if

$$1 - F(x) \geqq 1 - G(x), \quad \text{for every } x. \tag{3.2}$$

For $0 < u < 1$, let $F^{-1}(u) = \inf\{x : F(x) \geqq u\}$ and G^{-1} defined similarly. Then (3.2) is equivalent to

$$F^{-1}(u) \geqq G^{-1}(u), \quad \text{for } u \in (0,1). \tag{3.4}$$

Using the fact that G is a nondecreasing function, from (3.4) it follows that $F^{-1}(u) \geqq G^{-1}(v)$, $0 < v \leqq u < 1$. In other words, F is stochastically larger than G

$\iff S_{F,G} \supset \{(u,v) : 0 < v \leqq u < 1\}$ where

$S_{F,G} = \{(u,v) : F^{-1}(u) \geqq G^{-1}(v)\}$.

An advantage of the above definition is that $S_{F,G}$ provides a picture of "stochastic difference" between F,G. Larger the set $S_{F,G}$, "more apart" are the distributions with F "ahead of G". In the extreme case when $S_{F,G}$ is the same as open unit square, the support of F except for the boundary is disjoint from that of G. Thus partial ordering of the subsets of the unit square produces stochastic ordering between pairs. Suppose now F,G is a pair of bivariate distributions and F_x, G_x denote conditional distribution functions given the first component. Then the distribution F is said to have larger (positive) regression dependence than G if for every $x_2 > x_1$, $S_{F_{x_2}, F_{x_1}} \supset S_{G_{x_2}, G_{x_1}}$. Many of the properties of this ordering become transparent. For example, if H_1 and H_2 are marginals of a bivariate distribution function H and $L(x,y) = H_1(x)H_2(y)$. Then clearly, conditional distributions corresponding to L are identical so that H has larger regression dependence than L if and only if H is in B_4.

The ordered family of bivariate distributions provides a class nonparametric alternatives to the hypothesis of independence.

4. ASSOCIATION AND MULTIVARIATE DEPENDENCE CONCEPTS.

In order to consider the dependence concepts for multivariate distributions, condition (1.2') seems to be unsuitable. One could define "positive orthant dependence" by similar inequalities, however, it does not provide any useful results other than those characterizing independence [for example, see Jogdeo and Patil (1967) and Jogdeo (1968)]. Condition (1.3) has natural extension and was studied by Esary, Proschan and Walkup (1967). Let $\underset{\sim}{X} = (X_1,\ldots,X_k)$ be a k-vector of real random variables. Then $X_1,\ldots,X_k$ are said to be associated if

$$\text{Cov}[f(\underset{\sim}{X}),g(\underset{\sim}{X})] \geqq 0, \tag{4.1}$$

where f,g are real functions, nondecreasing in each of their k arguments. We list the properties which can be shown by the following simple relation,

$$\text{Cov}[X,Y] = \text{Cov}\{E(X|Z),E(Y|Z)\} + E\{\text{Cov}(X,Y)|Z\}. \tag{4.2}$$

Properties. (i) A single real random variable is associated. (This is immediate from Tchebyshev's inequality.) (ii) If $X_1,\ldots,X_k$ are independent then they are associated. (iii) The union of independent sets of associated random variables comprised of associated random variables. (iv) Nondecreasing functions of associated random variables are associated. (v) If $X_1,\ldots,X_k$ are associated then

$$P[X_i \geqq c_i,\ i = 1,\ldots,k] \underset{(\leqq)}{\geqq} \underset{(\leqq)}{P[X_j \geqq c_j,\ j \in A]}\,\underset{(\leqq)}{P[X_j \geqq c_j,\ j \in B]},$$

where A and B are subsets of K = 1,2,...,k such that $A \cup B = K$.

Properties (ii), (iv) and (v) are useful in several applications and Esary, Proschan and Walkup (1967) have mainly exploited these. A natural question would be whether a result analogous to Theorem 2.1 holds or not. The author (1974) has shown that with a more general definition of concordance, the concordant functions constructed from independent k-vectors of associated random variables inherit association. We will label this result as property (vi).

5. MULTIVARIATE NORMAL AND OTHER DISTRIBUTIONS. Motivated mainly by applications to conservative confidence interval estimation, the following probability inequality was established by Sidák (1967). Let $X_1,\ldots,X_k$ be jointly normally distributed random variables with zero means. Then

$$P[|X_i| \leqq c_i,\ i = 1,\ldots,k] \geqq \prod_{i=1}^{k} P[|X_i| \leqq c_i], \tag{5.1}$$

where $c_1, \ldots, c_k$ are constants. A more detailed result was later given by Sidák (1968) [see Jogdeo (1970) for a simpler proof], which asserts that if correlation coefficients of one component X_1 with others are changed simultaneously by a factor λ (that is, keeping everything else same, correlation coefficients are expressed as $\lambda\rho_{12}, \lambda\rho_{13}, \ldots, \lambda\rho_{1k}$) then the left side of (5.1) is a nondecreasing function of λ. For $\lambda = 0$, the minimum is achieved and thus

$$P[|X_i| \leqq c_i,\ i = 1, \ldots, k] \geqq P[|X_1| \leqq c_1] P[X_i \leqq c_i,\ 1 = 2, \ldots, k], \tag{5.2}$$

from which (5.1) follows by iteration. Sidák (1971), however, shows that in general one cannot replace $\leqq c_i$ by $\geqq c_i$ in above inequalities. In view of property (v) of Section 4, it follows that in general, $|X_1|, \ldots, |X_k|$ are not associated. One of the problems which is still open is whether (5.2) holds when X_i is replaced by student statistic t_i. Sidák (1971) shows that if correlations have a special structure,

$$\rho_{ij} = \lambda_i \lambda_j, \quad i \neq j, \tag{5.3}$$

then the inequality (5.2) holds with $\leqq c_i$ replaced by $\geqq c_i$ and also X_i replaced by t_i. In Jogdeo (1974), the author has shown that with (5.3), $|X_i|$, $i = 1, \ldots, k$ are in fact associated. This has been shown to hold for a class of distributions much wider than normal.

The model which generates this class may be labeled as "contaminated independence model". Suppose instead of observing independent Y_i, $i = 1, \ldots, k$; one observes $X_i = Y_i + \lambda_i Z$, where Z is a contaminating random variable, assumed to be independent of Y observations. If Y_i have symmetric unimodal distribution then it is shown that $|X_i|$ are associated.

Recently, Sidák (1972) has shown that if the k components of multivariate normal distribution satisfying (5.3) is split into m disjoint subsets then the joint probability that i^{th} subset is in a centrally symmetric convex set A_i, $i = 1, \ldots, k$, is larger than or equal to the product of corresponding probabilities. This generalization of (5.1) can be shown to hold for the above model.

One of the useful tools for proving these results has been an inequality due to Anderson (1955) which is based on a definition of multivariate unimodality. Roughly speaking, Anderson's definition calls for convex contours for the density surface. Das Gupta, Eaton and others (1970) have obtained inequalities for the distributions having elliptically contoured densities.

Although the literature devoted to positive dependence is extensive there are some studies of negative dependence in multivariate distributions. For example, Mallows (1968) has shown that for a multinomial distribution and for every set of constants $(c_1,\dots,c_k)$.

$$P[X_i \geqq c_i,\ i = 1,\dots,k] \leqq \prod_{i=1}^{k} P[X_i \geqq c_i]. \tag{5.4}$$

Recently Patil and the author (1974) have given a simple method by which several well known multivariate distributions are shown to obey (5.4).

Finally, we invite the reader to solve the following puzzle. A person at a fair wants to shoot a circular target of radius R. He is given a choice between two guns, A and B, having the following characteristics. Assuming one aims at the center, the first gun, A, is known to hit a point (X,Y) where (X,Y) is has circular bivariate normal distribution (the mean vector at the center of target, variances unity and cov(X,Y) = 0). On the other hand, the gun B is known to concentrate hits on the 45° line and (X,Y) has a singular normal distribution ($X \equiv Y$, var X = 1). In order to maximize the probability of hitting the target which gun should the person choose?

[A sketch of the solution: Let P(A) and P(B) be the probabilities of hitting the target with guns A and B and $z = R/\sqrt{2}$. Then

$P(B) - P(A) = 2\Phi(z) + e^{-z^2} - 2 = g(z)$, say. Then

$$\lim_{z\to 0} g(z) = 0 = \lim_{z\to\infty} g(z) \tag{5.5}$$

$$g'(z) = (2/\pi)^{1/2} e^{-z^2} \{e^{z^2/2} - z\sqrt{2}\}, \tag{5.6}$$

which implies $g'(0) > 0$. It follows now that for some $z_1 < z_2$,

$$g'(z) \lesseqgtr 0 \text{ according as } \begin{cases} z \in (z_1,z_2) \\ z = z_1 \text{ or } z_2 \\ z \notin [z_1,z_2]. \end{cases}$$

Thus $g(z) \gtreqless 0$ if $z \lesseqgtr z^*$, where z^* is the unique solution if

$$2\Phi(z) + e^{-z^2} = 2.$$

This example also throws some light on difficulties on comparing peakedness in multivariate distributions, recently studied by Mudholkar (1972).

REFERENCES

[1] Anderson, T. W. (1955). Proc. Amer. Math. Soc. 6, 170-176.
[2] Das Gupta, S., Eaton, M. L. and others. (1970). Sixth Berkeley Symposium, Vol. II, 241-265.
[3] Esary, J. D., Proschan, F. and Walkup, D. W. (1967). Ann. Math. Statist. 38, 1466-1474.
[4] Halperin, M. (1967). J. Amer. Statist. Assoc. 62, 603-606.
[5] Hoeffding, Wassily. (1940). Schriften. Math. Inst. Univ. Berlin 5, 181-233.
[6] Jogdeo, K. (1968). Ann. Math. Statist. 39, 433-441.
[7] Jogdeo, K. (1970). Ann. Math. Statist. 41, 1357-1359.
[8] Jogdeo, K. (1974). Association and probability inequalities. Submitted.
[9] Jogdeo, K. and Patil, G. P. (1967). Bull. Int. Statist. Inst. 42, 313-316.
[10] Jogdeo, K. and Patil, G. P. (1974). Probability inequalities for certain multivariate discrete distributions. To appear in Sankhyā.
[11] Lehmann, E. L. (1966). Ann. Math. Statist. 37, 1137-1153.
[12] Mallows, C. L. (1968). Biometrika 55, 422-424.
[13] Mudholkar, G. S. (1972). Ann. Inst. Statist. Math. 24, 127-135.
[14] Sidák, Z. (1967). J. Amer. Statist. Assoc. 62, 626-633.
[15] Sidák, Z. (1968). Ann. Math. Statist. 39, 1425-1432.
[16] Sidák, Z. (1971). Ann. Math. Statist. 42, 169-175.
[17] Yanagimoto, T. and Okamoto, M. (1969). Ann. Inst. Statist. Math. 21, 489-506.
[18] Yanagimoto, T. (1972). Ann. Inst. Statist. Math. 24, 559-573.
[19] Yanagimoto, T. and Sibuya, M. (1972). Ann. Inst. Statist. Math. 24, 259-269.
[20] Yanagimoto, T. and Sibuya, M. (1972). Ann. Inst. Statist. Math. 24, 423-434.

NEW FAMILIES OF MULTIVARIATE DISTRIBUTIONS

J. J. J. Roux

Department of Mathematical Statistics, University of South Africa, Pretoria, South Africa

SUMMARY. This lecture gives a review of different techniques in deriving and characterizing generalized multivariate distributions. Some properties and uses of these distributions are given.

KEY WORDS. Random matrices, matrix-variate hypergeometric distribution, multivariate beta, zonal polynomials, generalized special functions, symmetrized distributions.

1. INTRODUCTION. The multivariate distributions of random matrices discussed here can be regarded as generalizations of the univariate distributions. Some of these forms are not in use at present and are not likely to have useful applications. This does not mean that these forms may not become of greater importance at a later stage. From these generalized multivariate distributions most of the known multivariate distributions are obtained as special cases. Some of these known distributions, for example the Wishart and multivariate Beta, play an important part in deriving a number of test-statistics in multivariate analysis. In this review of generalized multivariate distributions we use expansions involving zonal polynomials and generalized special functions with matrix argument. Some special and interesting properties enjoyed by these distributions are also given.

G. P. Patil et al. (eds.), Statistical Distributions in Scientific Work, Vol. 1, 281-297.

2. A CHARACTERIZATION OF A MULTIVARIATE DISTRIBUTION. Seshadri and Patil (1964) derived and applied a sufficient condition for the unique existence of a bivariate distribution. A similar approach is used by Roux (1971a) for sets of $k(\geq 2)$ variables to derive a sufficient condition for the unique existence of a multivariate distribution. By using logical extensions in notations, the theorems are generalized to include random matrices. In these papers, some results on characterization of a multivariate distribution are obtained essentially by the use of the generalized Laplace transform.

We have shown that given $f_{1,k-1}(X_1,\ldots,X_{k-1})$, the marginal probability density function (p.d.f.) of $X_1:p\times r,\ldots,X_{k-1}:p\times r$, and $f(X_1,\ldots,X_{k-1}|X_k)$, the conditional p.d.f. of $X_1:p\times r,\ldots,X_{k-1}:p\times r$ on $X_k:p\times r$, a sufficient condition for $f_k(X_k)$, and hence for $f(X_1,\ldots,X_k)$ to be unique is that the conditional p.d.f. is of the exponential form

$$f(X_1,\ldots,X_{k-1}|X_k) = \exp[\mathrm{tr}[X_k'g(X_1,\ldots,X_{k-1}) + h(X_1,\ldots,X_{k-1}) + t(X_k)]] \tag{2.1}$$

where $g(\cdot)$ and $h(\cdot)$ are functions of $X_1,\ldots,X_{k-1}$, $t(\cdot)$ a function of X_k and the multiplication of matrices holds. This also holds when a function of X_k, namely $s(X_k)$, is substituted for X_k under certain conditions. We consider the following example.

Let

$$f_{1,k-1}(X_1,\ldots,X_{k-1}) = \frac{\Gamma_p\left(\sum_{i=1}^{k-1} n_i+m\right)}{\prod_{i=1}^{k-1} \Gamma_p(n_i)\Gamma_p(m)} \prod_{i=1}^{k-1} |X_i|^{n_i-\frac{1}{2}(p+1)} \cdot \left|I+\sum_{i=1}^{k-1} X_i\right|^{-\left(\sum_{i=1}^{k-1} n_i+m\right)}, \quad X_i > 0 \tag{2.2}$$

be a multivariate Dirichlet type II distribution as given by Tan (1969) and let

$$f(X_1,\ldots,X_{k-1}|X_k) = \frac{1}{\prod_{i=1}^{k-1} \Gamma_p(n_i)} \exp\left(-\mathrm{tr}X_k \sum_{i=1}^{k-1} X_i\right) \tag{2.3}$$

$$\cdot \prod_{i=1}^{k-1} |X_i|^{n_i - \frac{1}{2}(p+1)} |X_k|^{\sum_{i=1}^{k-1} n_i}, \quad X_k > 0.$$

Roux (1970) obtained that

$$f(X_1,\ldots,X_k) = \frac{1}{\prod_{i=1}^{k-1} \Gamma(n_i)\Gamma_p(m)} \exp\left(-\mathrm{tr}X_k \left(I + \sum_{i=1}^{k-1} X_i\right)\right) \tag{2.4}$$

$$\cdot |X_k|^{\sum_{i=1}^{k-1} n_i + m - \frac{1}{2}(p+1)} \prod_{i=1}^{k-1} |X_i|^{n_i - \frac{1}{2}(p+1)}, \quad X_i > 0$$

is the only joint distribution of multivariate exponential-type. For $p = 1$ we termed it a Dirichlet-gamma distribution. We derive a few multivariate distributions and their moment generating functions (m.g.f.'s).

A similar case where the m.g.f.'s are known and where the m.g.f. of the conditional p.d.f. is of the form

$$M(T_1,\ldots,T_{k-1};X_k) = \exp[\mathrm{tr}[s(X_k)'g(T_1,\ldots,T_{k-1}) \tag{2.5}$$

$$+ h(T_1,\ldots,T_{k-1}) + t(X_k)]]$$

where $g(\cdot)$ and $h(\cdot)$ are functions of $T_1,\ldots,T_{k-1}$. This is applied to characterize the multivariate normal distribution and Wishart distribution.

3. A GENERALIZED MATRIX-VARIATE HYPERGEOMETRIC DISTRIBUTION.

The generalized hypergeometric distribution or a general family of statistical probability distributions was introduced by Mathai and Saxena (1966), from which almost all the classical probability distributions are obtained as special cases. A similar multivariate probability distribution is considered by Van der Merwe and Roux (1974) from which the multivariate Beta and Wishart distributions are obtained as special cases, as are certain functions of the multivariate normal matrices.

Let $X{:}p\times p$ be a random matrix whose p.d.f. is,

$$f(\bar{X}) = \begin{cases} C|X|^{n-\frac{1}{2}(p+1)} {}_2F_1(\alpha,\beta;\theta;-AX), \; X > 0 \\ 0 \text{ elsewhere} \end{cases} \tag{3.1}$$

where ${}_2F_1(\cdot)$ is a hypergeometric function with matrix argument [Constantine (1963)] and

$$C = \frac{|A|^n\Gamma_p(\alpha)\Gamma_p(\beta)\Gamma_p(\theta-n)}{\Gamma_p(n)\Gamma_p(\theta)\Gamma_p(\alpha-n)\Gamma_p(\beta-n)} \tag{3.2}$$

for $\mathrm{Re}(\theta-n) > \frac{1}{2}(p-1)$, $\mathrm{Re}(\alpha-n) > \frac{1}{2}(p-1)$ and $\mathrm{Re}(\beta-n) > \frac{1}{2}(p-1)$. The parameters α, β and $\theta$are restricted to take values for which $f(X)$ is positive.

In (3.1) replace $A{:}p\times p$ by $\alpha^{-1}A$ and take the limit when α tends to infinity, then using the formula [James (1964]

$$\lim_{\alpha\to\infty} {}_2F_1(\alpha,\beta;\theta;-\alpha^{-1}AX) = {}_1F_1(\beta;\theta;-AX) \tag{3.3}$$

and the fact that

$$\lim_{\alpha\to\infty}|\alpha^{-1}A|^n \frac{\Gamma_p(\alpha)}{\Gamma_p(\alpha-n)} = |A|^n \tag{3.4}$$

we obtain

$$f_2(X) = \begin{cases} \dfrac{|A|^n\Gamma_p(\beta)\Gamma_p(\theta-n)}{\Gamma_p(n)\Gamma_p(\theta)\Gamma_p(\beta-n)} |X|^{n-\frac{1}{2}(p+1)} {}_1F_1(\beta;\theta;-AX), \; X > 0 \\ 0 \text{ elsewhere.} \end{cases} \tag{3.5}$$

Similarly by replacing $A{:}p\times p$ by $\beta^{-1}A$ in (3.5) and taking the limit when β tends to infinity, we obtain

$$f_3(X) = \begin{cases} \dfrac{|A|^n \Gamma_p(\theta-n)}{\Gamma_p(n)\Gamma_p(\theta)} |X|^{n-\frac{1}{2}(p+1)} {}_0F_1(\theta;-AX), & X > 0 \\ 0 \text{ elsewhere} \end{cases} \tag{3.6}$$

If however we replace $A{:}p\times p$ by θA in (3.5) and take the limit when θ tends to infinity we obtain

$$f_4(X) = \begin{cases} \dfrac{|A|^n \Gamma_p(\beta)}{\Gamma_p(n)\Gamma_p(\beta-n)} |X|^{n-\frac{1}{2}(p+1)} {}_1F_0(\beta;-AX), & X > 0 \\ 0 \text{ elsewhere.} \end{cases} \tag{3.7}$$

In (3.7) replace A by $\beta^{-1}A$ and let β tend to infinity, then we obtain

$$f_5(X) = \begin{cases} \dfrac{|A|^n}{\Gamma_p(n)} |X|^{n-\frac{1}{2}(p+1)} {}_0F_0(-AX), & X > 0 \\ 0 \text{ elsewhere.} \end{cases} \tag{3.8}$$

The functions $f_i(x)$, $i = 1,2,3,4,5$ satisfy the condition $\int_{X>0} f_i(X)dX = 1$.

Let $V{:}p\times m$ be a random matrix with p.d.f.

$$f(V) = \begin{cases} C \dfrac{\Gamma_p(\frac{1}{2}m)}{\pi^{\frac{1}{2}mp}} |VV'|^{n-\frac{1}{2}m} {}_2F_1(\alpha,\beta;\theta;-AVV'), & V: p\times m \\ 0 & \text{elsewhere.} \end{cases} \tag{3.9}$$

The following special cases are obtained.

(i) The Wishart distribution.

In (3.5) put $\beta = \theta$, $n = \frac{1}{2}m$ and $A = \frac{1}{2}\Sigma^{-1}$ or in (3.8) put $A = \frac{1}{2}\Sigma^{-1}$.

(ii) The multivariate Beta distribution.

In (3.1) put $\beta = \theta$ and $A = I_p$ or in (3.7) put $A = I_p$.

(iii) The multivariate folded normal distribution.
In (3.9) put $\beta = \theta$, $A = \frac{1}{2}\Sigma^{-1}$, $n = \frac{1}{2}m$, replace $A:p\times p$ by $\alpha^{-1}A$, and take the limit when α tends to infinity.

(iv) The special multivariate t distribution.
In (3.9) put $\beta = \theta$ and $n = \frac{1}{2}m$.

(v) The general family or univariate distributions.
In (3.1) put $p = 1$. Thus for $p = 1$ we find that almost all classical statistical univariate distributions are contained in these multivariate distributions. [Mathai and Saxena (1966)].

Some of these special central multivariate distributions are studied by Anderson (1958), Roy (1957), Khatri (1959), and Rubin (1964) and Dickey (1967) because of the importance and the practical application of these in testing certain statistical hypotheses. The multivariate t distribution has also been studied by Juritz (1973). The following properties are investigated.

When $S:p\times p$ has the distribution (3.1), the incomplete distribution function is given by

$$P(S < X) = C \frac{\Gamma_p(n)\Gamma_p(\frac{1}{2}(p+1))}{\Gamma_p(n+\frac{1}{2}(p+1))}|X|^n {}_3F_2(\alpha,\beta,n;n+\frac{1}{2}(p+1),\theta;-AX) \quad (3.10)$$

where for $p \geq 2$, $P(S < X) \neq 1 - P(S > X)$. [Constantine (1963)].

The joint distributions of the ordered characteristic roots of certain matrices have been determined by Anderson (1958). If $X:p\times p$ has a density function (3.1) the joint density of the characteristic roots of $X:p\times p$ can easily be found as

$$g(\lambda_1,\ldots,\lambda_p) = C\frac{\pi^{\frac{1}{2}p^2}}{\Gamma_p(\frac{1}{2}p)} \prod_{i=1}^{p} \lambda_i^{n-\frac{1}{2}(p+1)} {}_2F_1(\alpha,\beta;\theta;-A,D_\lambda) \prod_{i<j}^{p} (\lambda_i-\lambda_j),$$

$$\lambda_1 >\ldots> \lambda_p > 0 \quad (3.11)$$

where $D_\lambda = \text{diag}(\lambda_1,\ldots,\lambda_p)$. The hypergeometric function ${}_2F_1(\cdot)$ with two matrix arguments is defined by James (1964). If $\lambda_1 >\ldots> \lambda_p > 0$ are the roots of $|X - \lambda I| = 0$, then by a transforming of variables and integration the distribution of the largest root of $X:p\times p$ is found to be

$$f(\lambda_1) = C \frac{\Gamma_p(n)\Gamma_p(\frac{1}{2}(p+1))}{\Gamma_p(n+\frac{1}{2}(p+1))} \sum_{k=0}^{\infty} \sum_{K} \lambda_1^{np+k-1}(np+k)\frac{(\alpha)_K(\beta)_K(n)_K}{(\theta)_K(n+\frac{1}{2}(p+1))_K}$$

$$\cdot \frac{C_K(-A)}{k!}, 0 < \lambda_1 < 1. \quad (3.12)$$

The distribution of the smallest root of matrix $X:p\times p$ follows similarly.

Let $X_i:p\times p$, $i = 1,2$ be random matrices independently distributed with the following density function

$$f(X_i) = \begin{cases} C_i|X_i|^{n_i-\frac{1}{2}(p+1)} {}_2F_1(\alpha_i,\beta_i;\theta_i,-A_iX_i), & X_i > 0 \\ 0 & \text{elsewhere} \end{cases} \quad (3.13)$$

where

$$C_i = \frac{|A_i|^{n_i}\Gamma_p(\alpha_i)\Gamma_p(\beta_i)\Gamma_p(\theta_i-n_i)}{\Gamma_p(n_i)\Gamma_p(\theta_i)\Gamma_p(\alpha_i-n_i)\Gamma_p(\beta_i-n_i)}, \quad i = 1,2. \tag{3.14}$$

Consider the h-th order moment of $v = |X_1| \; |X_2|^{-1}$ and use the inverse Mellin transform [Erdelyi et al. (1954)], then we obtain the density function of the random variable v as

$$f(v) = \prod_{i=1}^{2} \left\{ \frac{\Gamma_p(\theta_i-n_i)}{\Gamma_p(n_i)\Gamma_p(\alpha_i-n_i)\Gamma_p(\beta_i-n_i)} \right\} \pi^{p(p-1)} v^{-1} \tag{3.15}$$

$$G^{3p\;3p}_{4p\;4p}[\cdot]\left[\frac{|A_1|}{|A_2|}\, v \middle| \begin{matrix} a_1,\ldots,a_{4p} \\ b_1,\ldots,b_{4p} \end{matrix}\right]$$

where a_j and $\mathbf{b}_j$ are function of the parameters and the general Meijer G-function is equal to

$$G^{r\;s}_{p\;q}\left[x \middle| \begin{matrix} a_1,\ldots,a_p \\ b_1,\ldots,b_q \end{matrix}\right] = (2\pi i)^{-1} \int_{c-i\infty}^{c+i\infty} \frac{\prod_{j=1}^{r}\Gamma(b_j+h)\prod_{j=1}^{s}\Gamma(1-a_j-h)}{\prod_{j=r+1}^{q}\Gamma(1-b_j-h)\prod_{j=s+1}^{p}\Gamma(a_j+h)}$$

$$\cdot\, x^{-h} dh \qquad \text{[Mathai (1971)].} \tag{3.16}$$

Van der Merwe (1974) has obtained the exact distribution of the product of two independent matrices taken from samples with p.d.f.'s given by (i) (3.5) for $A = I_p$, (ii) a multivariate beta type II and (iii) a Wishart distribution.

Distributions of the products of two independent random determinants are also investigated for the above three cases using the technique based on their h-th order moments and the Mellin inversion theorem. The results arrived at are in the most general form that is in terms of Meijer G-functions.

4. GENERALIZED MULTIVARIATE DISTRIBUTIONS. Many distributions in multivariate analysis can be expressed in a form involving the generalized hypergeometric functions with matrix argument as considered before. Roux (1971b) defined the following generalized multivariate distributions.

(i) A distribution which includes the central [Anderson (1958)] and noncentral [Constantine (1963)] Wishart distributions with the p.d.f.

$$f(X) = \begin{cases} \dfrac{|B|^n}{\Gamma_p(n)\, {}_{r+1}F_q(a_1,\ldots,a_r,n;b_1,\ldots,b_q;BR)} \exp(-\mathrm{tr}BX)\,|X|^{n-\frac{1}{2}(p+1)} \\ \cdot\ {}_rF_q(a_1,\ldots,a_r;b_1,\ldots,b_q;BRBX),\ X > 0 \\ 0 \quad \text{elsewhere} \end{cases} \tag{4.1}$$

where $R{:}p\times p$ and $B{:}p\times p$ are positive definite matrices. The parameters a_i, b_j; $i = 1,2,\ldots,r$, $j = 1,2,\ldots,q$ are restricted to take those values for which $f(X)$ is positive.

(ii) A distribution which includes the central and non-central multivariate Beta distributions with the p.d.f.'s

$$f(X) = \begin{cases} \dfrac{\Gamma_p(m+n)}{\Gamma_p(m)\Gamma_p(n)\, {}_{r+1}F_{q+1}(a_1,\ldots,a_r,m;b_1,\ldots,b_q,m+n;R)} \\ \cdot\ |X|^{m-\frac{1}{2}(p+1)}\,|I-X|^{n-\frac{1}{2}(p+1)}\, {}_rF_q(a_1,\ldots,a_r,b_1,\ldots,b_q;RX), \\ \qquad 0 < X < I \\ 0 \quad \text{elsewhere} \end{cases} \tag{4.2}$$

and

$$f(Y) = \begin{cases} \dfrac{\Gamma_p(m+n)}{\Gamma_p(m)\Gamma_p(n)\,{}_{r+1}F_{q+1}(a_1,\ldots,a_r,n;b_1,\ldots,b_q,m+n;R)} \\ \cdot\ |I+Y|^{-(m+n)}|Y|^{m-\frac{1}{2}(p+1)}\,{}_rF_q(a_1,\ldots,a_r;b_1,\ldots,b_q; \\ \qquad R(I+Y)^{-1}),\ Y > 0 \\ 0 \quad \text{elsewhere} \end{cases} \tag{4.3}$$

the parameters are restricted to take those values for which f(X) and f(Y) are positive. The distributions with p.d.f.'s (4.2) and (4.3) include under certain conditions the central [Anderson (1958) and Olkin and Rubin (1964)] and noncentral [De Waal (1968), (1969)] multivariate Beta distributions.

(iii) A distribution which includes under certain conditions the central [Olkin and Rubin (1964) and Troskie (1966)] and noncentral [De Waal (1968)] multivariate Dirichlet distributions with p.d.f.'s and extension of (ii).

Some special and interesting properties namely, the h-th moments of the determinants and the distributions of the trace of matrices enjoyed by these generalized multivariate distributions are given.

5. MULTIVARIATE DISTRIBUTIONS HAVING PROPERTIES USUALLY ASSOCIATED WITH THE WISHART DISTRIBUTION. In this section Roux and Van der Merwe (1974) introduced two families of multivariate distributions which in many respects could be considered as analogues of the univariate distributions [Mathai and Saxena (1968), (1969)]. We considered some general families of multivariate p.d.f.'s with the same properties. We shall, however, restrict ourselves to the following two properties namely the Wishart and the ratio property.

It is well-known that the Wishart distribution has the property that

$$[M(\tfrac{T}{\alpha};\ \Sigma^{-1},\ n)]^{\alpha} = M(T;\alpha\Sigma^{-1},\ \alpha n) \tag{5.1}$$

where M(•) denotes the common m.g.f., T a p×p matrix and α the size of a sample of random matrices. From the above m.g.f. it thus follows, by using the inverse Laplace transform [Constantine (1963)], that the distribution of the sample mean for a simple random sample from a Wishart distribution again has a Wishart distribution with the parameters scaled by the sample size. We shall point out that this is not a property unique to the Wishart, but that one can find a family of p.d.f.'s other that the Wishart with this property. The Wishart property may in general be defined as

$$[M(\frac{T}{\alpha};n_1,\ldots,n_k)]^{\alpha} = M(T;\alpha n_1,\ldots,\alpha n_k), \tag{5.2}$$

where $n_1,\ldots,n_k$ are parameters.

We show that if $X:p\times p$ has the p.d.f.

$$f(X;P,B,D,v,m)=\begin{cases} \dfrac{|P-B|^{2v-2m}}{\Gamma_p(2v-2m)\,{}_1F_0(2v;(D-B)(P-B)^{-1})} \\ \cdot\exp(-\mathrm{tr}X(P-B))\,|X|^{2v-2m-\frac{1}{2}(p+1)} \\ \cdot\ {}_1F_1(2v;2v-2m;(D-B)X),\ X>0 \\ 0 \quad \text{elsewhere} \end{cases} \tag{5.3}$$

where $\mathrm{Re}(u-m) > \frac{1}{2}(p-1)$, then the sample mean has the distribution $f(X;\alpha P,\alpha B,\alpha D,\alpha v,\alpha m)$ where $P:p\times p$, $B:p\times p$ and $D:p\times p$ are positive definite matrix parameters. Also if $X:p\times p$ has the p.d.f.

$$f(X;P,B,D,t) = \begin{cases} \dfrac{|P|^t}{\Gamma_p(t)}\exp(-\mathrm{tr}\ PB^{-\frac{1}{2}}DB^{-\frac{1}{2}})\exp(-\mathrm{tr}\ PX)\,|X|^{t-\frac{1}{2}(p+1)} \\ \cdot\,{}_1F_0(-t+\frac{1}{2}(p+1);-B^{-\frac{1}{2}}DB^{-\frac{1}{2}}X^{-1}),\ X>0 \\ 0 \quad \text{elsewhere} \end{cases} \tag{5.4}$$

where $\mathrm{Re}(t) > k_1 + \frac{1}{2}(p-1)$, then the sample mean has the distribution $f(X;\alpha P,\alpha B,\alpha D,\alpha t)$. Positivity of the above function is assumed since this can be easily achieved by selecting the parameters properly.

If $X_1:p\times p$ and $X_2:p\times p$ are independent random matrices both drawn from a Wishart distribution, then it is known that the distribution of $X_2^{-\frac{1}{2}}X_1X_2^{-\frac{1}{2}}$ has a Beta type II distribution [Olkin and Rubin (1964)]. The question now arises that if X_1:p×p and X_2:p×p are independent identically distributed matrices drawn from a certain distribution and $X_2^{-\frac{1}{2}}X_1X_2^{-\frac{1}{2}}$ has a Beta type II distribution, does it necessarily mean that X_1 and X_2 come from a Wishart distribution? We shall show that this need not be the case, by considering a p.d.f. [Van der Merwe and Roux (1974)] different from the Wishart for which the ratio turns out to be a Beta type II distribution. This property does not characterize the Wishart distribution uniquely. We shall state one such distribution here.

If X_1:p×p and X_2:p×p are independent random matrices with p.d.f.'s given by $f_{a_i}(X_i)$ for i = 1,2, where

$$f_{a_i}(X_i) = \begin{cases} \dfrac{\Gamma_p(a_i+\frac{1}{2}\gamma+\frac{1}{4}(p+1))}{\Gamma_p(a_i)\Gamma_p(\gamma+\frac{1}{2}(p+1))}|X_i|^{\frac{1}{2}\gamma-\frac{1}{4}(p+1)} \\ \cdot {}_1F_1(a_i+\frac{1}{2}\gamma+\frac{1}{4}(p+1);\gamma+\frac{1}{2}(p+1);-X_i),\ X_i>0 \\ 0 \text{ elsewhere} \end{cases} \tag{5.5}$$

then $X_2^{-\frac{1}{2}}X_1X_2^{-\frac{1}{2}}$ has a Beta type II distribution for all values of γ such that $f_{a_i}(X_i)$ exists. By applying a property of the Hankel transform [Herz (1955)] one can evaluate certain infinite integrals to prove the ratio property.

6. GENERALIZED LAGUERRE SERIES FORMS OF WISHART DISTRIBUTIONS.

The Laguerre series expansion of the noncentral Wishart is developed by Roux and Raath (1973) in terms of Laguerre polynomials, namely

$$f^*(X) = f(X)\sum_{k=0}^{\infty}\sum_K \alpha_K L_K^{\frac{1}{2}(n-p-1)}(\tfrac{1}{2}\Sigma^{-1}X) \tag{6.1}$$

where f(X) is the p.d.f. of the central Wishart distribution

[Anderson (1958)], $L_K^{\frac{1}{2}(n-p-1)}(\frac{1}{2}\Sigma^{-1}X)$ are generalized Laguerre polynomials [Constantine (1966)] and

$$\alpha_K = \frac{1}{k!}\sum_{j=0}^{k}\sum_{J} a_{K,J}\frac{(-1)^j}{(\frac{1}{2}n)_J C_J(I_p)}\exp(-\mathrm{tr}\Omega)\sum_{v=0}^{\infty}\sum_{V}\frac{1}{(\frac{1}{2}n)_V v!}$$

$$\cdot\frac{C_V(\Omega)}{C_V(I_p)}\sum_{\delta} g_{V,J}^{\delta}(\tfrac{1}{2}n)_{\delta}C_{\delta}(I_p) \tag{6.2}$$

are functions of the noncentral parameter $\Omega = \frac{1}{2}\mu\mu'\Sigma^{-1}$.

In (6.1) we can also replace the central Wishart distribution by the approximated noncentral Wishart distribution, derived by Steyn and Roux (1972), with density function

$$f(X) = \frac{|X|^{\frac{1}{2}(n-p-1)}\exp(-\frac{1}{2}\mathrm{tr}(\Sigma+\frac{1}{n}\mu\mu')X)}{2^{\frac{1}{2}np}\pi^{\frac{1}{4}p(p-1)}|\Sigma+\frac{1}{n}\mu\mu'|^{\frac{1}{2}n}\prod_{i=1}^{p}\Gamma[\frac{1}{2}(n+1-i)]}. \tag{6.3}$$

The density function of a joint Wishart distribution can be formally expanded as a sum of generalized Laguerre polynomials, namely

$$f(X_1,X_2) = f_{n_1}(X_1)f_{n_2}(X_2)\sum_{k=0}^{\infty}\sum_{K}\beta_K L_K^{\frac{1}{2}(n_1-p-1)}(\tfrac{1}{2}\Sigma_1^{-1}X_1)$$

$$\cdot L_K^{\frac{1}{2}(n_2-p-1)}(\tfrac{1}{2}\Sigma_2^{-1}X_2) \tag{6.4}$$

where $f_{n_i}(X_i)$ is the p.d.f. of a Wishart distribution and

$$\beta_k = \frac{E[L_K^{\frac{1}{2}(n_1-p-1)}(\frac{1}{2}\Sigma_1^{-1}X_1)L_K^{\frac{1}{2}(n_2-p-1)}(\frac{1}{2}\Sigma_2^{-1}X_2)]}{(C_K(I_p))^2(\frac{1}{2}n_1)_K(\frac{1}{2}n_2)_K(k!)^2} \tag{6.5}$$

are constants.

The results obtained appear important in giving a simple method of approximating the expressions for large samples and to derive the distributions of some statistics.

7. SYMMETRIZED MULTIVARIATE DISTRIBUTIONS. In multivariate analysis there exists a large class of important hypothesis testing problems all of which may be tested by a set of criteria that depend functionally on a matrix variate with a Beta distribution. Examples of such hypotheses are the following.

(i) Hypotheses of independence of two sets of variates, considering the second set fixed.

(ii) Linear hypotheses about regression coefficients.

(iii) General linear hypotheses in multivariate analysis of variance.

These may be tested using one of the following test criteria.

(i) Wilks' likelihood ratio criterion [Wilks (1932), Anderson (1958)].

(ii) Roy's largest and smallest root criteria [Roy (1957)].

(iii) Hotelling's T_0^2 or Pillai's $U^{(p)}$ criterion [Pillai (1955)].

(iv) Pillai's V and Q criteria [Pillai (1955)].

More specifically, these criteria are all functions of the characteristic roots of a noncentral Beta type I matrix [De Waal (1968)]. A first step in the investigation of the non-null distributions of these criteria is thus to derive the noncentral Beta type I density function and the joint density of the roots of a noncentral Beta type I matrix.

Greenacre (1973) defined the symmetrized multivariate density of a positive definite matrix $X:p\times p$ as

$$f_s(X) = \int_{O(p)} f(HXH')dH \tag{7.1}$$

where $X{:}p\times p$ has the density function $f(X)$. The symmetrized density clearly satisfies the conditions of a density function and has the important property that the distribution of the characteristic roots of $X{:}p\times p$ is the same for $f_s(X)$ and $f(X)$.

Consider the form of the actual noncentral Beta type I density with an unsolved integral namely [De Waal (1968)]

$$f(L) = \frac{1}{\Gamma_p(\frac{1}{2}m)\Gamma_p(\frac{1}{2}n)|2\Sigma|^{\frac{1}{2}(m+n)}} \exp(-\tfrac{1}{2}\mathrm{tr}\Omega)|L|^{\frac{1}{2}(n-p-1)}$$

$$\cdot|I-L|^{\frac{1}{2}(m-p-1)} \int_{S>0} \exp(-\tfrac{1}{2}\mathrm{tr}\Sigma^{-1}S)|S|^{\frac{1}{2}(m+n-p-1)} \tag{7.2}$$

$$\cdot{}_0F_1(\tfrac{1}{2}n;\tfrac{1}{4}\Omega\Sigma^{-1}S^{\frac{1}{2}}LS^{\frac{1}{2}})dS,\ 0 < L < I.$$

The symmetrized equivalent is solved and is in the form of the central density multiplied by a weighting function

$$f_s(L) = \frac{\Gamma_p(\frac{1}{2}(m+n))}{\Gamma_p(\frac{1}{2}m)\Gamma_p(\frac{1}{2}n)} |L|^{\frac{1}{2}(n-p-1)} |I-L|^{\frac{1}{2}(m-p-1)} \tag{7.3}$$

$$\cdot\exp(-\tfrac{1}{2}\mathrm{tr}\Omega)\,{}_1F_1(\tfrac{1}{2}(m+n);\tfrac{1}{2}n;\tfrac{1}{2}\Omega,L).$$

Similar results of other multivariate Beta densities and the density of the generalized multiple correlation matrix, will be illustrated [Greenacre (1972)].

Thus using the symmetrized density of $X{:}p\times p$ the densities of all functions of $X{:}p\times p$ which are invariant under congruence transformations of $X{:}p\times p$ by the orthogonal group $O(p)$ can be obtained. Amongst these functions are $|X|$, trace X, and in fact all functions which depend on X through its characteristic roots only.

Roux and Raath (1974) introduced a new class of m.g.f.'s with desirable properties, that are in certain cases equivalent to the actual m.g.f. and may in fact be used in its place. The symmetrized m.g.f. of the positive definite matrix $X:p\times p$ is defined as

$$M_s(T) = \int_{0(p)} M(HTH')dH \tag{7.4}$$

where $M(T)$ is the ordinary m.g.f. of the density function $f(X)$. The symmetrized m.g.f. satisfies the conditions of a m.g.f. and has the important property that the symmetrized m.g.f. $M_s(T)$ is the m.g.f. of the symmetrized function $f_s(X)$. We mention only as examples the Wishart and noncentral Wishart m.g.f.'s.

8. COMPOUND MULTIVARIATE DISTRIBUTIONS. Let $X:p\times p$ be a random matrix with distribution function $f_i(X;P)$ (conditional on $P:p\times p$) and the prior distribution function of the parameter $P:p\times p$ given by $h_i(P)$. Then the compound distribution (posterior) is given by

$$g_i(X) = \int f_i(X;P)h_i(P)dP \ . \tag{8.1}$$

It is possible to build up a family of multivariate distributions by changing the conditional distribution or the distribution of the parameter [Roux (1971b), Van der Merwe and Roux (1974) and Roux and Becker (1974)].

REFERENCES

[1] Anderson, T. W. (1958). An Introduction to Multivariate Statistical Analysis. Wiley, New York.

[2] Constantine, A. G. (1963). Ann. Math. Statist. 34, 1270-1285.

[3] Constantine, A. G. (1966). Ann. Math. Statist. 37, 215-225.

[4] De Waal, D. J. (1968). Nie-sentrale meerveranderlike Beta-verdelings. Ph.D. Thesis. University of Cape Town, Cape Town, South Africa.

[5] De Waal, D. J. (1969). S. Afr. Statist. J. 3, 101-108.

[6] Dickey, J. M. (1967). Ann. Math. Statist. 38, 511-518.

[7] Erdélyi, A., et al. (1954). Tables of Integral Transforms, 1. McGraw-Hill, New York.

[8] Greenacre, M. (1972). Some noncentral distributions and applications in multivariate analysis. M.Sc. dissertation. University of South Africa, Pretoria.

[9] Greenacre, M. (1973). S. Afr. Statist. J. 7, 95-101.

[10] Herz, C. S. (1955). Ann. Math. 61, 474-523.
[11] James, A. T. (1964). Ann. Math. Statist. 35, 475-497.
[12] Jaritz, J. M. (1973). Aspects of noncentral multivariate t distributions. Ph.D. Thesis. University of Cape Town, Cape Town, South Africa.
[13] Khatri, C. G. (1959). Ann. Math. Statist. 30, 1258-1262.
[14] Mathai, A. M. (1971). S. Afr. Statist. J. 5, 71-90.
[15] Mathai, A. M. and Saxena, R. K. (1966). Metrika 11, 127-132.
[16] Mathai, A. M. and Saxena, R. K. (1968). Metrika 13, 10-15.
[17] Mathai, A. M. and Saxena, R. K. (1969). S. Afr. Statist. J. 3, 27-34.
[18] Olkin, I. and Rubin, H. (1964). Ann. Math. Statist. 35, 261-269.
[19] Pillai, K. C. S. (1955). Ann. Math. Statist. 26, 117-121.
[20] Roux, J. J. J. (1970). Oor die gebruik van voortbringende funksies in die veld van meerveranderlike verdelingsteorie. Ph.D. Thesis, University of South Africa, Pret
[21] Roux, J. J. J. (1971a). S. Afr. Statist. J. 5, 27-36.
[22] Roux, J. J. J. (1971b). S. Afr. Statist. J. 5, 91-100.
[23] Roux, J. J. J. and Raath, E. L. (1973). S. Afr. Statist. J. 7, 23-34.
[24] Roux, J. J. J. and Raath, E. L. (1974). Some extensions of the Wishart moment generating function. Submitted for publication in S. Afr. Statist. J.
[25] Roux, J. J. J. and Van der Merwe, G. J. (1974). Families of multivariate distributions having properties usually associated with the Wishart distribution. S. Afr. Statist. J. 8, No. 2.
[26] Roux, J. J. J. and Becker, P. J. (1974). Gemodifiseerde Bessel-funksiemodelle in lewenstoetsing. Research lecture.
[27] Roy, S. N. (1957). Some Aspects of Multivariate Analysis. Wiley, New York.
[28] Seshadri, V. and Patil, G. P. (1964). Inst. Statist. Math. Ann. 15, 215-221.
[29] Steyn, H. S. and Roux, J. J. J. (1972). S. Afr. Statist. J. 6, 165-173.
[30] Tan, W. Y. (1969). J. Amer. Statist. Assoc. 64, 230-241.
[31] Troskie, C. G. (1966). Tydskrif vir Natuurwetenskappe 6, 58-71.
[32] Van der Merwe, G. J. (1974). Exact distributions of the product of certain independent matrices. S. Afr. Statist. J. 8, No. 2.
[33] Van der Merwe, G. J. and Roux, J. J. J. (1974). On a generalized matrix-variate hypergeometric distribution. S. Afr. Statist. J. 8, No. 1.
[34] Wilks, S. S. (1932). Biometrika 24, 471-494.

ASYMPTOTIC EXPANSIONS FOR THE NONNULL DISTRIBUTIONS OF THE MULTIVARIATE TEST STATISTICS

Minoru Siotani

Department of Statistics, Kansas State University,
Manhattan, Kansas, U.S.A.

SUMMARY. Three methods are discussed for obtaining the asymptotic expansions for the nonnull distributions of multivariate test statistics; (i) the method by the Taylor expansion including the studentization method, (ii) the method by the expansion of exact characteristic functions using the formulas for weighted sums of zonal polynomials, and (iii) the method based on the system of partial differential equations satisfied by the hypergeometric functions. A short note on some other useful methods is also given. The emphasis in the discussion is put on the basic idea in each method.

KEY WORDS. Asymptotic expansion, nonnull distribution, Taylor expansion, studentization, characteristic function, zonal polynomial, hypergeometric function with matrix argument, partial differential equations.

1. INTRODUCTION. The power functions of various multivariate tests were known very little before the works by A. T. James and A. G. Constantine who developed the method for deriving general or nonnull distributions of the test statistics in terms of zonal polynomials of a positive definite symmetric matrix. At the present time we also have the expressions in terms of the generalized Laguerre polynomials and the generalized Hermite polynomials for the distributions of some test statistics. These exact formulas are unfortunately not so convenient for the numerical investigation on powers of the tests. By this reason, asymptotic expansions for the nonnull distributions of test criteria have been studied from the various points of view.

G. P. Patil et al. (eds.), Statistical Distributions in Scientific Work, Vol. 1, 299-317.

Recently two different approaches to the power investigation, one from the exact consideration and one from the asymptotic consideration, have been successfully connected by obtaining the useful formulas for deriving the asymptotic expansions from the exact expressions of the density functions, which will be seen in Section 3.

The main aim of this paper is to explain the basic idea of the methods recently developed for deriving the asymptotic expansions for the nonnull distributions in the multivariate analysis. However, owing to the restriction of given number of pages, we shall only give the exposition on the following three methods: (i) method by the Taylor expansion including the studentization (Section 2), (ii) Derivation from the exact expression of characteristic function (Section 3), and (iii) Method based on a system of partial differential equations (Section 4). In the last section, a short note on some other methods will be given.

We are not here going to discuss the exact distributions, but readers are recommended to refer to Johnson and Kotz's book (1972) which gives a survey on the continuous multivariate distributions.

2. METHOD BY TAYLOR EXPANSION: STUDENTIZED STATISTICS.

2.1 General description of the method

In the multivariate normal theory, many test statistics come out as a function of Wishart matrices. Let $V = nS \sim W_p(\Sigma, n)$ and Z be another random matrix. Suppose that we are interested in a statistic $g(Z,S)$ and its sampling distribution. Suppose that $g(Z,S)$ converges in probability to $g(Z,\Sigma)$ as $n \to \infty$ and the limiting distribution of $g(Z,S)$ is known as the distribution of $g(Z,\Sigma)$, but exact one is not known or very complicated for practical uses. The method in this section is aimed primarily to meet such a situation by applying the Taylor expansion to desired probability distribution or the characteristic function (ch. f.) of $g(Z,S)$ with respect to S about Σ. The method can be easily extended so as to enable us to treat the joint distribution of several statistics having S in common.

<u>Lemma 2.1</u> (James (1954)). Let $f(S)$ be a function of S where $nS \sim W_p(\Sigma,n)$. Suppose that $f(S)$ has the Taylor expansion about $S = \Sigma$. Then

$$E\{f(S)\} = \frac{n^{pn/2}}{\Gamma_p(n/2)|2\Sigma|^{n/2}} \int_{S>0} f(S)|S|^{(n-p-1)/2} \operatorname{etr}(-\tfrac{n}{2}\Sigma^{-1}S)dS \tag{2.1}$$

$$= \Theta \cdot f(\Lambda)\Big|_{\Lambda=\Sigma}$$

where $\Gamma_p(a) = \pi^{p(p-1)/4} \Pi_{i=1}^{p} \Gamma(a-(i-1)/2)$ for $a > (p-1)/2$, $\mathrm{etr}(A) = \exp(\mathrm{tr}\, A)$,

$$\Theta = \left|I - \frac{2}{n}\Sigma\partial\right|^{-n/2} \mathrm{etr}(-\Sigma\partial) \tag{2.2}$$

$$\begin{aligned} &= 1 + \frac{1}{n}\mathrm{tr}(\Sigma\partial)^2 + \frac{1}{n^2}[\tfrac{4}{3}\mathrm{tr}(\Sigma\partial)^3 + \tfrac{1}{2}\{\mathrm{tr}(\Sigma\partial)^2\}^2] \\ &\quad + \frac{1}{n^3}[2\mathrm{tr}(\Sigma\partial)^4 + \tfrac{4}{3}\mathrm{tr}(\Sigma\partial)^2\,\mathrm{tr}(\Sigma\partial)^3 + \tfrac{1}{6}\{\mathrm{tr}(\Sigma\partial)^2\}^3] \\ &\quad + O(n^{-4}) \end{aligned} \tag{2.3}$$

and $\partial: p\times p = ((1/2)(1 + \delta_{ij})\partial/\partial\lambda_{ij})$, δ_{ij} being the Kronecker symbol.

Lemma 2.2. Let $nS \sim W_p(\Sigma,n)$ and $Z: p\times m$ be a random matrix distributed independently of S. Suppose a function of Z and S, $g(Z,S)$, has the Taylor expansion about $S = \Sigma$. Then

$$\Pr\{g(Z,S) \le c\} = \Theta\cdot\Pr\{g(Z,\Lambda) \le c\}\Big|_{\Lambda=\Sigma}, \tag{2.4}$$

where Θ is the same as (2.2) or (2.3).

Lemma 2.3. Let $\phi(t)$ be the ch. f. of $g(Z,S)$ and let $\phi(t;\Lambda) = E_Z[\exp\{itg(Z,\Lambda)\}]$. Then

$$\phi(t) = \Theta\cdot\phi(t;\Lambda)\Big|_{\Lambda=\Sigma}. \tag{2.5}$$

Lemmas 2.1 - 2.3 can be proved first by fixing S, secondly by expanding f(S), conditional probability function and ch. f. about $S = \Sigma$ using the Taylor expansion formula, respectively, and finally by taking the expectation with respect to S.

Those lemmas can be easily extended to the noncentral Wishart matrix.

In order to calculate the explicit expansion for the distribution of $g(Z,S)$ according to Lemma 2.2 or Lemma 2.3, we need to evaluate the various derivatives in those formulas. The following idea is essentially due to Welch (1947). Consider

$$J = \Pr\{g(Z,\Sigma + \varepsilon \leq c\} \tag{2.6}$$

where $\varepsilon : p\times p = (\varepsilon_{jk})$ is a symmetric matrix consisting of small increments ε_{jk} to σ_{jk} such that $\Sigma + \varepsilon$ is still positive definite (p. d.). Then by the Taylor expansion

$$J = [1 + \Sigma\, \varepsilon_{rs}\partial_{rs} + \frac{1}{2!}\Sigma\, \varepsilon_{rs}\varepsilon_{tu}\partial_{rs}\partial_{tu} + \ldots]\{\Pr g(Z,\Lambda) \leq c\}\Big|_{\Lambda=\Sigma} \tag{2.7}$$

On the other hand, J can be expressed as

$$J = \int_{D^*} f(Z)dZ \tag{2.8}$$

where $D^* = \{Z\colon g(Z,\Sigma + \varepsilon) \leq c\}$ and $f(Z)$ is the p. d. f. of Z. If we can expand J of (2.8) in power series of ε_{jk}'s, then we can obtain the desired derivatives by comparing those with (2.7) and hence an asymptotic expansion (asymp. expan.) for the distribution of $g(Z,S)$. In particular, if $Z:p\times m \sim N_p(M;\Sigma)$, then

$$J = (2\pi)^{-pm/2}\,|\Sigma|^{-m/2}\int_{D^*} \mathrm{etr}\{-\tfrac{1}{2}\Sigma^{-1}(Z-M)(Z-M)'\}dZ$$

$$= (2\pi)^{-pm/2}\,|I - D_\xi|^{m/2}\int_{D^{**}} \mathrm{etr} - \tfrac{1}{2}(Y-H)'(I-D_\xi)(Y-H)\}dY \tag{2.9}$$

by making the nonsingular transformation $Z = LY$ with L such that $L'(\Sigma+\varepsilon)^{-1}L = I$, $L'\Sigma^{-1}L = I-D_\xi$, where $D_\xi = \mathrm{diag}(\xi_1,\ldots,\xi_p)$ with $|\xi_j| < 1$ for all j, $H = L^{-1}M$, and $D^{**} = \{Y\colon g(LY,LL') \leq c\}$. If $g(LY,LL')$ is independent of L such as $\mathrm{tr}(\Sigma+\varepsilon)^{-1}ZZ' = \mathrm{tr}(YY')$, the integral in (2.9) might be evaluated explicitly.

2.2. Some examples

(i) Nonnull distribution of Hotelling's generalized T_o^2. Let $Z:p\times m \sim N_p(M;\Sigma)$, $nS:p\times p \sim W_p(\Sigma,n)$, and Z and S be independent. Then the statistic $T_o^2 = g(Z,S) = \mathrm{tr}(S^{-1}ZZ')$ was proposed by Hotelling (1951) for testing the hypothesis $H\colon M = 0$ against $K\colon M \neq 0$. Since $g(Z,\Sigma) = \mathrm{tr}(\Sigma^{-1}ZZ')$ is distributed as the noncentral χ^2 with pm d. f. and noncentrality $\omega^2 = tr(\Sigma^{-1}MM')$. Since $g(LY,LL') = tr(YY')$, the integral J of (2.9) can be evaluated and expressed in the following form:

$$J = \frac{|I - eD_\xi|^{m/2}}{|I - D_\xi|} \, etr(-\tfrac{1}{2}\Omega) etr\{\tfrac{1}{2}(I-D_\xi)(I-eD_\xi)^{-1}(I-D_\xi)HH'e\}$$

$$\chi^2_{pm}(c;0)$$

$$= |I - \Delta X|^{-m/2} \, etr\{-\tfrac{1}{2}\Omega + \tfrac{1}{2}e(I - \Delta X)^{-1}\Omega\} \cdot \chi^2_{pm}(c;0) \quad (2.10)$$

where e is an operator such that $e^r\chi^2_f(c;\omega^2) = \chi^2_{f+2r}(c;\omega^2)$, $r=0,1,2,\ldots$, $^2_f(c;\omega^2)$ is the cumulative distribution function (c. d. f.) of the noncentral chi square distribution with f d. f. and noncentrality ω^2, and

$$\Delta = e - 1, \qquad \Omega = \Sigma^{-1}MM', \qquad X = (\Sigma + \varepsilon)^{-1}\Sigma - I.$$

If the derivation is based on the ch. f. of T_0^2, we have

$$\phi(t;\Sigma + \varepsilon) = |I - \Delta(t)X|^{-m/2} \, etr\{-\tfrac{1}{2}\Omega + \tfrac{1}{2}e(t)(I - \Delta(t)X)^{-1}\Omega\}$$

$$\cdot \, \psi_{pm}(t;0) \quad (2.11)$$

where Ω, X are the same as before and

$$e(t) = (1-2it)^{-1}, \qquad \Delta(t) = e(t) - 1 = 2it(1 - 2it)^{-1}$$

$$\psi_{pm}(t;\omega^2) = (1 - 2it)^{-pm/2}\exp\{\tfrac{1}{2}\,\omega^2\,\Delta(t)\}$$ = the ch. f. of the noncentral chi square distribution with pm d. f. and noncentrality ω^2. J of (2.10) and $\phi(t;\Sigma + \varepsilon)$ of (2.11) can be expanded as a series in powers of ε_{rs} and hence we can obtain the desired derivatives to get an asymp. expan. of the distribution of T_0^2. For details of the calculation and the final form of the expansion were given in Siotani (1968, 1971, b).

It is noted here that several authors have obtained asymp. expan. for the noncentral distribution of T_0^2 by different methods; for example, Fujikoshi (1970), Hayakawa (1972), Ito (1960), Muirhead (1972.b), and Yoong-Sin Lee (1971). Chattopadhyay (1972) has considered an asymp. expan. for the distribution of T_0^2 by the same method in the case where the covariance matrices Σ_1 of Z and Σ_2 of $W_p(\Sigma_2, n)$ are different.

(ii) Asymptotic expansion for the distribution of a characteristic root of a Wishart matrix. Let $nS = n(s_{jk}) \sim W_p(\Sigma, n)$ and $c_\alpha(S)$ be the αth largest ch. r. of S. We may assume that $\Sigma = D_\lambda = \mathrm{diag}(\lambda_1, \ldots, \lambda_p)$ with $\lambda_1 \geq \ldots \geq \lambda_p > 0$. Using the expanded form of $c_\alpha(S)$ given by Lawley (1956) for sufficiently large n and for the case of simple λ_α, the ch. f. of $\sqrt{n}(c_\alpha(S) - \lambda_\alpha)$ can be expressed as

$$\phi(t) = E[\exp\{it\sqrt{n}(s_{\alpha\alpha} - \lambda_\alpha)\} \cdot \{1 + \frac{it}{\sqrt{n}} q_1(S) + \frac{it}{n} q_2(S) + \frac{(it)^2}{2n} q_1^2(S)\}] + O(n^{-3/2}) \tag{2.12}$$

where

$$q_1(S) = n \sum_{j \neq \alpha} \frac{1}{\lambda_\alpha - \lambda_j} s_{\alpha j}^2,$$

$$q_2(S) = n^{3/2} \sum_{j \neq \alpha} \frac{1}{(\lambda_\alpha - \lambda_j)^2} s_{\alpha j}^2 (s_{jj} - \lambda_j) - \sum_{j \neq \alpha} \frac{1}{(\lambda_\alpha - \lambda_j)^2} s_{\alpha j}^2 (s_{\alpha\alpha} - \lambda_\alpha) + \sum_{\substack{j<k \\ j,k \neq \alpha}} \frac{2}{(\lambda_\alpha - \lambda_j)(\lambda_\alpha - \lambda_k)} s_{\alpha j} s_{\alpha k} s_{jk} .$$

In order to evaluate the expectation, consider $E[\{\mathrm{etr}(itAU)\} f(S)]$, where A:pxp is any symmetric matrix and $U = \sqrt{n}(S - \Sigma)$. Since the Taylor expansion of f(S) about $S = \Sigma$ is $f(X) = \{\mathrm{etr}(S - \Sigma)\partial\} f(\Lambda)|_{\Lambda=\Sigma}$, we obtain, in the same was as in Section 2.1,

$$E[\{etr(itAU)\}f(S)] = \left|I - \frac{2it}{\sqrt{n}} A - \frac{2}{n}\Sigma\partial\right|^{-n/2}$$

$$etr(-it\sqrt{n}A\Sigma - \Sigma\partial)\cdot f(\Lambda)\Big|_{\Lambda=\Sigma}$$

$$= etr\{-t^2(A\Sigma)^2\}[1 + \frac{1}{\sqrt{n}}\{2it\cdot tr(\Sigma A\Sigma\partial) + \frac{4}{3}(it)^3 tr(\Sigma A)^3\}$$

$$+ \frac{1}{n}\{tr(\Sigma\partial)^2 + 4(it)^2 tr(\Sigma A)^2\Sigma\partial + 2(it)^2(tr(\Sigma A\Sigma\partial))^2 \tag{2.13}$$

$$+ 2(it)^4 tr(\Sigma A)^4 + \frac{8}{3}(it)^4 tr(\Sigma A)^3 tr(\Sigma A\Sigma\partial) + \frac{8}{9}(it)^6(tr(\Sigma A)^3)^2$$

$$+ O(n^{-3/2})]\ f(\Lambda)\Big|_{\Lambda=\Sigma}.$$

Now put $\Sigma = D_\lambda$ ang $A = diag(0,\ldots,0, \underset{\alpha}{1}, 0,\ldots, 0)$ in (2.13).
Then (2.12) can be written as

$$\phi(t) = E[etr(itAU)] + \frac{it}{\sqrt{n}} E[\{etr(itAU)\}q_1(S)]$$

$$+ \frac{it}{n} E[\{etr(itAU)\}q_2(S)] \tag{2.14}$$

$$+ \frac{(it)^2}{2n} E[\{etr(itAU)\}q_1^2(S)] + O(n^{-3/2}).$$

Thus we obtain an asymp. expan. for $\phi(t)$ by evaluating each expectation in (2.14) according to (2.13). The above discussion is due to Sugiura (1973, b).

(iii) Chou and Siotani (1972) have treated the ratio of two quasi-independent T_0^2 statistics; $F_0 = tr(S^{-1}Z_1Z_1')/tr(S^{-1}Z_2Z_2')$, where $nS \sim W_p(\Sigma,n)$, $Z_1 \sim N_p(M;\Sigma)$, and $Z_2 \sim N_p(0;\Sigma)$, and Z_1, Z_2 are independent. Chattopadhyay and Pillai (1971) have considered and given an asymp. expan. for the distribution of $F^* = (m_1/n_1)tr(S_1S_4^{-1})/tr(S_3S_2^{-1})$ by the same technique as in this section, where m_1S_1, m_2S_2, n_1S_3 and n_2S_4 are independently distributed Wishart matrices with m_1, m_2, n_1, and n_2 d. f. respectively and each of pairs (S_1, S_2) and (S_3, S_4) has a common covariance matrix.

Siotani (1959) has obtained an asymp. expan. for $\Pr(T_1^2 \leq c_1^2, T_2^2 \leq c_2^2)$ where $T_j^2 = Z_j' S^{-1} Z_j$ $(j = 1,2)$ and $Z_j \sim N_p(0,\Sigma)$, Z_j's being correlated in a certain way and nS is a $W_p(\Sigma, n)$ matrix independent of Z_j's.

3. DERIVATION FROM THE EXACT EXPRESSION: THE EXPANSION OF CHARACTERISTIC FUNCTIONS.

3.1. Some useful formulas for weighted sums of zonal polynomials.

As mentioned in the introduction, we have at the present time many exact mathematical expressions for p. d. f. or ch. f. of the nonnull distributions of test statistics in terms of zonal polynomials, the generalized Laguerre polynomials, and the generalized Hermite polynomials. But the explanation in this section will be mainly given for the zonal polynomials. For the definition and the properties of zonal polynomials, see James (1961, a; 1964) and Constantine (1963).

Lemma 3.1 [Fujikoshi (1970)]. Let $C_x(\Sigma)$ be a zonal polynomial corresponding to the partition $x = (k_1, \ldots, k_p)$ of k $(k_1 \geq \ldots \geq k_p \geq 0, k_1 + \ldots + k_p = k)$ and let $D_\lambda = \mathrm{diag}(\lambda_1, \ldots, \lambda_p)$ where λ_j's are ch. roots of Σ:pxp. Put

$$a_1(x) = \prod_{\alpha=1}^{p} k_\alpha(k_\alpha - \alpha), \quad a_2(x) = \prod_{\alpha=1}^{p} k_\alpha(4k_\alpha^2 - 6k_\alpha\alpha + 3\alpha^2). \tag{3.1}$$

Then the following relations hold:

$$a_1(x) C_x(\Sigma) = \mathrm{tr}(D_\lambda \partial)^2 \, C_x(\Lambda)\Big|_{\Lambda=D_\lambda}, \tag{3.2}$$

$$\{3a_1^2(x) - a_2(x) + k\} \, C_x(\Sigma) = [3\{\mathrm{tr}(D_\lambda \partial)^2\}^2 + 8\mathrm{tr}(D_\lambda \partial)^3] \, C_x(\Lambda)\Big|_{\Lambda=D_\lambda} \tag{3.3}$$

where ∂:pxp is the matrix of differential operators defined in the last section.

Lemma 3.2. For large n, we have

$$(n)_x = n^k[1 + \frac{1}{2n} a_1(x) + \frac{1}{24n^2} \{3a_1^2(x) - a_2(x) + k\} + O(n^{-3})], \tag{3.4}$$

where $(n)_x$ is the notation defined by

$$(n)_x = \prod_{j=1}^{p} (n - \frac{j-1}{2})_{k_j}, \quad (c)_m = c(c+1)\ldots(c+m-1), \quad (c)_0 = 1 \tag{3.5}$$

for the partition $x = (k_1, \ldots, k_p)$ of the integer k.

Lemma 3.1 is proved as follows: let $f(S) = C_x(S)$ in (2.1). Then

$$C_x(\Lambda)|_{\Lambda=\Sigma} = [\Gamma_p(\tfrac{n}{2})]^{-1}|2\Sigma|^{-n/2} \int_{S>0} C_x(S)|S|^{(n-p-1)/2} \operatorname{etr}(-\tfrac{n}{2}\Sigma^{-1}S)dS \tag{3.6}$$

$$= (2/n)^k(n/2)_x C_x(\Sigma) \quad \text{(by formula (20) in Constantine (1963)).}$$

$$= [1 + \frac{1}{n} a_1(x) + \frac{1}{6n^2} \{3a_1^2(x) - a_2(x) + k\} + O(n^{-3})]C_x(\Sigma)$$

by the formula in Lemma 3.2. On the other hand, since the integral in (3.6) is invariant under the orthogonal transformation; H'SH and H'ΣH, we obtain

$$C_x(\Lambda)|_{\Lambda=\Sigma} = [1 + \frac{1}{n}\operatorname{tr}(D_\lambda\partial)^2 + \frac{1}{6n^2}\{3(\operatorname{tr}(D_\lambda\partial)^2)^2 + 8\operatorname{tr}(D_\lambda\partial)^3\} + O(n^{-3})]\ C_x(\Lambda)|_{\Lambda=D_\lambda}.$$

Comparing coefficients of $1/n$ and $1/n^2$, we have (3.2) and (3.3).

Lemma 3.3 [Sugiura and Fujikoshi (1969)].

$$\sum_{k=r}^{\infty} \sum_{(x)} C_x(Z)/(k - r)! = (\operatorname{tr} Z)^4 \operatorname{etr} Z, \tag{3.7}$$

$$\sum_{k=0}^{\infty} \sum_{(x)} C_x(Z)a_1(x)/k! = (\mathrm{tr}\, Z^2)\mathrm{etr}\, Z, \tag{3.8}$$

$$\sum_{k=1}^{\infty} \sum_{(x)} C_x(Z)a_1(x)/(k-1)! = \{2\mathrm{tr}\, Z^2 + (\mathrm{tr}\, Z)\mathrm{tr}\, Z^2\}\ \mathrm{etr}\, Z, \tag{3.9}$$

$$\sum_{k=0}^{\infty} \sum_{(x)} C_x(Z)a_1^2(x)/k! = \{(\mathrm{tr}\, Z)^2 + \mathrm{tr}\, Z^2 + 4\mathrm{tr}\, Z^3 + (\mathrm{tr}\, Z^2)^2\}\ \mathrm{etr}\, Z, \tag{3.10}$$

$$\sum_{k=0}^{\infty} \sum_{(x)} C_x(Z)a_2(x)/k! = \{\mathrm{tr}\, Z + 3(\mathrm{tr}\, Z)^2 + 3\mathrm{tr}\, Z^2 + 4\mathrm{tr}\, Z^3\}\ \mathrm{etr}\, Z, \tag{3.11}$$

where $a_1(x)$ and $a_2(x)$ are defined by (3.1).

<u>Lemma 3.4</u> [Fujikoshi (1970)]. Let Z be any symmetric p. d. matrix whose characteristic roots have the absolute values less than one and let $V = Z(I - Z)^{-1}$. Then

$$\sum_{k=1}^{\infty} \sum_{(x)} (b)_x C_x(Z)/(k-1)! = b(\mathrm{tr}\, V)|I - Z|^{-b}, \tag{3.12}$$

$$\sum_{k=2}^{\infty} \sum_{(x)} (b)_x C_x(Z)/(k-2)! = b\{b(\mathrm{tr}\, V)^2 + \mathrm{tr}\, V^2\}\ |I - Z|^{-b}, \tag{3.13}$$

$$\sum_{k=0}^{\infty} \sum_{(x)} (b)_x C_x(Z)a_1(x)/k! = (b/2)\{(\mathrm{tr}\, V)^2 + (2b + 1)\mathrm{tr}\, V^2\}\ |I - Z|^{-b}, \tag{3.14}$$

$$\sum_{k=1}^{\infty} \sum_{(x)} (b)_x C_x(Z)a_1(x)/(k-1)! = (b/2)\{2(\mathrm{tr}\, V)^2 + 2(2b+1)\mathrm{tr}\, V^2 + b(\mathrm{tr}\, V)^3 + (2b^2+b+2)(\mathrm{tr}\, V)\mathrm{tr}\, V^2 + 2(2b+1)\mathrm{tr}\, V^3\}|I - Z|^{-b}, \tag{3.15}$$

$$\sum_{k=0}^{\infty} \sum_{(x)} (b)_x C_x(Z)a_1^2(x)/k! = (b/4)\{2(2b+1)(\mathrm{tr}\, V)^2 + 2(2b+3)\mathrm{tr}\, V^2 + 4(\mathrm{tr}\, V)^3 + 12(2b+1)(\mathrm{tr}\, V)\mathrm{tr}\, V^2$$

$$+ 8(2b^2+3b+2)\operatorname{tr} V^3 + b(\operatorname{tr} V)^4 + 2(2b^2+b+2)\cdot$$

$$\cdot(\operatorname{tr} V)^2 \operatorname{tr} V^2 + (2b+1)(2b^2+b+2)(\operatorname{tr} V^2)^2$$

$$+ 8(2b+1)(\operatorname{tr} V)\operatorname{tr} V^3 + 2(8b^2+10b+5)\operatorname{tr} V^4\}|I - Z|^{-b} \qquad (3.16)$$

$$\sum_{k=0}^{\infty} \sum_{(x)} (b)_x C_x(Z) a_2(x)/k! = (b/2)\{2\operatorname{tr} V + 3(2b+1)(\operatorname{tr} V)^2$$

$$+ 3(2b+3)\operatorname{tr} V^2 + 2(\operatorname{tr} V)^3 + 6(2b+1)(\operatorname{tr} V)\operatorname{tr} V^2$$

$$+ 4(2b^2+3b+2)\operatorname{tr} V^3\}|I - Z|^{-b}. \qquad (3.17)$$

Lemmas 3.3 and 3.4 can be proved based on Lemma 3.1 and the fundamental expansions

$$(\operatorname{tr} Z)^k = \sum_{(x)} C_x(Z), \qquad \operatorname{etr} Z = \sum_{k=0}^{\infty} \sum_{(x)} C_x(Z)/k! \,. \qquad (3.18)$$

The symbol $\sum_{(x)}$ in the above formulas stands for the summation with respect to partitions of k into not more than p parts.

Sugiura (1971) has given a set of the formulas stronger than those in Lemma 3.3. A set of similar formulas for Laguerre polynomials $L_x^{(n-p-1)/2}(Z)$ has been given by Fujikoshi (1970), and see also Hayakawa (1972).

It is possible to extend the formulas in Lemmas 3.3 and 3.4 by writing the explicit relations for coefficients of n^{-3}, n^{-4},... in the expansions of $(2/n)^k(n/2)_x C_x(\Sigma)$ and $\Theta C_x(\Lambda)|_{\Lambda=\Sigma}$ which were used to get Lemma 3.1. Actually Fujikoshi (1972) has given more identities of those type.

3.2. Some Applications of the Formulae.

(i) The likelihood ratio criterion for multivariate linear hypothesis. Let V_e:p×p and V_h:p×p be the matrices of sums of squares and products due to error and due to the hypothesis in MANOVA respectively. Let $V_e \sim W_p(\Sigma, N_e)$ and $V_h \sim W_p(\Sigma, m, \Omega)$. The LRC for H: $\Omega = 0$ is $\lambda = \{|V_e|/|V_h+V_e|\}^{N/2}$, where N is the total sample size. Put $n = N_e\frac{1}{2}(m-p-1)$ and $\rho = n/N$. The exact ch. f. of $-2\rho \log \lambda$ can be obtained from the formula for the moments of $|V_e|/|V_h+V_e|$ given by Constantine (1963) as

$$\phi(t) = \phi_o(t) \quad {}_1F_1(-itn; \frac{n}{2}(1-2it)+\frac{1}{4}(m+p+1); -\Omega) \tag{3.19}$$

where $\phi_o(t)$ is the ch.f. in the null case, whose asymptotic expansion is known; see Box (1949) or Anderson (1958), Section 8.6.1. ${}_1F_1(a; b; Z)$ is a special case of the hypergeometric functions defined by Constantine (1963) and is given by

$${}_1F_1(a;\, b;\, Z) = \sum_{k=0}^{\infty} \sum_{(x)} \frac{(a)_x}{(b)_x} \frac{C_x(Z)}{k!} . \tag{3.20}$$

Hence an asymp. expan. for ${}_1F_1$ in (3.9) can be obtained by using (3.4) and then by simplifying the resultant by the formulae in Lemma 3.3. In this way we obtain an asymp. expan. for $\phi(t)$, and hence we finally obtain an asymp. expan. for $\Pr\{-2\rho \log \lambda \leq x\}$ by inverting the expansion for $\phi(t)$. The detail of the computation and the final result are given by Sugiura and Fujikoshi (1969).

(ii) Fujikoshi (1970) obtained an asymp. expan. for the distribution of the trace criterion for multivariate linear hypothesis. Sugiura and Fujikoshi (1969) gave an asymp. expan. for the nonnull distribution of the LRC for the independence between two sets of variates.

4. DERIVATION FROM THE EXACT EXPRESSION: PARTIAL DIFFERENTIAL EQUATIONS FOR ${}_gF_h$ FUNCTIONS.

4.1. An outline of the expansion procedure.

A different approach from the last section for obtaining asymp. expan. for the nonnull distributions starting with the exact expressions has been given by Muirhead (1970 a, b). He has proved that hypergeometric functions ${}_2F_1(a, b; c; S)$, ${}_1F_1(a; c; S)$, and ${}_0F_1(c; S)$ satisfy certain systems of partial differential equations.

Let us explain the expansion procedure for ${}_1F_1(a;\, b-\tfrac{1}{2}\varepsilon n;\, \tfrac{1}{2}nR)$ (we call this ${}_1F_1$ of type A) with respect to n^{-1}, where $\varepsilon = 1$ or -1. From the system of partial differential equations (part. diff. eq.) given by Muirhead (1970, a) for ${}_1F_1(a; c; S)$, it follows that

$$r_j \frac{\partial^2 F}{\partial r_j^2} + \{b - \frac{1}{2}\varepsilon n - \frac{1}{2}(p-1) - \frac{1}{2}nr_j + \frac{1}{2}\sum_{\substack{k=1\\k\neq j}}^{p} \frac{r_j}{r_j - r_k}\} \frac{\partial F}{\partial r_j}$$

$$- \frac{1}{2}\sum_{\substack{k=1\\k\neq j}}^{p} \frac{r_k}{r_j - r_k} \frac{\partial F}{\partial r_k} = \frac{1}{2}\, anF \qquad (j = 1,\ldots,p) \tag{4.1}$$

Where $F = {}_1F_1(a;\, b - \tfrac{1}{2}\varepsilon n;\, \tfrac{1}{2}nR)$, and r_j, $j=1,\ldots,p$ are ch. roots of a pxp symmetric matrix R, subject to the conditions that (i) F is symmetric in $r_1,\ldots,r_p$ and (ii) F is analytic about $R = 0$ and $F(0) = 1$. Because of the symmetry of F in $r_1,\ldots,r_p$, it is sufficient to perform the actual computation only on the first equation. Let $W = 1 - (I + \varepsilon R)^{-1}$, so that F is 1 at $W = 0$. Then from (4.1) with $j = 1$, F satisfies

$$w_1(1-w_1)^2 \frac{\partial^2 F}{\partial w_1^2} + \{b - \frac{1}{2}(p-1) - 2w_1 - \frac{1}{2}n\varepsilon(1-w_1)^{-1}$$

$$+ \frac{1}{2}\sum_{k=2}^{p} \frac{w_1(1-w_k)}{w_1 - w_k}\} \cdot (1-w_1)\frac{\partial F}{\partial w_1} \tag{4.2}$$

$$- \frac{1}{2}\sum_{k=2}^{p} \frac{w_k(1-w_k)^2}{w_1 - w_k} \frac{\partial F}{\partial w_k} = \frac{1}{2}\varepsilon an(1-w_1)^{-1}F$$

where $w_1,\ldots,w_p$ are ch. roots of W.

Using the formula (3.4), it can be shown that ${}_1F_1$ of type A has the limit $|I - \varepsilon R|^{-a}$ or $|I - W|^a$ as $n \to \infty$. Hence we put

$$F = |I - W|^a \cdot G(W) \tag{4.3}$$

and consider $\log F = a \log|I - W| + H(W)$, where $H(W) = \log G(W)$, so that $H(0)=0$. It follows from (4.2) that H satisfies the part. diff. eq.

$$w_1(1-w_1)^2\{\frac{\partial^2 H}{\partial w_1^2} + (\frac{\partial H}{\partial w_1})^2\} + \{b - \frac{1}{2}(p-1) - \frac{1}{2}\varepsilon n - w_1[b - \frac{1}{2}(p-5) + 2a]$$

$$+ 2w_1^2(a+1) + \frac{1}{2}w_1(1-w_1)\sum_{k=2}^{p}\frac{1-w_k}{w_1-w_k}\}\frac{\partial H}{\partial w_1} \tag{4.4}$$

$$-\frac{1}{2}\sum_{k=2}^{p}\frac{w_k(1-w_k)^2}{w_1-w_k}\frac{\partial H}{\partial w_k} = ab - w_1 a(a+1) - \frac{1}{2}a\sum_{k=2}^{p} w_k .$$

Now we set the form of a solution of (4.4) as $H(W) = \Sigma_{k=1}^{\infty} q_k(W) n^{-k}$ where $q_k(0) = 0$ for all k, substitute this into (4.4) and equate coefficients of same powers of n^{-1} on both sides. Then we obtain a set of successive part. diff. eq. for $q_k(W)$ with respect to w_1. For example, we have

$$\frac{\partial q_1}{\partial w_1} = a\ \varepsilon\{\sum_{k=2}^{p} w_k + 2w_1(a+1) - 2b\} \tag{4.5}$$

$$\frac{\partial q_2}{\partial w_1} = 2\varepsilon[w_1(1-w_1)^2\frac{\partial^2 q_1}{\partial w_1^2} + \{b - \frac{1}{2}(p-1) - w_1(b-\tfrac{1}{2}(p-5)+2a)$$

$$+ 2w_1^2(a+1) + \frac{1}{2}w_1(1-w_1)\sum_{k=2}^{p}\frac{1-w_k}{w_1-w_k}\}\frac{\partial q_1}{\partial w_1} - \frac{1}{2}\sum_{k=2}^{p}\frac{w_k(1-w_k)^2}{w_1-w_k}$$

$$\cdot\frac{\partial q_1}{\partial w_k}] . \tag{4.6}$$

Solving those equations successively in taking care of conditions; (i) q_k are symmetric in $w_1,\ldots,w_p$, and (ii) $q_k(0) = 0$, we obtain an asymp. expan. for the distribution of statistics whose exact density functions involve ${}_1F_1$ function of type A. The final form of the asymp. expan. for ${}_1F_1$ of type A will be

$$|I - W|^a \exp\{\frac{1}{n} q_1(W) + \frac{1}{n^2} q_2(W) + O(n^{-3})\} \tag{4.7}$$

where $W = I - (I + \varepsilon R)^{-1}$. Muirhead (1970, b) has given the general expressions of $q_k(W)$, $k = 1, 2, 3$.

Muirhead (1970, b) has shown that

$$\Psi(a, c; R) = [1/\Gamma_p(a)]\int_{V>0} \operatorname{etr}(-RV)|V|^{a-(p+1)/2}$$

$$|I+V|^{c-a-(p+1)/2} dV \tag{4.8}$$

satisfies the same system of part. diff. eq. as for ${}_1F_1(a; c; R)$, and hence the asymp. expan. for $\psi(a, b-\frac{1}{2}\varepsilon n; \frac{1}{2}nR)$ has the same form as (4.7).

In exactly same way, Sugiura (1972) has given asymp. expan. for

$${}_1F_1(a_1\sqrt{n} + a_0; b_2n + b_1\sqrt{n} + b_0; (c_2n + c_1\sqrt{n} + c_0)R)$$

– ${}_1F_1$ of type B, and ${}_1F_1(a_1n + a_0; b_1n + b_0; \sqrt{n}tR)$

– ${}_1F_1$ of type C

in the following form:

$${}_1F_1 \text{ of type B} = |I-W|^{a_1\sqrt{n}+a_0} \exp\{Q_0(W) + \frac{1}{\sqrt{n}} Q_1(W)$$

$$+ \frac{1}{n} Q_2(W) + O(n^{-3/2})\}, \tag{4.9}$$

where $W = I - (I - (c_2/b_2)R)^{-1}$, and

$${}_1F_1 \text{ of type C} = \exp\{\sqrt{n}Q^*(R) + Q_0(R) + n^{-\frac{1}{2}}Q_1(R)$$

$$+ n^{-1}Q_2(R) + O(n^{-3/2})\}, \tag{4.10}$$

respectively.

Constantine and Muirhead (1972) have extended the systems of part. diff. eq. to the case of hypergeometric functions ${}_2F_1$, ${}_1F_1$, ${}_1F_0$, ${}_0F_1$, and ${}_0F_0$ with two argument matrices.

4.2. Some examples.

(i) The largest ch. root of a Wishart matrix. Let nS: $p \times p \sim W_p(\Sigma, n)$ and let r_{max} be the largest ch. root of S. Then the c. d. f. of r_{max} is

$$\Pr\{r_{max} \leq r\} = \left(\frac{nr}{2}\right)^{pn/2} |\Sigma|^{-n/2} \frac{\Gamma_p\left(\frac{p+1}{2}\right)}{\Gamma_p\left(\frac{n}{2} + \frac{p+1}{2}\right)} \operatorname{etr}\left(-\frac{1}{2} nr\Sigma^{-1}\right)$$

$$\cdot {}_1F_1\left(\frac{p+1}{2}; \frac{n}{2} + \frac{p+1}{2}; \frac{nr}{2}\Sigma^{-1}\right) \tag{4.11}$$

[Constantine (1963), Sugiyama (1967)]. The hypergeometric function in (4.11) is an ${}_1F_1$ of type A with $a = (p+1)/2$, $b = (p+1)/2$, $\varepsilon = -1$ and $R = r\Sigma^{-1}$. Hence an asymp. expan. for the c. d. f. of r_{max} is obtained by combining the expansion for the ${}_1F_1$ with the expansion for other factors including gamma functions. See Muirhead (1970, b).

(ii) Muirhead (1972, a) has treated the nonnull distribution of the LRC for the test of independence given (3.26) by the technique in this section. He (1972, b) has also treated the nonnull distribution of Hotelling's generalized T_0^2 defined in Section 2.2 (i) in the similar way and given the asymp. expan. up to term of order n^{-3}.

5. A SHORT NOTE ON SOME OTHER TOPICS.

5.1. Method based on the representation of an orthogonal matrix by a skew-symmetric matrix.

By James (1960), the joint p. d. f. of sample ch. roots $r_1 \geq r_2 \geq \ldots \geq r_p \geq 0$ of V: $p \times p \sim W_p(\Sigma, n)$ can be written as

$$\frac{\pi^{p^2/2}(n/2)^{pn/2}}{\Gamma_p(n/2)\Gamma_p(p/2)} \left(\prod_{j=1}^{p} r_j\right)^{(n-p-1)/2} \prod_{j<k} (r_j - r_k) |\Sigma|^{-n/2}$$

$$\cdot \int_{O(p)} \operatorname{etr}\left(-\frac{n}{2} \Sigma^{-1} H D_r H'\right) d(H) \tag{5.1}$$

where $O(p)$ is the orthogonal group of order p, $d(H)$ is the normalized invariant measure over $O(p)$, and $D_r = \mathrm{diag}(r_1,\ldots,r_p)$. Anderson (1965) has obtained an asymp. expan. for (5.6) by making the transformation $H = \exp(S)$, where S: pxp is a skew-symmetric matrix, by expanding $d(H)$ and $\mathrm{tr}(\Sigma^{-1}HD_rH')$ in powers of S, and by evaluating the integrals with respect to S term by term.

Anderson (1970) has also treated the case of the noncentral Wishart distribution using the similar method.

5.2. Method by an expansion of test statistics in terms of normal variates.

Let us consider a simple case where the test statistic is a function of a Wishart matrix, V: $p\mathrm{x}p \sim W_p(\Sigma,n)$, i.e., $f = f(V)$. Let $m = O(n)$. Then it is easy to see that the limiting distribution of the random matrix Y: $p\mathrm{x}p = (\Sigma^{-\frac{1}{2}}V\Sigma^{-\frac{1}{2}} - mI)/\sqrt{2m}$ is the $p(p+1)/2$ variate normal with mean zero matrix and covariance matrix $\mathrm{diag}(1,\ldots,1, 1/2,\ldots,1/2)$ whose c. d. f. is const. $\mathrm{etr}(-Y^2/2)$. In this method, first the statistic $f(V)$ is expanded with respect to Y in an appropriate way; secondly based on this expansion, the ch. f. of f is evaluated, and finally an asymp. expan. for the distribution of f is obtained by inverting the ch. f. thus obtained.

6. ACKNOWLEDGMENT. This research was partially supported by Agricultural Experiment Station, Kansas State University (Contribution Number 234) and also by Technogy, Incorporated under contract F33615-73-C-4155.

The author wishes to express his thanks to Dr. P. R. Krishnaiah, Aerospace Research Laboratories, W-P A. F. B. for his helpful comment and to the referee for his helpful suggestion to make the paper concise. He also thanks Mrs. Vickie J. Claar, Kansas State University for her help in the preparation of this paper.

REFERENCES.

[1] Anderson, G. A. (1965). Ann. Math. Statist. 36, 1153-1166.
[2] Anderson, G. A. (1970). Ann. Math. Statist. 41, 1700-1707.
[3] Anderson, T. W. (1946). Ann. Math. Statist. 17, 409-431.
[4] Anderson, T. W. (1951). Psychometrika 16, 31-50.
[5] Anderson, T. W. (1958). An Introduction to Multivariate Statistical Analysis. Wiley, New York.

[6] Box, G. E. P. (1949). Biometrika 36, 317-346.
[7] Chattopadhyay, A. K., and Pillai, K. C. S. (1971). J. Multivariate Anal. 1, 215-231.
[8] Chattopadhyay, A. K. (1972). Nonnull distribution of Hotelling's generalized T_0^2 statistic. ARL 72-0033, Aerospace Research Laboratories, W-P AFB, Ohio.
[9] Chou, C., and Siotani, M. (1974). Asymptotic expansion of the nonnull distribution of the ratio of two conditionally independent Hotelling's T_0^2-statistics. Accepted for Ann. Inst. Statist. Math. 26.
[10] Constantine, A. G. (1963). Ann. Math. Statist. 34, 1270-1285.
[11] Constantine, A. G., and Muirhead, R. J. (1972). J. Multivariate Anal. 2, 332-338.
[12] Fujikoshi, Y. (1968). J. Sci. Hiroshima Univ. Ser A-I 32, 293-299.
[13] Fujikoshi, Y. (1970). J. Sci. Hiroshima Univ. Ser A-I 34, 73-144.
[14] Fujikoshi, Y. (1972). New formulae for the weighted sums of zonal polynomials. Submitted to Ann. Statist.
[15] Hayakawa, T. (1972). Ann. Inst. Statist. Math. 24, 19-32.
[16] Hayakawa, T. (1974). Asymptotic expansion of the distribution of the likelihood ratio criterion for homogeneity of parameters under local alternatives. Research Memorandum No. 59, Inst. Statist. Math., Tokyo.
[17] Hotelling, H. (1951). In Proc. 2nd Berkeley Symp. on Math. Stat. and Prob., J. Neyman (ed.), University of California Press, Berkeley, Calif., 23-42.
[18] Ito, K. (1960). Ann. Math. Statist. 31, 1148-1153.
[19] James, A. T. (1960). Ann. Math. Statist. 31, 151-158.
[20] James, A. T. (1961, a). Ann. Math. 74, 456-469.
[21] James, A. T. (1961, b). Ann. Math. Statist. 32, 874-882.
[22] James, A. T. (1964). Ann. Math. Statist. 35, 475-501.
[23] James, G. S. (1954). Biometrika 41, 19-43.
[24] Johnson, N. L., and Kotz, S. (1972). Distribution in Statistics: Continuous Multivariate Distributions. Wiley, New York.
[25] Khatri, C. G., and Srivastava, M. S. (1974). Ann. Statist. 2, 109-117.
[26] Krishnaiah, P. R. (1965). Ann. Inst. Statist. Math. 17, 35-53.
[27] Lawley, D. N. (1956). Biometrika 43, 128-136.
[28] Lee, Yoong-Sin (1971). Ann. Math. Statist. 42, 526-537.
[29] Muirhead, R. J. (1970, a). Ann. Math. Statist. 41, 991-1001.
[30] Muirhead, R. J. (1970, b). Ann. Math. Statist. 41, 1002-1010.
[31] Muirhead, R. J. (1972, a). Ann. Math. Statist. 43, 1491-1497.
[32] Muirhead, R. J. (1972, b). Ann. Math. Statist. 43, 1671-1677.
[33] Nagao, H. (1970). J. Sci. Hiroshima Univ. Ser A-I 34, 153-247.

[34] Nagao, H. (1972). Ann. Inst. Statist. Math. 24, 67-79.
[35] Nagao, H. (1973, a). Ann. Statist. 1, 700-709.
[36] Nagao, H. (1973, b). J. Multivariate Anal. 3, 435-444.
[37] Okamoto, M. (1963). Ann. Math. Statist. 34, 1286-1301; Correction, 39, (1968), 1358-1359.
[38] Olkin, I., and Siotani, M. (1964). Asymptotic distribution of functions of a correlation matrix. Technical Report No. 6, Dept. of Statistics, Stanford University.
[39] Pillai, K. C. S. (1968). Ann. Math. Statist. 39, 877-880.
[40] Siotani, M. (1959). Ann. Inst. Statist. Math. 10, 183-208.
[41] Siotani, M., and Hayakawa, T. (1964). Proc. Inst. Statist. Math. 12, 191-198.
[42] Siotani, M. (1968). Some methods for asymptotic distributions in the multivariate analysis. Mimeo Ser. 595, Inst. Stat., Univ. North Carolina, Chapel Hill, N. C.
[43] Siotani, M. (1971, a). Simultaneous confidence intervals relating to the multivariate regression matrices. Tech. Report No. 15, Dept. Statist., Kansas State University.
[44] Siotani, M. (1971, b). Ann. Math. Statist. 42, 560-571.
[45] Sugiura, N., and Fujikoshi, Y. (1969). Ann. Math. Statist. 40, 942-952.
[46] Sugiura, N. (1971). Ann. Math. Statist. 42, 768-772.
[47] Sugiura, N. (1972). Ann. Inst. Statist. Math. 24, 517-524.
[48] Sugiura, N. (1973, a). Ann. Statist. 1, 718-728.
[49] Sugiura, N. (1973, b). Comm. in Statist. 1, 393-417.
[50] Sugiyama, T. (1967). Ann. Math. Statist. 38, 1148-1151.
[51] Welch, B. L. (1947). Biometrika 34, 28-35.

A MULTIVARIATE GAMMA TYPE DISTRIBUTION WHOSE MARGINAL LAWS ARE GAMMA, AND WHICH HAS A PROPERTY SIMILAR TO A CHARACTERISTIC PROPERTY OF THE NORMAL CASE

A. Dussauchoy and R. Berland

Maître de Conférences Maîtrise M.I.A.G., Université Lyon I, Villeurbanne and Maître de Conferences I.U.T. de Limoges, Limoges, France

SUMMARY. Reasoning by analogy with a characteristic property of the multivariate normal distribution, we give here a distribution with marginal laws which have the same property as the characteristic property of the normal law. This distribution has one dimensional marginal laws which are gamma laws.

KEY WORDS. Multivariate gamma, multivariate normal.

1. INTRODUCTION. It seems that two approaches have been used to define multivariate gamma distributions:

(a) The first approach is that of Krisnamoorthy and Parthasarathy (1951) whose use results of Kibble (1945) on series expansion of bivariate frequency function using orthogonal functions on the marginal distributions. In the bivariate case for marginal gamma laws the characteristic function is Griffiths (1969):

$$\psi(u,v) = (1 - iu)^{-e_1}(1 - iv)^{-e_2}[1 + z\ u\ v/(1 - iu)(1 - iv)]^{-e_3}$$

$e_1, e_2, e_3 > 0$ and $0 \leq z \leq 1$

(b) The second approach is that of David and Fix (1961) who use three independent gamma random variables to define two non-independent gamma random variables by linear combination of the first variables. The characteristic function is

G. P. Patil et al. (eds.), Statistical Distributions in Scientific Work, Vol. 1, 319-328.

$$\psi(u,v) = (1 - iu)^{-e_1} (1 - iv)^{-e_2} [1 - i(u + v)]^{-e_3}$$

In Dussauchoy and Berland (1972) we have given the distribution for two dependent gamma random variables X and Y with the property that Y - βX and X are independent.

$$\psi(u,v) = (1 - iva_2^{-1})^{-e_2}(1 - i\beta va_1^{-1})^{+e_1}(1 - i(u + \beta v)a_1^{-1})^{-e_1}$$

This characteristic function is similar to the characteristic function in (b) but comparing the results given in David and Fix (1965) and in Dussauchoy and Berland (1972), it is clear that the distributions are different.

In the second approach it is impossible to find β which makes Y - βX independent of X.

Generalizing to the multivariate case the property of independence given here we derive the law of an n dimensional random gamma-type vector whose marginal laws have the same property as a characteristic property of the normal laws.

2. PRELIMINARY RESULTS. Let Z be a random vector with p + q dimensions composed of two random vectors X and Y having respectively p and q dimensions and such that there exists a matrix B with q rows and p columns and such that the vector Y - BX is independent of vector X.

Let $X' = [X_1,\ldots,X_p]$
$Y' = [X_{p+1},\ldots,X_{p+q}]$
$Z' = [X' \vdots Y']$

where X' stands for the transposed vector of X. Let $U \in R^p$ and $V \in R^q$ with $U' = [u_1,\ldots,u_p]$ $V' = [u_{p+1},\ldots,u_{p+q}]$. Let then the vector $W \in R^{p+q}$ be such that $W' = [U' \vdots V']$.

Theorem 1. With the above mentioned hypothesis, the characteristic function ψ_Z of Z can be obtained from the characteristic function ψ_X and ψ_Y of respectively X and Y, by the relation

$$\psi_Z (U, V) = \psi_Y (V) \frac{\psi_X (U + B' V)}{\psi_X (B' V)} \tag{1}$$

where the second part of (1) is effectively a characteristic function.

Proof. From the definition of the characteristic function of a random vector

$$\psi_Z (W) = \psi_Z (U, V) = E [\exp (i W' Z)] = E [\exp(i U' X + i V' Y)].$$

Besides it is clear that

$$\psi_Z (U, V) = E [\exp\{i U' X + i V' B X + i V' (Y - B X)\}] =$$

$$E [\exp \{i (U' + V' B) X\}] \; E [\exp \{i V' (Y - B X)\}] .$$

Since Y - BX and X are independent

$$\psi_Z (U, V) = \psi_X (U + B' V) \cdot \psi_Z (-B' V, V) .$$

But

$$\psi_Y (V) = \psi_Z (0, V) = \psi_X (B' V) \cdot \psi_Z (-B' V, V).$$

Hence the theorem is proved.

3. APPLICATION TO THE NORMAL CASE.

Lemma. If Z is a normal random vector,

(i) of n dimensions, composed of two vectors X and Y respectively of p and q dimensions,

(ii) with the variance-covariance matrix

$$A = \left[\begin{array}{c|c} A_{11} & {}^tA_{21} \\ \hline A_{21} & A_{22} \end{array}\right]$$

where A_{11} [resp. A_{22}] is the variance-covariance matrix of X[resp. of Y].

Then if $p \geq 1$ and $q \geq 1$ such that $p + q = n$ and A_{11} is regular, the matrix $B = A_{21} A_{11}^{-1}$ is such that vector Y - B X is independent of the vector X.

Proof. Let C be a real matrix n x n, then the vector $Z_1 = C\,Z$ has the variance-covariance matrix $A_1 = C\,A\,C'$.

If $C = \left[\begin{array}{c|c} I & 0 \\ \hline B & I \end{array}\right]$ then $Z_1 = \left[\begin{array}{c} X \\ \hline y - BX \end{array}\right]$ [Cramer (1963)].

One can easily verify that if $B = A_{21}\,A_{11}^{-1}$, then

$$A_1 = \left[\begin{array}{c|c} A_{11} & 0 \\ \hline 0 & A_{22} - A_{21}\,A_{11}^{-1}\,A'_{21} \end{array}\right]$$

Hence the lemma is proved.

Theorem 2. If Z is a normal random vector composed n random variables with strictly positive variances, then

$$\psi_Z\,(u_1,u_2,\ldots,u_n) = \prod_{j=1}^{n} \frac{\psi_{X_j}\,(u_j + a_{jj}^{-1} \sum_{k=j+1}^{n} a_{jk}\,u_k)}{\psi_{X_j}\,(a_{jj}^{-1} \sum_{k=j+1}^{n} a_{jk}\,u_k)}, \qquad (2)$$

Where ψ_Z [resp. ψ_{X_j}] is the characteristic function of Z [resp. of the component X_j of Z] and where $A = ||a_{ij}||$ is the variance-covariance matrix of Z.

Proof.

(i) For n = 2 the result is obvious since,

(a) from the lemma, if $[X_{n-1}, X_n]$ is a normal random vector, the random variable $X_n - a_{n,n-1}\,a_{n-1,n-1}^{-1}\,X_{n-1}$ is independent of X_{n-1} provided $a_{n-1,n-1} \neq 0$.

(b) According to Theorem 1

$$\psi_{[X_{n-1},X_n]}\,(u_{n-1},u_n) = \psi_{X_n}\,(u_n) \cdot \frac{\psi_{X_{n-1}}\,(u_{n-1} + a_{n-1,n-1}^{-1} \cdot a_{n,n-1} u_n)}{\psi_{X_{n-1}}\,(a_{n-1,n-1}^{-1} \cdot a_{n,n-1} u_n)}$$

and it is easy to check that the second member is indeed a characteristic function.

(ii) Assuming that the result is true for a normal random vector Y such as $Y' = (X_2, X_3, \ldots, X_n)$ then by applying the Theorem 1 to the vector $Z = [X_1, Y']$, we obtain the result (2) since with B' given by

$$B' = [A_{21} A_{11}^{-1}]' = [a_{21} a_{11}^{-1}, \ldots\ldots, a_{n1} a_{11}^{-1}]$$

the lemma shows that Y - B X is independent of X.

4. A MULTIVARIATE GAMMA-TYPE DISTRIBUTION.

Definition. We shall say that a n dimensional random vector Z with components $X_1, X_2 ---, X_n$ is a gamma type random vector if the characteristic function of this vector is

$$\psi_Z (u_1, u_2, \ldots u_n) = \prod_{j=1}^{n} \frac{\psi_{X_j} (u_j + \sum_{k=j+1}^{n} \beta_{jk} u_k)}{\psi_{X_j} (\sum_{k=j+1}^{n} \beta_{jk} u_k)} \tag{3}$$

where:

(a) $\psi_{X_j} (u_j) = (1 - iu_j a_j^{-1})^{-e_j}$ for all $\forall\, j = 1, \ldots, n$

is the characteristic function of the component X_j, e.g. X_j is a gamma random variable with parameters (a_j, e_j)

(b) $\beta_{jk} \geq 0$; $a_j \geq \beta_{jk} a_k > 0$ $\forall j < k \in [1, 2, \ldots, n]$

$0 < e_1 \leq e_2 \leq \ldots\ldots \leq e_n$

This definition is obtained recursively:

(i) We define a two dimensional gamma vector [as in Dussauchoy and Berland (1972)] $[X_{n-1}, X_n]$ such that $X_n - \beta_{n-1,n} X_{n-1}$ is independent of X_{n-1} and such that the marginal laws of

X_{n-1} and X_n are gamma laws with parameters respectively $[a_{n-1}, e_{n-1}]$ and $[a_n, e_n]$ with $\beta_{n-1,n} \geq 0$; $0 \leq \beta_{n-1,n} \cdot a_n < a_{n-1}$ and $0 < e_{n-1} \leq e_n$.

(ii) We define now a three dimensional gamma vector $[X_{n-2}, X_{n-1}, X_n]$

(a) $[X_{n-1}, X_n]$ is a two dimensional gamma vector defined as in (i)

(b) X_{n-2} is a gamma random variable such that

$X_n - \beta_{n-2,n} X_{n-2}$ is independent of X_{n-2} and

$X_{n-1} - \beta_{n-2,n-1} \cdot X_{n-2}$ is independent of X_{n-2} and so on. . . . .

Theorem 3. If Z is a gamma type random vector with components $[X_1,\ldots,X_j,\ldots,X_k,\ldots,X_n]$, satisfying the definition above there exists β_{jk} with $j < k$ for any $[X_j, X_k]$ such as $X_k - \beta_{jk} \cdot X_j$ be independent of X_j.

All the marginal laws of the vector Z are laws of a gamma type random vector.

It is possible to construct recursively the probability density function of Z.

Proof.

(i) According to (3) the characteristic function of the pair $[X_j, X_k]$ with $j < k$ is:

$$\psi_{[X_j,\, X_k]}(u_j, u_k) = \psi_Z\,(0,\ldots,0,\, u_j,\ldots,u_k,\, 0,\ldots,0)$$

$$= \frac{\psi_{X_j}\,(u_j + \beta_{jk}\, u_k) \cdot \psi_{X_k}\,(u_k)}{\psi_{X_j}\,(\beta_{jk}\, u_k)}$$

Then using the result of Dussauchoy and Berland (1972), we obtain (i) of the Theorem 3. It is to be noted that all the properties demonstrated in Dussauchoy and Berland (1972) apply to any pair $[X_j, X_k]$ with $j < k$.

(ii) Using the expression of the characteristic function of the marginal laws of Z in function of ψ_Z it is clear that all this characteristic function have the form (iii) of the definition.

(iii) In Dussauchoy and Berland (1972) we give the probability density function of all the pairs $[X_j, X_k]$ of Z and in particular that of $[X_1, X_2]$.

Let $\psi_{X_3/x_1,x_2}(v_3; x_1, x_2) = E[\exp\{iv_3 (X_3|X_1 = x_1 \cap X_2 = x_2\}]$

(This is the characteristic function of the random variable X_3 given that $X_1 = x_1 \cap X_2 = x_2$.)

Then:

$$\psi_{X_3|x_1,x_2}(v_3;x_1,x_2) = \exp(iv_3\beta_{23}x_2)\ E[\exp\{iv_3(X_3|x_1,x_2)-\beta_{23}x_2)\}]$$

$$= \exp(iv_3\beta_{23}x_2)\ E[\exp\{iv_3[(X_3-\beta_{23}X_2)/X_1=x_1\cap X_2=x_2]\}]$$

$$= \exp(iv_3\beta_{23}x_2)\ E[\exp\{iv_3[(X_3-\beta_{23}X_2)/X_1=x_1]\}] \qquad (4)$$

Since $X_3 - \beta_{23} X_2$ is independent of X_2, it is easy to show from (4) in the same manner that:

$$\psi_{X_3|x_1,x_2}(v_3;v_1,x_2) = \exp(ix_2\beta_{23}v_3)\ \exp[ix_1(\beta_{13}-\beta_{12}\beta_{23})v_3]$$

$$\cdot\ E[\exp\{iv_3([X_3 - \beta_{13}X_1) - \beta_{23}(X_2 - \beta_{12}X_1)] \mid X_1 = x_1)\}]$$

$$= \exp(ix_1 (\beta_{13} - \beta_{12} \beta_{23}) v_3) \exp(ix_2 \beta_{23} v_3)$$

$$\cdot E[\exp\{iv_3 [(X_3 - \beta_{13} X_1) - \beta_{23} (X_2 - \beta_{12} X_1)]\}]$$

$$= \exp(ix_1 (\beta_{13} - \beta_{12} \beta_{23}) v_3) \cdot \exp(ix_2 \beta_{23} v_3)$$

$$\cdot \psi_{[X_1, X_2, X_3]} [(-\beta_{13} + \beta_{12} \cdot \beta_{23}) v_3, - \beta_{23} v_3, v_3] , \qquad (5)$$

where the characteristic function $\psi_{[X_1, X_2, X_3]}$ is given by (3).

Then the third term of the second member of (5) is

$$\frac{\psi_{X_3} (v_3)}{\psi_{X_1} [(\beta_{13} - \beta_{12} \beta_{23}) v_3] \cdot \psi_{X_2} (\beta_{23} v_3)} .$$

Thus

$$\psi_{X_3|X_1=x_1} (v_3;x_1,x_2) = \exp(ix_1(\beta_{13}-\beta_{12} \beta_{23})v_3)$$

$$\cdot \left(1 - \frac{i(\beta_{13}-\beta_{12}\beta_{23}) v_3}{a_1}\right)^{e_1} \cdot \psi_{X_3|X_2=x_2} (v_3; x_2) . \qquad (6)$$

However,

$$\psi_{X_3|X_2=x_2} (v_3;x_2) = \exp(ix_2\beta_{23} v_3) \left(1 - \frac{i\beta_{23} v_3}{a_2}\right)^{e_2} \psi_{X_3} (v_3)$$

[see Dussauchoy and Berland (1972) and Cramer (1963)].

The inverse Fourier transform of (7) is known [Dussauchoy and Berland (1972)].

Using the inverse Fourier transform of (6) it is possible to obtain when it exists, the probability density function of the random variable $X_3 | X_2 = x_2 \cap X_1 = x_1$. As we know the probability density function of $[X_1, X_2, X_3]$.

It is easy to iterate this process for cases of more than three dimensions.

5. CONCLUSION. We have given a model of the probability law for a random vector, all the marginal laws of which belong, to the same family. This model can be applied in some processes where the independence hypothesis of the space separating the successive occurrences of the phenomen is not satisfied [see Berland and Dussauchoy (1973)].

Let's note that the extensions of the results of this paper can be obtained in various ways: using the maximum likelihood methods of estimation of the parameters as in Bruce and Al (1972); using a generalization of the gamma distribution [Stacy (1962)]; by means of generalization of results previously given [Bondesson (1973), Mathai (1972)] on random vector composed of independent gamma random variables.

REFERENCES

[1] Berland, R. and Dussauchoy, A. (1973). Aspects statistiques des régimes de micro-décharges électriques entre électrodes métalliques placées dans le vide industriel. VACUUM.

[2] Bondesson, L. (1973). Z. Wahrschein Verw. Geb. 26, 335-344.

[3] Bruce, M., et al. (1972). Amer. Statist. Assoc. 67, 927-929.

[4] Cramer, H. (1963). Mathematical Methods of Statistics. Princeton University Press.

[5] David, F. N. and Fix, E. (1961). In Proceedings of the Fourth Berkeley Symposium, Vol. 1. University of California Press, pp. 177-197.

[6] Dussauchoy, A. and Berland R. (1972). Lois gamma à deux dimensions. CRAS Paris T 274 série A, 1946-1949.

[7] Griffiths, P. C. (1969). The canonical correlation coefficients of bivariate gamma distribuionns. Ann. Math. Statist. 40.

[8] Kibble, W. F. (1945). An extension of theorem of Mehler on Hermite polynomials. Proc. Cambridge Phil. Soc. 41.

[9] Krishnamoorthy, A. S. and Parthasarathy, M. (1951). Ann. Math. Statist. 22, 549-557.

[10] Mathai, A. M. (1972). Products and ratios of generalized gamma variate. Skandin. Aktuarietidskrift, pp. 194-198.

[11] Stacy, E. W. (1962). Ann. Math. Statist. 33, 1187-1192.

THE BIVARIATE BURR DISTRIBUTION

Frederick C. Durling

University of Waikato, Hamilton, New Zealand

SUMMARY. The bivariate Burr distribution,

$$F(x,y) = 1 - (1+x^{b_1})^{-p} - (1+y^{b_2})^{-p} + (1+x^{b_1}+y^{b_2}+rx^{b_1}y^{b_2})^{-p};$$

$$x,\ y \geq 0,\ 0 \leq r \leq p + 1;\ F(x,y) = 0 \quad \text{elsewhere}$$

is developed and investigated. Two special cases of the distribution occur when the parameter $r = 0$ and 1 respectively. For the limiting case $r = 0$, $F(x,y)$ reduces to the bivariate case of the multivariate Burr distribution developed by Takahasi (1965). When $r = 1$, $F(x,y) = F(x)\cdot F(y)$, the independent case. The relationship of the bivariate Burr distribution and its marginals to the Pearson curves is discussed.

KEYWORDS. Burr distribution; general system of distributions; Pearson system of frequency curves; bivariate Burr distribution.

1. INTRODUCTION. The general system of distribution referred to here was first given by Burr (1942). Using a simple algebraic form as an expression for the distribution function

$$F(x) = \begin{cases} 1 - (1+x^b)^{-p} & x \geq 0;\ b,\ p > 0 \\ 0 & x < 0 \end{cases}, \tag{1}$$

G. P. Patil et al. (eds.), Statistical Distributions in Scientific Work, Vol. 1, 329-335.

F(x) covers an important region of the standardized third and fourth central moments in a way analogous to the Pearson system of distributions. The Burr system of distributions as given by F(x) covers the curve shape characteristics for the normal, logistic and exponential (Pearson Type X) distributions as well as a significant portion of the curve shape characteristics for Pearson Types I (beta), II, III (gamma), V, VII, IX and XII. Thus, the Burr system of distributions is quite general.

2. THE TAKAHASI BIVARIATE BURR DISTRIBUTION. Takahasi (1965) developed a multivariate Burr distribution by using the fact that a Burr distribution is a compound Weibull distribution with a gamma distribution as a compounder. That is, if

$$w(x;\, b,\, \theta) = \begin{cases} \theta b x^{b-1} e^{-\theta x^{b}} & x \geq 0 \\ 0 & x < 0 \end{cases}, \tag{2}$$

and θ is a random variable such that

$$g(\theta;\, p,\, 1) = \begin{cases} \theta^{p-1} e^{-\theta} & \theta \geq 0 \\ 0 & \theta < 0 \end{cases}, \tag{3}$$

then the resultant probability density function is Burr. The special case of the bivariate density is

$$f(x_1, x_2) = \begin{cases} \dfrac{\Gamma(p+2)}{\Gamma(p)}\, b_1 b_2 a_1 a_2 x_1^{b_1-1} x_2^{b_2-1} (1+a_1 x_1^{b_1}+a_2 x_2^{b_2})^{-(p+2)} & \\ \qquad x_i > 0,\ (i = 1,2) & \\ 0 \qquad \text{elsewhere} . \end{cases} \tag{4}$$

The bivariate distribution is

$$F(x_1,x_2) = \begin{cases} 1 - (1+a_1x_1^{b_1})^{-p} - (1+a_2x_2^{b_2})^{-p} + (1+a_1x_2^{b_2}+a_2x_2^{b_2})^{-p} , & \\ \qquad\qquad x_i \geqq 0 & \\ 0 \qquad \text{elsewhere} . & \end{cases} \tag{5}$$

It should be noted that the a_i's are equal to one in the Burr distribution as given by Burr (1942). If the transformation $x_i = D_i(z_i + c_i)$ is made, it is easily seen that the a_i's are redundant. In addition, if the b_i's and p are held constant, e.g., the third and fourth moments can be set equal to those of the normal distribution by proper choice of the b_i's and p, then the correlation coefficient is restricted to a point defined by a function of the b_i's and p. A generalization of the bivariate Burr distribution, such that the correlation coefficient would not be restricted to a point under these conditions, would be of considerable practical value.

3. THE GENERALIZED BIVARIATE BURR DISTRIBUTION. Comparison of the mathematical forms of the Takahasi bivariate Burr distribution with the a_i's set equal to one and the product of two independent Burr distributions led to the investigation of the ensuing mathematical form. The bivariate distr bution is

$$F(x_1,x_2) = \begin{cases} 1 - (1+x_1^{b_1})^{-p} - (1+x_2^{b_2})^{-p} + (1+x_1^{b_1}+x_2^{b_2}+rx_1^{b_1}x_2^{b_2})^{-p} , & \\ \qquad x_i \geqq 0,\ b_i > 0 & \\ \qquad 0 \leqq r \leqq p + 1 & \\ 0 \qquad \text{elsewhere} . & \end{cases} \tag{6}$$

The bivariate density is

$$f(x_1,x_2) = \begin{cases} p(p+1)(1+rx_2^{b_2})(1+rx_1^{b_1})b_1b_2x_1^{b_1-1}x_2^{b_2-1}(1+x_1^{b_1}+x_2^{b_2} \\ \quad +rx_1^{b_1}x_2^{b_2})^{-(p+2)} - prb_1b_2x_1^{b_1-1}x_2^{b_2-1}(1+x_1^{b_1}+x_2^{b_2} \\ \quad + rx_1^{b_1}x_2^{b_2})^{-(p+1)} \quad x_i \geq 0,\ b_i > 0 \\ \quad 0 \leq r \leq p+1 \\ 0 \qquad \text{elsewhere.} \end{cases} \tag{7}$$

The marginals are of the form given in equation (1) by Burr (1942). $F(x_1,x_2)$ in equation (6) reduces to the Takahasi bivariate distribution for the limiting case of $r = 0$. For $r = 1$, $F(x_1,x_2)$ becomes the product of two independent Burr distributions.

The conditional distribution of x_i given x_j, $i \neq j$, is

$$F(x_i|x_j) = \begin{cases} 1 - r\left\{\dfrac{1+x_j^{b_j}}{1+rx_j^{b_j}}\right\}\left[1 + \left\{\dfrac{1+rx_j^{b_j}}{1+x_j^{b_j}}\right\}x_i^{b_i}\right]^{-p} \\ \quad + \left\{\dfrac{r-1}{1+rx_j^{b_j}}\right\}\left[1 + \left\{\dfrac{1+rx_j^{b_j}}{1+x_j^{b_j}}\right\}x_i^{b_i}\right]^{-(p+1)}, \\ \qquad x_i \geq 0,\ b_i > 0 \\ 0 \quad \text{elsewhere} \quad . \end{cases} \tag{8}$$

The conditional density of x_i given x_j is

$$f(x_i\ x_j) = \begin{cases} (p+1)(1+rx_i^{b_i}) \left\{\dfrac{1+rx_j^{b_j}}{1+x_j^{b_j}}\right\} b_i x_i^{b_i-1} \left[1 + \left\{\dfrac{1+rx_j^{b_j}}{1+x_j^{b_j}}\right\} x_i^{b_i}\right]^{-(p+2)} \\ \quad - rb_i x_i^{b_i-1} \left[1 + \left\{\dfrac{1+rx_j^{b_j}}{1+x_j^{b_j}}\right\} x_i^{b_i}\right]^{-(p+1)}, \\ \qquad x_i \geq 0,\ b_i > 0 \\ 0 \quad \text{elsewhere.} \end{cases} \quad (9)$$

The correlation coefficient is

$$\rho_{x_1x_2} = \Gamma(p - \frac{1}{b_1})\Gamma(p - \frac{1}{b_2})\Gamma(1 + \frac{1}{b_1})\Gamma(1 + \frac{1}{b_2}).$$

$$\{(1 - \frac{1}{pb_1})\,{}_2F_1(p + 1 - \frac{1}{b_1}, 1 + \frac{1}{b_2}; p + 1; 1 - r)$$

$$+ \frac{r}{pb_1}\,{}_2F_1(p - \frac{1}{b_1}, 1 + \frac{1}{b_2}; p + 1; 1 - r)\}/ \quad (10)$$

$$\{[\Gamma(1 + \frac{2}{b_1})\Gamma(p - \frac{2}{b_1})\Gamma(p) - \Gamma^2(1 + \frac{1}{b_1})\Gamma^2(p - \frac{1}{b_1})].$$

$$[\Gamma(1 + \frac{2}{b_2})\Gamma(p - \frac{2}{b_2})\Gamma(p) - \Gamma^2(1 + \frac{1}{b_2})\Gamma^2(p - \frac{1}{b_2})]\}^{\frac{1}{2}},$$

where ${}_2F_1(a, b; c, z)$ is Gauss' hypergeometric function. If $r = 1$, then $\rho_{x_1 x_2} = 0$. The correlation coefficient is a function of the b_i's, p, and r, which does effectively generalize the bivariate Burr distribution in the desired way.

4. APPLICATIONS. Examples of the use of the Burr system of distributions which take advantage of the flexibility of the curve shape characteristics include the development of non-normal control charts [Burr, (1967a)], simulations of approximations to the normal [Burr (1967b)], simulation of approximations to various non-normal distributions in a study of tests on variances [Wheeler (1970)], estimation and confidence interval construction for quantal response or sensitivity data [Seibert (1970)], and estimation for quantal response or sensitivity data arising from a bivariate situation [Durling (1969)]. The ease with which the Burr system of distributions can be used in many applications should be facilitated considerably by the recent publication [Burr (1973)] of b and p values for a wide grid of the standardized third and fourth central moments.

Future applications may well arise in the study of data from such fields as Psychology, Sociology, Economics, etc., in which there is often reason to study data which is distinctly non-normal in character. The physical sciences may well find applications in the fitting of hydrological, meteorological or life testing data which require distributions which have heavier tails than a gamma tail [Bryson (1974)]. Probably the potential applications are far greater than any suggestions that could be made at this point in time.

REFERENCES

[1] Bryson, M. C. (1974). Technometrics 16, 61-68.
[2] Burr, I. W. (1942). Ann. of Math. Statist. 13, 215-232.
[3] Burr, I. W. (1967a). Industrial Quality Control 23, 563-569.
[4] Burr, I. W. (1967b). Technometrics 9, 647-651.
[5] Burr, I. W. (1973). Comm. Statist. 2, 1-21.
[6] Burr, I. W. and Cisiak, P. J. (1968). J. Amer. Statist. Assoc. 63, 627-643.
[7] Craig, C. C. (1936). Ann. Math. Statist. 7, 16-28.
[8] Durling, F. C. (1969). Bivariate Probit, Logit and Burrit Analysis. Themis Signal Analysis Statistics Research Program, Tech. Rept. 41, Department of Statistics, Southern Methodist University, Dallas.

[9] Durling, F. C., Owen, D. B., and Drane, J. W. (1970). Ann. Math. Statist. 41, 1135.

[10] Gradshteyn, I. S. and Ryshik, I. M. (1965). Table of Integrals, Series and Products. Academic Press, New York and London.

[11] Hatke, Sister M. A. (1949). Ann. Math. Statist. 20, 461-463.

[12] Johnson, N. L. and Kotz, S. (1972). Distributions in Statistics: Continuous Multivariate Distributions. John Wiley and Sons, Inc., New York and London.

[13] Pearson, E. S. and Hartley, H. O. (1966). Biometrika Tables for Statisticians, Vol. I, 3rd ed. Cambridge University Press, Cambridge.

[14] Seibert, G. B., Jr. (1970). Estimation and Confidence Intervals for Quantal Response or Sensitivity Data. Themis Signal Analysis Statistics Research Program, Tech. Rept. 66, Department of Statistics, Southern Methodist University, Dallas.

[15] Takahasi, K. (1965). Ann. Inst. Statist. Math. (Tokyo) 17, 257-260.

[16] Wheeler, D. J. (1970). An Alternative to an F-Test on Variances. Themis Signal Analysis Statistics Research Program, Tech. Rept. 76, Department of Statistics, Southern Methodist University, Dallas.

MULTIVARIATE BETA DISTRIBUTION

R. P. Gupta

Mathematics Department, Dalhousie University, Halifax,
Nova Scotia, Canada

SUMMARY. Multivariate beta distribution (central and non-central) has been defined by many authors [e.g. Khatri (1959), Kshirsagar (1961) and Das Gupta (1972)]. We have tried to clarify this definition as given by these authors. Several properties and its decomposition is given.

KEYWORDS. Multivariate beta distribution, Wishart distribution, non-central, decomposition.

1. INTRODUCTION. Let A and B be $(p \times p)$ symmetric positive definite matrices having Wishart density $W_p(A|\Sigma|n_1)$ and $W_p(B|\Sigma|n_2)$ respectively, where

$$W(A|\Sigma|n_1) = C_1 \exp\{-\frac{1}{2}\operatorname{tr}\Sigma^{-1}A\} \quad |A|^{\frac{n_1-p-1}{2}} \tag{1}$$

and $C_1^{-1} = 2^{\frac{n_1 p}{2}} \pi^{p(p-p)/4} \prod_{i=1}^{p} \Gamma(\frac{n_1+1-i}{2})$.

There exists a lower triangular matrix C with positive diagonal elements such that

$$\left.\begin{aligned} A + B &= CC' \\ \text{Define L by} \quad A &= CLC' \end{aligned}\right\} \tag{2}$$

G. P. Patil et al. (eds.), Statistical Distributions in Scientific Work, Vol. 1, 337-344.

Khatri (1959) has shown that L and C are independent and he derived the density of L

$$f(L) = C_2 \; |L|^{\frac{1}{2}(n_1-p-1)} \; |I-L|^{\frac{1}{2}(n_2-p-1)} \qquad (3)$$

$$\text{where } C_2 = \frac{\Pi \; \Gamma(\frac{n_1+n_2-i+1}{2})}{p(p-1)/4 \; \prod_{i=1}^{p} \Gamma(\frac{n_1-i+1}{2})\Gamma(\frac{n_2-i+1}{2})} \; .$$

The range of ℓij's are restricted by the condition that L and I-L both are positive definite. Kshirsagar (1961) called this distribution of L as multivariate beta distribution. We will denote (3) by $B_p(\frac{n_1}{2}, \frac{n_2}{2})$.

Note that the density function of L does not depend upon Σ, hence there is no loss of generality in taking $\Sigma = I$.

If B has a non-central (linear) Wishart distribution

$$W(B/I/n_2) \; e^{-\lambda^2/2} \Sigma \; \frac{(\lambda^2/2)^r}{r!} \; \frac{b_{11}^r}{2^r} \; \frac{\Gamma(\frac{n_2}{2})}{\Gamma(\frac{n_2}{2}+r)}$$

then the non-central (linear) multivariate beta density of L is given by, see Kshirsagar (1961).

$$\text{Const.} \; |L|^{\frac{1}{2}(n_1-p-1)} \; |I-L|^{\frac{1}{2}(n_2-p-1)} \; e^{-\lambda^2/2} \; {}_1F_1[\tfrac{1}{2}(n_1+n_2), \; \tfrac{1}{2} n_2, \; \tfrac{1}{2} \lambda^2(1-\ell_{11}^2)] \qquad (4)$$

In case B has a non-central (planar) Wishart distribution

$$W(B|I|n_2)e^{-\frac{\lambda_1^2+\lambda_2^2}{2}}\Gamma(\tfrac{n_2}{2})\Gamma(\tfrac{n_2}{2}-1)$$

$$\cdot \Sigma\Sigma \frac{(\lambda_1^2\lambda_2^2)^{\alpha}}{2^{4\alpha+2\beta}} \frac{(b_{11}b_{22}-b_{12})^{\alpha}(\lambda_1^2 b_{11}+\lambda_2^2 b_{22})^{\beta}}{\alpha!\beta!\Gamma(\frac{1}{2}n_2-1+\alpha)\Gamma(\frac{n_2}{2}+2\alpha+\beta)}$$

the density function of L is

$$f(L) = \text{const.}\ |L|^{\frac{1}{2}(n_1-p-1)}|I-L|^{\frac{1}{2}(n_2-p-1)}\int \pi c_{ii}^{n_1+n_2-i}\ e^{-\frac{1}{2}\mathrm{tr}CC'}$$

$$\times\ {}_0F_1[\tfrac{1}{2}n_2,\ \tfrac{1}{2}c_{11}(I-L_{11})c_{11}']dL\ dc\ldots$$

Gupta and Kabe (1970) integrated the above, but the result involves five summations and seems to be of no practical significance. Radcliffe (1968) noted that the density of L will involve the elements of L_{11} only where

$$L = \begin{Vmatrix} L_{11} & L_{12} \\ L_{21} & L_{22} \end{Vmatrix}.$$ Hence

$$f(L) = \text{const.}\ |L|^{\frac{1}{2}(n-p-1)}|I-L|^{\frac{1}{2}(n_2-p-1)}\ \phi(L_{11}) \qquad (6)$$

He used this to factorize the Wilk's Λ. We denote the non-central multivariate beta density of L by $B_p(\frac{n_1}{2}, \frac{n_2}{2}, \Delta)$ where Δ is the non-centrality parameter.

2. DENSITY FREE APPROACH TO THE MATRIX VARIATE BETA DISTRIBUTION

Mitra (1970) introduced the multivariate beta density in the following distribution free approach.

Let A and B be independent Wishart matrices $A \sim W_p(A|\Sigma|n_1)$, $B \sim W_p(B|\Sigma|n_2)$, assume $n_1 + n_2 \geq p$. Define

$L = (A+B)^{-\frac{1}{2}} A \left((A+B)^{-\frac{1}{2}}\right)'$. It can be easily seen that the density of L is free of Σ. He reffered it as a matrix variate beta distribution $B_p(\frac{n_1}{2}, \frac{n_2}{2})$.

Khatri (1970) defined the matrix beta distribution with a slightly different angle. His results are the same as that of Mitra (1970), but his method seems to be more useful. Let the column vectors of X: $(p\times n_1)$ and Y: $(p\times n_2)$ be independently and identically distributed as $N(0,\Sigma)$.

Let $n = n_1 + n_2 \geq p$ and Σ is non-singular. Define $L = ZZ'$ where $Z = S^{-\frac{1}{2}}X$, $S = XX' + YY'$ and $n_1 < p$, $n_2 < p$ or both. Then it is easy to see that L is matrix beta variate.

Das Gupta (1972) has also defined the matrix beta variate as $L = Z'Z$.

3. PROPERTIES

(i) (a) If $L \sim B_p(\frac{n_1}{2}, \frac{n_2}{2})$, then $(I-L) \sim B_p(\frac{n_1}{2}, \frac{n_2}{2})$

(b) If $L \sim B_p(\frac{n_1}{2}, \frac{n_2}{2}, \Delta)$ then $I - L \sim B_p(\frac{n_1}{2}, \frac{n_2}{2}, \Delta)$

(ii) If $L \sim B_p(\frac{n_1}{2}, \frac{n_2}{2})$, then for each fixed non-null matrix A, $\frac{A'LA}{A'A} \sim B_p(\frac{n_1}{2}, \frac{n_2}{2})$. This implies that the diagonal elements of L are distributed as $B(\frac{n_1}{2}, \frac{n_2}{2})$.

(iii) If $S \sim W_p(S|\Sigma|n)$ and is distributed independently of L, and if $n_1 + n_2 = n$ then $T_1 = S^{\frac{1}{2}}L(S^{\frac{1}{2}})'$ and $T_2 = S^{\frac{1}{2}}(I-L)(S^{\frac{1}{2}})'$ are independently distributed. $T_1 \sim W_p(T_1|\Sigma|n_1)$ and $T_2 \sim W_p(T_2|\Sigma|n_2)$

(iv) If $L \sim B_p(\frac{n_1}{2}, \frac{n_2}{2})$, $L_* \sim B_p(\frac{n_1+n_2}{2}, \frac{n_3}{2})$, L and L_* are independently distributed then

$$L_*^{\frac{1}{2}} L (L_*^{\frac{1}{2}})' \sim B_p(\frac{n_1}{2}, \frac{n_2+n_3}{2})$$

(v) (a) If $L = \begin{Vmatrix} L_{11} & L_{12} \\ L_{21} & L_{22} \end{Vmatrix} \sim B_p(\frac{n_1}{2}, \frac{n_2}{2})$, $n_1 \geq p - s$

then $L_{22.1} = L_{22} - L_{21}L_{11}^{-1}L_{12} \sim B_s(\frac{n_1-p+s}{2}, \frac{n_2}{2})$.

(b) If $L \sim B_p(\frac{n_1}{2}, \frac{n_2}{2}, \Delta)$ and partitioning Δ in the same way as L, then $L_{22.1} \sim B_s(\frac{n_1-p+s}{2}, \frac{n_2}{2}, \Delta)$.

(vi) (a) $L_{11} \sim B_{p-s}(\frac{n_1}{2}, \frac{n_2}{2})$ (b) L_{11} and $L_{22.1}$ are independent.

(vii) If $L \sim B_p(\frac{n_1}{2}, \frac{n_2}{2})$ and $n_1 \geq p$, $L_{11} = |L|_1, |L|_2/|L|_1$, $|L|_3/|L|_2, \ldots, |L|_p/|L|_{p-1}$ are independently distributed as $B(\frac{n_1}{2}, \frac{n_2}{2})$, $B(\frac{n_1-1}{2}, \frac{n_2}{2}), \ldots,$ and $B(\frac{n_1-p+1}{2}, \frac{n_2}{2})$ respectively

(viii) (a) If $L \sim B_p(\frac{n_1}{2}, \frac{n_2}{2})$, then for every fixed non-null vector a, $\frac{a'a}{a'L^{-1}a} \sim B(\frac{n_1-p+1}{2}, \frac{n_2}{2})$

(b) If $L \sim B_p(\frac{n_1}{2}, \frac{n_2}{2}, \Delta)$ then $\frac{a'a}{A'L^{-1}a} \sim B_p(\frac{n_1-p+1}{2}, \frac{n_2}{2}, a'\Delta a$.

(ix) (a) If $L \sim B_p(\frac{n_1}{2}, \frac{n_2}{2})$ and P: (s×p) is independently distributed of L and PP' = I then $PLP' \sim B(\frac{n_1}{2}, \frac{n_2}{2})$.

(b) If $n_1 > p-s$, then $(PL^+P')^+ \sim B_s(\frac{n_1-p+s}{2}, \frac{n_2}{2})$ where

P^+ is a Moore-Penrose inverse of P.

(x) Let L and L_1 be independently distributed as $B_p(\frac{n_1}{2}, \frac{n_2}{2})$ and $B_p(\frac{n_2}{2}, \frac{n_1}{2})$ respectively. Then, $L^{\frac{1}{2}}L_1(L^{\frac{1}{2}})'$ and $L_1^{\frac{1}{2}}L(L_1^{\frac{1}{2}})'$ are identically distributed.

(xi) Let Y_1 and Y_2 be two (p×p) random matrices such that $Y_1 \sim B_p(\frac{n_1}{2}, \frac{n_2}{2}, \Delta_1)$ and the conditional distribution of Y_2 given Y_1 is $B(\frac{n_1-n_2}{2}, \frac{s-r}{2}, Y_1^{\frac{1}{2}}(\Delta-\Delta_1)Y_1^{\frac{1}{2}\prime})$ then the distribution of Y is $B_p(\frac{n_1}{2}, \frac{n_2}{2}, \Delta)$ where $Y = Y_1^{\frac{1}{2}}\, Y_2\, Y_1^{\frac{1}{2}}$.

4. DECOMPOSITION. The density function of L from (6) is

$$f(L) = K|L|^{(n_1-p-1)/2}|I-L|^{(n_2-p-1)/2}\phi(L_{11}) .$$

$$\text{Where } L = \begin{Vmatrix} L_{11} & L_{12} \\ L_{21} & L_{22} \end{Vmatrix} .$$

Writing this as

$$f(L_{11},L_{21},L_{22}) = K \cdot |L_{22}-L_{21}L_{11}^{-1}L_{21}'|^{(n_1-p-1)/2}|L_{11}|^{(n_1-p-1)/2}$$

$$\cdot |I-L_{22}-L_{21}(I-L_{11})^{-1}L_{12}|^{(n_2-p-1)/2}$$

$$\cdot |I-L_{11}|^{(n_2-p-1)/2}\phi(L_{11}) \tag{7}$$

Setting $L_{22} = D + L_{21}L_{11}^{-1}L_{12}$, $L_{21} = V[(I - L_{11})L_{11}]^{\frac{1}{2}}$ the density of D, V and L_{11} is

$$f(D,V,L_{11}) = K\cdot|D|^{(n_1-p-1)/2}|I-VV'-D|^{(n_2-p-1)/2}|L_{11}|^{(n_1-p-s-1)/2}$$

$$\cdot|I-L_{11}|^{(n_2-s-1)/2}\phi(L_{11})$$

Again setting $D = (I-VV')^{\frac{1}{2}}R(I-VV')^{\frac{1}{2}}$ and

$$Z = (I-D)^{-\frac{1}{2}}V \tag{8}$$

$$f(D,L_{11},Z) = K|D|^{(n_1-p-1)/2}|I-D|^{(n_2-p-1)/2}|I-ZZ'|^{(n_1-s-1)/2}$$

$$\psi(L_{11}) \tag{9}$$

where $\psi(L_{11}) = |L_{11}|^{(n_1-s-1)/2}|I-L_{11}|^{(n_2-s-1)/2}\phi(L_{11})$.

Obviously the densities of D, Z and L_{11} are independent. The decomposition of L(central), given by Khatri and Pillai (1965) is as follows.

Let $L = (\ell ij)$ then the central part of the multivariate beta density may be decomposed in terms of p beta variates $Z_1,Z_2,\ldots,Z_p$ and $(p-1)Y_i$ vector variates having the joint density

$$f(Z_1,Z_2,\ldots,Z_p,Y_1,Y_2,\ldots,Y_{p-1}) = K\prod_{i=1}^{p} Z_i^{(n_1-p-i-2)/}\prod_{i=1}^{p-1}(1-Y_iY_i') .$$

Hence

$$\ell ij = Z_i + \ell_{(i)}\, L_{ii}^{-1}\, \ell'_{(i)} \qquad i = 1,2,\ldots,p-1$$

$$\ell_{pp} = Z_p$$

$$\ell_{(i)} = (\ell_i, \ell_{i+1},\ldots,\ell_{ip})$$

Where L_{ii} is obtained by omitting the first i rows and i columns of L.

Note that $Z_1, Z_2, \ldots, Z_p = |L|$. Further note that all idependent factors of L are expressible in terms of Z's. p Z's and $\frac{p(p-1)}{2}$ elements of Y_i's account for $\frac{p(p+1)}{2}$ elements of L.

REFERENCES

[1] Anderson, T. W. (1958). An Introduction to Multivariate Statistical Analysis. Wiley, New York.
[2] Constantine, A. G. (1963). Ann. Math. Statist. 34, 1270-1285.
[3] Das Gupta, S. (1972). Sankhyā 34, 357-362.
[4] DeWeal, D. J. (1963). S. Afr. Statist. J. 2, 77-84.
[5] Gupta, R. P. and Kabe, D. G. (1970). Trab. Estadis XX, 61-67.
[6] Gupta, R. P. and Kabe, D. G. (1971). Ann. Inst. Statist. Math. 23, 97-103.
[7] James, A. T. (1964). Ann. Math. Statist. 35, 475-501.
[8] Khatri, C. G. (1959). Ann. Math. Statist. 30, 1258-1262.
[9] Khatri, C. G. and Pillai, K. C. S. (1965). Ann. Math. Statist. 36, 1511-1520.
[10] Khatri, C. G. (1970). Sankhyā Ser A 32, 311-318.
[11] Kshirasagar, A. M. (1961). Ann. Math. Statist. 23, 104-111.
[12] Kshirasagar, A. M. (1970). Ann. Inst. Statist. Math. 2, 295-306.
[13] Mitra, S. K. (1970). Sankhyā Ser A 32, 81-88.
[14] Olkin, I. and Rubin, H. (1964). Ann. Math. Statist. 35, 261-269.

DISTRIBUTION OF A QUADRATIC FORM IN NORMAL VECTORS
(Multivariate Non-Central Case)

C. G. Khatri

Gujarat University, Ahmedabad, India

SUMMARY. Let the column vectors of $\underset{\sim}{X}$: pxn be distributed as independent normals with common covariance matrix $\underset{\sim}{\Sigma}$. Then, the quadratic form in normal vectors is denoted by $\underset{\sim}{X}\underset{\sim}{A}\underset{\sim}{X}' = \underset{\sim}{S}$ where $\underset{\sim}{A}$: nxn is a symmetric matrix which is assumed to be positive definite. This paper deals with a series representation of the density function of $\underset{\sim}{S}$ when $E(\underset{\sim}{X}) \neq \underset{\sim}{0}$, extending the idea of the author (1971) and Kotz et al. (1967b) to the multivariate non-central case. It is pointed out that this method gives a result which is not easy to obtain directly by integrating over an orthogonal space in the sense of James (1964).

KEY WORDS. Quadratic form in normal vectors, series representation.

1. INTRODUCTION. Suppose that $\underset{\sim}{x}_i$: px1 , (i=1,2,...,n) are distributed normally with means $E(\underset{\sim}{x}_i)$ and $\mathrm{Cov}(\underset{\sim}{x}_i, \underset{\sim}{x}_{i'}) = v_{ii'} \underset{\sim}{\Sigma}$ for i, i' = 1,2,...,n, where $\underset{\sim}{V} = (v_{ii'})$ and $\underset{\sim}{\Sigma}$ are symmetric positive definite (p.d.) matrices. Then, a quadratic form in normal vectors is defined by

$$\underset{\sim}{S} = \sum_{i,i'} a_{ii'} \underset{\sim}{x}_i \underset{\sim}{x}_{i'}' = \underset{\sim}{X}\underset{\sim}{A}\underset{\sim}{X}'$$

G. P. Patil et al. (eds.), Statistical Distributions in Scientific Work, Vol. 1, 345-354.

with $\underset{\sim}{X} = (\underset{\sim}{x}_1,\ldots,\underset{\sim}{x}_n)$ and $\underset{\sim}{A} = (a_{ii'})$, $a_{ii'} = a_{i'i}$. By carrying out suitable linear transformations on $\underset{\sim}{X}$, one can easily show that the distribution of $\underset{\sim}{S}$ is the same as that of

$$\sum_{i=1}^{n} \underset{\sim}{y}_i \underset{\sim}{y}_i' = \underset{\sim}{Y}\underset{\sim}{Y}' \tag{1.1}$$

where $\underset{\sim}{Y} = (\underset{\sim}{y}_1,\ldots,\underset{\sim}{y}_n)$, $\underset{\sim}{y}_i = (i=1,2,\ldots,n)$ are distributed normally with $E(\underset{\sim}{y}_i) = \underset{\sim}{\delta}_i$, $\mathrm{Cov}(\underset{\sim}{y}_i, \underset{\sim}{y}_j) = \alpha_i^2 \underset{\sim}{\Sigma}$ if $i=j$, and $=\underset{\sim}{0}$ if $i \neq j$ $\alpha_1^2 \geq \alpha_2^2 \geq \ldots \geq \alpha_n^2 > 0$ are the characteristic roots of VA. It is well-known that if $\alpha_1^2 = \alpha_2^2 = \ldots = \alpha_n^2$, then the distribution of $\underset{\sim}{S}$ is non-central Wishart and is studied by Constantine (1963) and James (1964). When all $\underset{\sim}{\delta}_i = \underset{\sim}{0}$ $(i=1,2,\ldots,n)$, the solutions in various series representations are given by Khatri (1971, 1966), Hayakawa (1966) and Shah (1970). When $p = 1$, the problem is studied by various authors (see the references) in various ways. The purpose of this paper is to obtain a density function of $\underset{\sim}{S}$. in the non-central case and to point out that the direct method gives an integral over an orthogonal space, which is not known, and one does not know how to obtain a solution, while a modification of the expression for a series representation given by the author in the previous paper (1971) gives a solution.

2. INTEGRATION PROBLEM OVER AN ORTHOGONAL SPACE. The joint density of $\underset{\sim}{Y} = (\underset{\sim}{y}_1,\ldots,\underset{\sim}{y}_n)$ can be written as

$$(2\pi)^{-pn/2} |\underset{\sim}{\Sigma}|^{-n/2} |\underset{\sim}{D}_\alpha|^{-p} \exp[-\tfrac{1}{2} \operatorname{tr} \underset{\sim}{\Sigma}^{-1}(\underset{\sim}{Y}-\underset{\sim}{\Delta})\underset{\sim}{D}_\alpha^{-2}(\underset{\sim}{Y}-\underset{\sim}{\Delta})'] \tag{2.1}$$

where $\underset{\sim}{D}_\alpha = \mathrm{diag}\,(\alpha_1,\ldots,\alpha_p)$ and $\underset{\sim}{\Delta} = (\underset{\sim}{\delta}_1,\underset{\sim}{\delta}_2,\ldots,\underset{\sim}{\delta}_p)$. Notice that $\underset{\sim}{S} = \underset{\sim}{Y}\underset{\sim}{Y}'$ and it is invariant under an orthogonal transformation $\underset{\sim}{Y} \to \underset{\sim}{Y}\underset{\sim}{H}$ where $\underset{\sim}{H}$ is any orthogonal matrix. For obtaining the distribution of $\underset{\sim}{S}$, we can multiply (2.1) by an invariant unit Haar measure $d\underset{\sim}{H}$ and then use the transformation $\underset{\sim}{Y} \to \underset{\sim}{Y}\underset{\sim}{H}$. This shows that the density function of $\underset{\sim}{S}$ will be obtained provided $\underset{\sim}{Y}$ has the density function given by

$$c \int_{O(n)} \exp[-\frac{1}{2} \text{tr}\, \Sigma^{-1} Y H D_\alpha^{-2} H' Y' + \text{tr}\, \Sigma^{-1} Y H D_\alpha^{-2} \Delta'] \, dH \quad (2.2)$$

where $c^{-1} = (2\pi)^{pn/2} |\Sigma|^{n/2} |D_\alpha|^p \exp(-\frac{1}{2} \text{tr}\, \Sigma^{-1} \Delta D_\alpha^{-2} \Delta')$ and the integration is over an orthogonal space $O(n)$. Notice that the results of the integration (2.2) are available only in particular cases, namely (i) $D_\alpha^2 = \beta I$, (ii) $\Delta = 0$ and (iii) $p = 1$. In general, however, this value for (2.2) is unknown and it seems to be difficult to evaluate. Moreover (2.2) can be rewritten as

$$c \exp(-r\, \text{tr}\, \Sigma^{-1} S) \int_{O(n)} \exp[\text{tr}\, W H D_\beta H' + \text{tr}\, \Delta_1' H'] \, dH \quad (2.3)$$

where $W = Y' \Sigma^{-1} Y$, $\Delta_1 = D_\alpha^{-2} \Delta' \Sigma^{-1} Y$, $r > 0$ is arbitrary and $D_\beta = r I_n - \frac{1}{2} D_\alpha^{-2}$. From (2.3), it is not clear in general whether (2.3) is a function of S or not. However, in this case, after some simplifications, (2.3) can be shown to be a function of S [see (3.25)]. Hence, we shall have to adopt a different method which will yield the required results. For this purpose, we shall adopt the method utilized by Kotz et al. (1967a, 1967b) and Khatri (1971) with appropriate changes.

3. A SERIES REPRESENTATION. The method adopted by Kotz et al. (1967a, 1967b) and Khatri (1971) can be described as follows.

Notice that the matrix $S = (s_{ij})$ has $p(p+1)/2$ random elements and hence let us assume that the density function of S can be represented by

$$f(S) = \sum_{k=0}^{\infty} \sum_\kappa a_\kappa f_\kappa(S) \quad (3.1)$$

where $\kappa = \{k_{11}, k_{12}, \ldots, k_{1p}, k_{22}, k_{23}, \ldots, k_{2p}, \ldots, k_{pp}\}$ with $\sum_{i=1}^{p} \sum_{j=i}^{p} k_{ij} = k$ and k_{ij} being non-negative integers, Σ_κ denotes the summation over such k_{ij}'s such that $\Sigma k_{ij} = k$, a_κ is constant and $f_\kappa(S)$ is a suitable function of S. Notice that when $p = 1$, we have the same type of the series representation as Kotz et al. (1967a, 1967b) while for any $p > 1$, the series given by Khatri

(1971) is different in the sense that it uses p summations, while the present one is in $p(p+1)/2$ summations as the number of variables. We have to put in (3.1) restriction of convergence of the series; namely

$$|f(S)| \leq b \exp(\mathrm{tr}\, BS) \text{ for all } S > 0$$

while b is a positive constant and $(-B)$ is positive semi-definite. The problem is to choose the suitable functions $f_\kappa(S)$ in (3.1) so that the Laplace transform of $f(S)$ is the same as $L_0(Z) = E\exp(-\mathrm{tr}\, Z\, S)$. It is easy to see from (2.1) that the latter part is given by

$$L_0(Z) = E(\exp(-\mathrm{tr}\, ZS)$$

$$= \prod_{j=1}^{n} |I_p + 2\alpha_j^2 \Sigma\; Z|^{-1/2} \exp[-\sum_{j=1}^{n} \delta_j' Z(I_p + 2\alpha_j^2 \Sigma Z)^{-1}\delta_j]. \tag{3.2}$$

Since Σ is p.d., we can find a nonsingular matrix P such that $\Sigma = PP'$. Then, instead of obtaining the distribution of $S = YY'$, we can obtain the density function of $P^{-1}SP'^{-1} = S_1$, (say), and then we can find that of S. For this purpose, we shall assume without loss of generality that $\Sigma = I$ in (3.2). Then $E\exp(-\mathrm{tr}\, Z\, S)$ is

$$L_0(Z) = \prod_{j=1}^{n} |I_p + 2\alpha_j^2 Z|^{-1/2} \exp[-\sum_{j=1}^{n} \delta_j' Z(I_p + 2\alpha_j^2 Z)^{-1}\delta_j]. \tag{3.3}$$

Suppose that the Laplace transform of the series $f(S)$ given by (3.1) is

$$\hat{f}(Z) = \int_{S>0} \exp(-\mathrm{tr}\, ZS)\, f(S)\, dS. \tag{3.4}$$

Then, in order for $f(S)$ to be a density function of S, we must have

$$L_0(Z) = \hat{f}(Z) \text{ for all } Z \text{ with } \mathrm{Re}(Z) \geq 0\;. \tag{3.5}$$

Let us write in (3.3), $\theta = (\gamma I + Z)^{-1}$ with γ being an arbitrary constant greater than zero and $\beta_j = \gamma - (2\alpha_j^2)^{-1}$, then we can rewrite (3.3) as

$$L_0(\underset{\sim}{\theta}^{-1} - \gamma \underset{\sim}{I}) = a_0 |\underset{\sim}{\theta}|^{n/2} \left(\prod_{j=1}^{n} |\underset{\sim}{I} - \beta_j \underset{\sim}{\theta}|^{-1/2} \right)$$

$$\exp\left[\sum_{j=1}^{n} \underset{\sim}{\mu}_j' \underset{\sim}{\theta} (\underset{\sim}{I} - \beta_j \underset{\sim}{\theta})^{-1} \underset{\sim}{\mu}_j \right] \tag{3.6}$$

where $a_0 = [\prod_{j=1}^{n} (2\alpha_j^2)^{-p/2}] \exp(-\frac{1}{2} \sum_{j=1}^{n} \alpha_j^{-2} \underset{\sim}{\delta}_j' \underset{\sim}{\delta}_j)$ and

$\underset{\sim}{\mu}_j = \underset{\sim}{\delta}_j / 2\alpha_j^2$. Let us suppose that we can write the expansion of (3.6) as

$$\sum_{k=0}^{\infty} \sum_{\kappa} a_\kappa M_\kappa(\underset{\sim}{\theta}) = a_0 \left(\prod_{j=1}^{n} |\underset{\sim}{I} - \beta_j \underset{\sim}{\theta}|^{-1/2} \right) \exp\left[\sum_{j=1}^{n} \underset{\sim}{\mu}_j' \underset{\sim}{\theta} (\underset{\sim}{I} - \beta_j \underset{\sim}{\theta})^{-1} \underset{\sim}{\mu}_j \right]$$

$$= C(\underset{\sim}{\theta}) \text{ , (say) ,} \tag{3.7}$$

where

$$M_\kappa(\underset{\sim}{\theta}) = \prod_{i=1}^{p} \prod_{j=i}^{p} (\theta_{ij}^{k_{ij}}) \text{ , } \underset{\sim}{\theta} = (\theta_{ij}) \text{ with } \theta_{ij} = \theta_{ji}. \tag{3.8}$$

Thus, using (3.7) and (3.6) in (3.5), we shall have the Laplace transform of $f(\underset{\sim}{S})$ as

$$\hat{f}(\underset{\sim}{\theta}^{-1} - \gamma \underset{\sim}{I}) = |\underset{\sim}{\theta}|^{n/2} \sum_{k=0}^{\infty} \sum_{\kappa} a_\kappa M_\kappa (\underset{\sim}{\theta}) . \tag{3.9}$$

Now, we have to find the function $f_\kappa(\underset{\sim}{S})$ of $\underset{\sim}{S}$ such that its Laplace transform $\hat{f}_\kappa(\underset{\sim}{Z})$ is

$$\hat{f}_\kappa(\underset{\sim}{Z}) = |\underset{\sim}{I}\gamma + \underset{\sim}{Z}|^{-n/2} M_\kappa((\underset{\sim}{I}\gamma + \underset{\sim}{Z})^{-1}) . \tag{3.10}$$

To find such a function $f_\kappa(\underset{\sim}{S})$, let us consider a function

$$g_k(\underset{\sim}{S}, \underset{\sim}{W}) = \Sigma_\eta^{(p)}\{(n/2)_\eta\}^{-1} C_\eta(\underset{\sim}{W}\underset{\sim}{S})\ h(n/2, \underset{\sim}{S}, \gamma \underset{\sim}{I}_p) \tag{3.11}$$

where

$$h(n/2, \underset{\sim}{S}, \gamma \underset{\sim}{I}_p) = \{\Gamma_p(n/2)\}^{-1}|\underset{\sim}{S}|^{(n-p-1)/2}\exp(-\gamma \mathrm{tr}\ \underset{\sim}{S}), \tag{3.12}$$

$$\Gamma_p(n/2) = \pi^{p(p-1)/4}\prod_{i=1}^{p}\Gamma\left(\frac{n-i+1}{2}\right), \quad \underset{\sim}{\eta}=\{\eta_1,\eta_2,\ldots,\eta_p\},\ \eta_1\geq\ldots\geq\eta_p\geq 0$$

is a partition of k with $\sum_{i=1}^{p}\eta_i = k$ and $C_\eta(\underset{\sim}{S})$ is a zonal polynomial of degree k in $\underset{\sim}{S}$, [tabulated by James (1968, 1964) $\Sigma_\eta^{(p)}$ indicates the summation over such partitions. Then, the Laplace transform $\hat{g}_k(\underset{\sim}{Z}, \underset{\sim}{W})$ is

$$\hat{g}_k(\underset{\sim}{Z},\underset{\sim}{W}) = \int_{\underset{\sim}{S}>\underset{\sim}{0}} \exp(-\mathrm{tr}\ \underset{\sim}{Z}\ \underset{\sim}{W})\ g_k(\underset{\sim}{S},\underset{\sim}{W})\ d\underset{\sim}{S}$$

$$= \Sigma_\eta^{(p)}\ |\gamma\underset{\sim}{I} + \underset{\sim}{Z}|^{-n/2}\ C_\eta(\underset{\sim}{W}(\underset{\sim}{I}\gamma + \underset{\sim}{Z})^{-1})$$

$$= |\gamma\underset{\sim}{I} + \underset{\sim}{Z}|^{-n/2}\{\mathrm{tr}\ (\underset{\sim}{W}(\underset{\sim}{I}\gamma + \underset{\sim}{Z})^{-1})\}^k$$

$$= \Sigma_\kappa\ \hat{f}_\kappa(\underset{\sim}{Z})\ M_\kappa(\underset{\sim}{W})\ c_\kappa \tag{3.13}$$

where $\hat{f}_\kappa(\underset{\sim}{Z})$ is defined in (3.10), $M_\kappa(\underset{\sim}{W})$ is defined by (3.8) and

$$c_\kappa = k!\left(\prod_{i=1}^{p}\prod_{j=i}^{p}(k_{ij}!)^{-1}\right)2^{k-\sum_{i=1}^{p}k_{ii}}\ . \tag{3.14}$$

Hence, if

$$f_\kappa(\underset{\sim}{S}) = \text{Coefficient of } \{M_\kappa(\underset{\sim}{W})\ c_\kappa\} \text{ in the expansion of } g_\kappa(\underset{\sim}{S}, \underset{\sim}{W}), \tag{3.15}$$

then

$\hat{f}_K(\underset{\sim}{Z})$ = Coefficient of $\{M_K(\underset{\sim}{W})\ c_K\}$ in the expansion of $\hat{g}_K(\underset{\sim}{Z}, \underset{\sim}{W})$. Using the uniqueness property of the Laplace transform, we get the density function of $\underset{\sim}{S}$ as

$$f(\underset{\sim}{S}) = \sum_{k=0}^{\infty} \Sigma_K\ a_K\ f_K(\underset{\sim}{S}) \tag{3.16}$$

where $f_K(\underset{\sim}{S})$ is defined by (3.15) and a_K is defined by (3.7). Let us try to verify the convergence of the series (3.16). For this we have to verify the validity of the expansion of (3.7). (3.7) is valid if

$$\max_i \mid \text{(ith characteristic root (Ch.) of } \underset{\sim}{\theta} \mid < 1/\varepsilon \tag{3.17}$$

where $\varepsilon = \underset{j}{\text{maximum}} |\beta_j| = \underset{j}{\text{max.}} |\gamma - (2\alpha_j^2)^{-1}|$. If $\underset{\sim}{\theta}$ is a real positive definite matrix, then all Ch $(\underset{\sim}{\theta}) < 1/\varepsilon$.

Let us choose non-negative real numbers ζ_{ij}'s such that $|\theta_{ij}| \leq \zeta_{ij}$ and $\underset{\sim}{\zeta} = (\zeta_{ij})$ is positive semi-definite with maximum Ch. $(\underset{\sim}{\zeta}) < 1/\varepsilon$. Then, by using Cauchy's inequality in (3.7), we obtain

$$|a_K| M_K(\underset{\sim}{\zeta}) \leq \underset{C\{\bigcap_{i,j}(|\theta_{ij}| \leq \zeta_{ij})\}}{\text{maximum}} \{C(\underset{\sim}{\theta})\} = d, \text{ (say)} \tag{3.18}$$

where $C(A)$ is the Compliment of the set A. Observe that

$$d \leq a_0 \left(\prod_{i=1}^{n} |\underset{\sim}{I} - |\beta_j|\ \underset{\sim}{\zeta}\ |^{-1/2}\ \exp\left[\sum_{j=1}^{n} \underset{\sim}{\mu}_j'\ \underset{\sim}{\zeta}\ (\underset{\sim}{I} - |\beta_j|\ \underset{\sim}{\zeta})^{-1} \underset{\sim}{\mu}_j\right]\right. \tag{3.19}$$

or by taking $\zeta_{ij} = \zeta$ for all i,j in (3.19),

$$d \leq a_0 \prod_{j=1}^{n} (1-\zeta p|\beta_j|)^{-1/2} \exp\left[\sum_{j=1}^{n} (\underset{\sim}{\mu}_j'\ \underset{\sim}{1})^2\ \zeta/(1-p\zeta|\beta_j|)\right] \tag{3.20}$$

where $\underset{\sim}{1}$ is a column vector having p unit elements and $p\ \zeta < 1/\varepsilon$. From (3.20), we have

$$d \leq a_0(1-\varepsilon\ p\ \zeta)^{-n/2} \exp(\zeta\lambda/(1-\varepsilon\ p\ \zeta)) \tag{3.21}$$

where $\lambda = \sum_{j=1}^{n} (\underset{\sim}{\mu}_j' \underset{\sim}{1})^2$. Now using (3.18) and (3.21), we have

$$|\Sigma_\kappa a_\kappa f_\kappa(\underset{\sim}{S})| \leq (2/n)\ \Sigma_\kappa \{M_\kappa(\underset{\sim}{\zeta})\}^{-1} |M_\kappa(\underset{\sim}{S})|\ c_\kappa$$

$$\leq c\ (2/n)\ (\text{tr}\ \underset{\sim}{S}_0\ \underset{\sim}{\zeta}^{-1})^k/k!\ \text{for}\ k \geq 1 \tag{3.22}$$

where $\underset{\sim}{S}_0 = (|s_{ij}|)$ and $(\text{tr}\ \underset{\sim}{S}_0\zeta^{-1}) \leq (\text{tr}\ S)/\zeta$ provided $\zeta_{ij} = \zeta$. Using (3.22), we can obtain the maximum error after n terms, and easily establish the best choice of γ. This will be the same as mentioned by Kotz et el (1967a, 1967b) or by Khatri (1971) namely

$\gamma = (1/4\alpha_{(1)}^2) + (1/4\alpha_{(n)}^2)$ where $\alpha_{(1)}^2$ and $\alpha_{(n)}^2$ are the maximum and the minimum values of α_j^2 $(j=1,2,\ldots,n)$.

In order to use (3.16) in practical situations, we must determine the values of a_κ and $f_\kappa(\underset{\sim}{S})$. We observe from (3.15) that if

$$\Sigma_\kappa c_\kappa M_\kappa(\underset{\sim}{W})\ f_\kappa^{(1)}\ (\underset{\sim}{S}) = \Sigma_\eta^{(p)} \{(n/2)_\eta\}^{-1}\ C_\eta(\underset{\sim}{W}\underset{\sim}{S})\ ,$$

then

$$f_\kappa(\underset{\sim}{S}) = f_\kappa^{(1)}\ (\underset{\sim}{S})\ h(n/2,\ \underset{\sim}{S}\ ,\ \gamma\underset{\sim}{I})\ . \tag{3.23}$$

The values of $f_\kappa^{(1)}(\underset{\sim}{S})$ are given for k = 1,2, in the table. To obtain a_κ, we use (3.7) and note that

$$\log C(\underset{\sim}{\theta}) = \log a_0 + \sum_{i=0}^{\infty} \text{tr}(\underset{\sim}{\theta}^{i+1}\ \underset{\sim}{B}_i)$$

where $\underset{\sim}{B}_i = \underset{\sim}{A}_i + \gamma_{i+1}\ \underset{\sim}{I}_p/2$; $\gamma_{i+1} = \sum_{j=1}^{n} \beta_j^{i+1}/(i+1)$ and

$\underset{\sim}{A}_i = \sum_{j=1}^{n} \underset{\sim}{\mu}_j\ \underset{\sim}{\mu}_j'\ \beta_j^i$ for i = 0,1,2,... . Thus, (3.7) can be rewritten as

$$\sum_{k=0}^{\infty} \sum_{\kappa} a_{\kappa} M_{\kappa}(\underset{\sim}{\theta}) = a_0 \exp[\sum_{i=0}^{\infty} \text{tr}\, (\underset{\sim}{\theta}^{i+1} \underset{\sim}{B}_i)] \tag{3.24}$$

From this one can compute a_κ's. These are given for k = 2 in the table in terms of $B_i = (b_{jj'}^{(i)})$.

Remark 1. We can connect the above result with that of orthogonal integration given by (2.3), i.e., the following is to be satisfied:

$$a_0 \int_{O(n)} \exp[\text{tr}\, (\underset{\sim}{Y}'\underset{\sim}{Y}\, \underset{\sim}{H}\, \underset{\sim}{D}_\beta\, \underset{\sim}{H}') + \text{tr}(\underset{\sim}{Y}'\, \underset{\sim}{\Delta}\, \underset{\sim}{D}_\alpha^{-2}\, \underset{\sim}{H}')]\, d\underset{\sim}{H}$$

$$= \sum_{k=0}^{\infty} \sum_{\kappa} a_\kappa f_\kappa^{(1)}(\underset{\sim}{S}). \tag{3.25}$$

REFERENCES

[1] Constantine, A. G. (1963). Ann. Math. Statist. 34, 1270-1285.
[2] Gurland, J. (1955). Ann. Math. Statist. 26, 122-127. Corrections in (1962), Ann. Math. Statist. 33, 813.
[3] Hayakawa, T. (1966). Ann. Inst. Statist. Math. 18, 191-200.
[4] James, A. T. (1964). Ann. Math. Statist. 35, 475-501.
[5] Khatri, C. G. (1966). Ann. Math. Statist. 37, 468-479.
[6] Khatri, C. G. (1971). J. Multivariate Anal. 1, 199-214.
[7] Kotz, S., Johnson, N. L. and Boyd, D. W. (1967a). Ann. Math. Statist. 38, 823-837.
[8] Kotz, S., Johnson, N. L. and Boyd, D. W. (1967b). Ann. Math. Statist. 38, 838-848.
[9] Pachares, J. (1955). Ann. Math. Statist. 26, 128-131.
[10] Robbins, H. (1948). Ann. Math. Statist. 19, 266-270.
[11] Ruben, H. (1962). Ann. Math. Statist. 33, 542-570.
[12] Ruben, H. (1963). Ann. Math. Statist. 34, 1582-1584.
[13] Shah, B. K. (1963). Ann. Math. Statist. 34, 186-190.
[14] Shah, B. K. (1968). Ann. Math. Statist. 39, 18, 1090.
[15] Shah, B. K. (1970). Ann. Math. Statist. 41, 692-697.
[16] Shah, B. K. and Khatri, C. G. (1961). Ann. Math. Statist. 32, 883-887. Corrections in (1963), Ann. Math. Statist. 34, 673.

TABLE

Values of $f_K^{(1)}(\underset{\sim}{S})$ and a_K up to k = 2

k	$M_K(\underset{\sim}{W})$	$c_K f_K^{(1)}(\underset{\sim}{S})$	a_K/a_0	c_K
1	(a) w_{ii}	$(2/n)\ s_{ii}$	$b_{ii}^{(1)}$	1
	(b) w_{ij} $(i \neq j)$	$(4/n)\ s_{ij}$	$2\ b_{ij}^{(1)}$	2
2	(a) w_{ii}^2	$\{4/n(n+2)\}\ s_{ii}^2$	$b_{ii}^{(2)} + \frac{1}{2}(b_{ii}^{(1)})^2$	1
	(b) $w_{ii}\ w_{jj}$ $(i \neq j)$	$f_1\ s_{ii}\ s_{jj} - f_2\ s_{ij}^2$	$b_{ii}^{(1)}\ b_{jj}^{(1)}$	2
	(c) $w_{ii}\ w_{ij}$ $(i \neq j)$	$\{16/n(n+2)\}\ s_{ii}\ s_{ij}$	$2(b_{ij}^{(2)} + b_{ii}^{(1)}\ b_{ij}^{(1)})$	4
	(d) $w_{ii}\ w_{jk}$ $(i \neq j \neq k)$	$2f_1\ s_{ii}\ s_{jk} - 4f_2\ s_{ij}\ s_{ik}$	$2b_{ii}^{(1)}\ b_{jk}^{(1)}$	4
	(e) w_{ij}^2 $(i \neq j)$	$f_3\ s_{ij}^2 - 2f_2\ s_{ii}\ s_{jj}$	$2(b_{ij}^{(2)}) + 2(b_{ij}^{(1)})^2$	4
	(f) $w_{ij}\ w_{jk}$ $(i \neq j \neq k)$	$2f_3\ s_{ij}\ s_{jk} - 4f_2\ s_{jj}\ s_{ki}$	$2b_{ik}^{(2)} + 4b_{ij}^{(1)}\ b_{jk}^{(1)}$	8
	(g) $w_{ij}\ w_{k\ell}$ $(i \neq j \neq k \neq \ell)$	$4f_1\ s_{ij}\ s_{k\ell} - 2f_2(s_{ij}\ s_{\ell j} + s_{i\ell}\ s_{jk})$	$4b_{ij}^{(1)}\ b_{jk}^{(1)}$	8

where $f_2 = 8/n(n-1)(n+2)$; $f_1 = (n+1)f_2$; $f_3 = 16/(n-1)(n+2)$

BIVARIATE AND MULTIVARIATE EXTREME DISTRIBUTIONS

J. Tiago de Oliveira

Center of Applied Mathematics, Faculty of Sciences,
Lisbon, Portugal

SUMMARY. The paper discusses some characteristic properties of bivariate and multivariate extreme distributions.

KEY WORDS. Bivariate extreme distributions, stability postulate, Gumbel marginals.

1. INTRODUCTION. The purpose of this paper is to present a review of some characteristic properties of bivariate and multivariate extreme distributions.

The basic papers in which univariate extremes have been discussed are those of Dodd (1923), Frechet (1927), Fisher-Tippett (1928), von Mises (1936), Gumbel (1935) and Gnedenko (1943). Bivariate extremes have been studied by Finkelshteyn (1953), Geffroy (1958/59), Sibuya (1960) and Tiago de Oliveira (1958).

2. THE BASIC IDEAS. Let $X_1,\ldots,X_n,\ldots$ be a sequence of independent and identically distributed random variables with distribution function $F(x) = Pr\{X_i \leq x\}$.

The distribution function of $\max(X_1,\ldots,X_n)$ is $Pr\{\max(X_1,\ldots,X_n) \leq x\} = Pr\{X_1 \leq x,\ldots,X_n \leq x\} = F^n(x)$.

The distribution function of $\min(X_1,\ldots,X_n)$ is

G. P. Patil et al. (eds.), Statistical Distributions in Scientific Work, Vol. 1, 355-361.

$\Pr\{\min(X_1,\ldots,X_n)\leq x\} = 1-\Pr\{\min(X_1,\ldots,X_n)>x\} =$

$1-\Pr\{X_1>x,\ldots,X_n>x\} = 1-(1-F(x))^n$. Note that $\min(X_1,\ldots,X_n) =$

$-\max(-X_1,\ldots,-X_n)$.

This kind of "duality" allows reduction of minima results to maxima results. In what follows, we shall deal only with maxima case.

Evidently the manipulation of maxima (largest values) of samples depends on the full knowledge of F(x). In practical applications, chiefly when dealing with large samples--as done for the behavior of sums--we will resort to asymptotic results, that is, use those limiting distributions L(x) such that, for suitable coefficients λ_n and $\delta_n (> o)$ we have $F^n(\lambda_n+\delta_n x) \overset{s}{} L(x)$.

These limiting distributions L(x) are called, usually, the extreme distributions. The limiting distributions, without location and dispersion parameters, are of the three forms:

Fréchet: $\Phi_\alpha(x) = o \quad \text{if} \quad x<o$

$\qquad\qquad = \exp(-x^{-\alpha}) \quad \text{if} \quad x\geq o \;, \quad \alpha>o$

Weibull: $\Psi_\alpha(x) = \exp(-(-x)^{\alpha}) \quad \text{if} \quad x\leq o \;, \quad \alpha>o$

$\qquad\qquad = 1 \quad \text{if} \quad x>o$

Gumbel: $\Lambda(x) = \exp(-e^{-x})$,

except for the degenerate or unitary distribution

$H(x) = o \quad \text{if} \quad x<o$

$\qquad = 1 \quad \text{if} \quad x\geq o$

Note that Gumbel distribution is, generally, called the extreme distribution.

3. THE BIVARIATE EXTREME CASE. Let $(X_1, Y_1),\ldots$ be a sequence of independent and identically distributed random pairs with distribution function F(x,y).

The distribution function of $(\max X_i, \max Y_i)$ is $\Pr\{\max X_i \leq x, \max Y_i \leq y\} = F^n(x,y)$.

We can, also, search coefficients λ_n, $\delta_n > 0$, λ'_n, $\delta'_n > 0$ such that $F^n(\lambda_n+\delta_n x, \lambda'_n + \delta'_n y) \overset{s}{\to} L(x,y)$ and, using Khintchine's theorem [see Feller (1966)], we see that the distribution function $L(x,y)$ satisfies a stability relation $L^k(a_k + b_k x, a'_k + b'_k y) = L(x,y)$.

We can also assume that the marginals are Gumbel marginals $L(x, +\infty) = \Lambda(x)$, $L(+\infty,y) = \Lambda(y)$. Finally, using the extension from the natural k to rational r and then to real z we obtain $\Lambda^z(x,y) = \Lambda(x-\log z,\ y-\log z)$.

Taking $x = \log z$ we have $\Lambda(x,y) = \Lambda^{e^{-x}}(0,\ y-x)$. Denoting $\Lambda(0,w) = \exp(-(1 + e^{-w})\,k(w))$ we finally obtain that the limiting and stable distributions of bivariate maxima with Gumbel reduced marginals are given by $\Lambda(x,y) = \exp(-(e^{-x}+e^{y})k(y-x)) = \{\Lambda(x)\Lambda(y)\}^{k(y-x)}$.

It remains now to study the dependence function $k(w)$, which is obviously continuous and non-negative.

We show that $k(w)$ is a dependence function of a limiting distribution of bivariate maxima with Gumbel marginals if and only if $k(-\infty) = k(+\infty) = 1$; $(1 + e^{-w})k(w)$ is non-increasing, $(1 + e^{w})\,k(w)$ is non-decreasing, and $\Delta^2_{xy}[(e^{-x} + e^{-y})\,k\,(y-x)] \leq 0$.

It is well known that $\Lambda(x,y)$ is a bivariate distribution function with marginals $\Lambda(x)$ and $\Lambda(y)$ if and only if $\Lambda(x,+\infty) = \Lambda(x)$, $\Lambda(+\infty,y) = \Lambda(y)$ and $\Delta^2_{xy}\,\Lambda(x,y) \geq 0$.

The first two conditions are easily seen to be equivalent to $k(+\infty)=1$, $k(-\infty)=1$.

Let us now consider the non-negativity condition. It can be written as $\Lambda(\xi,\eta) + \Lambda(x,y) \geq \Lambda(\xi,y) + \Lambda(x,\eta)$, $\xi \geq x$, $\eta \geq y$.

Let $y \to -\infty$ and $x \to -\infty$. We get that $(1+e^{w})d(w)$ is a non-decreasing function, and $(1+e^{-w})k(w)$ is a non-increasing function.

Denote now by $\mu = \exp(-e^{-\xi})$, $A = (e^{-(x-\xi)} + e^{-(y-\xi)})k(y-x)$, $B = (1+e^{-(\eta-\xi)})k\,(\eta-\xi)$, $C = (e^{-(\eta-\xi)}+e^{-(\eta-\xi)})k(\eta-\xi)$ and $D = (1+e^{-(y-\xi)})k(y-\xi)$, where A,B,C,D being fixed provided $x-\xi$, $y-\xi$, $\eta-\xi$ are fixed; the non-negativity condition can be written $f(\mu) = \mu^A + \mu^B - \mu^C - \mu^D > 0$ for $\mu \in [0,1]$.

As $f(1) = 0$ we have $f'(1) \leq 0$ so that $A + B \leq C + D$ which leads to the last condition $\Delta^2_{x,y}\,[(e^{-x} + e^{-y})\,k(y-x)] \leq 0$.

Let us now prove the converse. The conditions on $(1+e^{w})k(w)$ and $(1+e^{-w})k(w)$, imply in the previous notation, that $A \geq D$, $A \geq C$, $D \geq B$, $C \geq B$, $B \geq 0$.

Since $A+B \leq C+D$, we have $f(\mu) = \mu^{A}+\mu^{B}-\mu^{C}-\mu^{D} \geq (\mu^{C}-\mu^{A})(\mu^{D}-\mu^{A})/\mu^{A} \geq 0$ which is equivalent to $\Delta^{2}_{x,y}\Lambda(x,y) \geq 0$.

From Fréchet (1951) we have the relation for any bivariate distribution function F(x,y) and their marginals as follows: $\max(0,F(x,+\infty)+F(+\infty,y)-1) \leq F(x,y) \leq \min(F(x,+\infty),F(+\infty,y))$. Applying to $\Lambda(x,y)$ and using the stability condition, we obtain

$$(1/2 \leq)\ \frac{\max(1,e^{w})}{1+e^{w}} \leq k(w) \leq 1.$$

Note that since $(1+e^{w})k(w)$ is non-decreasing, we have $(1+e^{w})k(w) \geq 1$ and since $(1+e^{-w})k(w)$ is non-increasing we obtain $(1+e^{-w})k(w) \geq 1$ which is equivalent to the lower bound obtained from Fréchet's inequalities.

Letting ξ, $\eta \to +\infty$ in the condition $\Delta^{2}_{x,y}[(e^{-x}+e^{-y})k(y-x)] \leq 0$ we obtain the upper bound $k(w) \leq 1$.

It must be noted that the bounds $k(w)=1$ and $k(w)=\max(1,e^{w})/(1+e^{w})$ correspond to actual bivariate extreme distribution functions; the upper bound leads to the independence case and the lower bound leads to the diagonal case where we have X=Y with probability one.

Note that in the last case the density function, does not exist. The cases where no density exists--known so far--are the so called biextremal model with $k(w/\theta) = \dfrac{1-\theta+\max(\theta,e^{w})}{1+e^{w}}$ $(0 \leq \theta \leq 1)$ and the Gumbel model $k(w/\theta)=1-\theta\dfrac{\min(1,e^{w})}{1+e^{w}}$, the latter being symmetrical $(k(w/\theta)= k(-w/\theta))$. Both include the extremes $\theta=0$ and $\theta=1$ corresponding to independence and the diagonal cases.

It is useful to remark that if $k_1(w)$ and $k_2(w)$ are dependence functions then $\theta k_1(w)+(1-\theta)k_2(w)\ (0 \leq \theta \leq 1)$ is, also, a dependence function; also if $k(w)$ is a dependence function then $k(-w)$ is such a function.

Directly or using the previous result we can show: If $\Lambda(x,y)$ possesses a planar density $\partial^2\Lambda/\partial x\partial y$ then $k(w)$ must satisfy the following conditions $k(-\infty)=k(+\infty)=1$; $[(1+e^w)k(w)]'\geq 0$; $[(1+e^{-w})k(w)]'\leq 0$; $(1+e^{-w})k''(w)+(1-e^{-w})k'(w)\geq 0$.

The two known models, with a planar density, are: the logistic model with $k(w/\theta)=(1+e^{-w/(1-\theta)})^{1-\theta}/(1+e^{1-w})$ $(0 \leq \theta \leq 1)$ and the mixed model with $k(w/\theta)= 1-\theta e^w/(1+e^w)^2$ $(0\leq\theta\leq 1)$; both are symmetrical and include the independence case for $\theta=0$; the diagonal case is only contained in the logistic model for $\theta=1$.

In the differentiable case, the following representation theorem for $k(w)$ is valid: $k(w)$ is a dependence function in the differentiable case if and only if it exists a non-negative function $Z(w)$ such that:

$$\int_{-\infty}^{+\infty} e^{\beta} Z(\beta)dw \leq 1 \text{ and } \int_{-\infty}^{+\infty} Z(\beta)d\beta \leq 1$$

and $k(w)$ is of the form $k(w)=1-(e^w\int_w^{+\infty}Z(\beta)d\beta+\int_{-\infty}^{w}e^{\beta}Z(\beta)d\beta)/(1+e^w)$; $Z(w)$ is uniquely defined for each twice-differentiable $k(w)$.

Observe that since $Z(w)\geq 0$, $k(w)=1$ (independence) if and only if for some $a(-\infty<a<+\infty)$ we have $k(a)=1$.

4. MULTIVARIATE EXTENSIONS. If the bivariate extremes area is largely open we can say that the field of multivariate extremes is almost completely open.

The following remarks may be useful for researchers and students of this important field.

In the same way as above we can show that $\Lambda(x_1,\ldots,x_m)$ is a multivariate extreme stable distribution with Gumbel marginals $\Lambda(x_i)$ if and only if $\Lambda(x_1,\ldots,x_m)= \exp\{-(e^{-x_1}+\ldots+e^{-x_2})k(x_2-x_1,\ldots,x_m-x_1)\} = \{\Lambda(x_1)\ldots\Lambda(x_m)\}^{k(x_2-x_1,\ldots,x_m-x_1)}$. (The dependence function k satisfies appropriate relations which assure that $\Lambda(x_1,\ldots,x_m)$ is a distribution function.) We also have the following basic inequalities $\Lambda(x_1)\ldots\Lambda(x_m)\Lambda\leq\Lambda(x_1,\ldots,x_m)$ $\leq \min[\Lambda(x_1),\ldots,\Lambda(x_m)]$, $[\prod_{i\neq j}\Lambda_{ij}(x_i,x_j)]\ 1/2(m-1) \leq$ $\Lambda(x_1,\ldots,x_m) \leq [\prod_{i\neq j}\Lambda(x_i,x_j)]^{1/2}/(\prod\Lambda(x_i))^{m-2}$, where $\Lambda(x_i)$ and

$\Lambda_{ij}(x_i,x_j)$ denote the univariate and bivariate marginals, respectively.

Indeed, from Fréchet (1940) we have
$\max(0, \Sigma\, \Lambda(x_i)-m+1) \leq \Lambda(x_1,\ldots,x_m) \leq \min[\Lambda(x_1),\ldots,\Lambda(x_m)]$ and from the stability can be written
$\Lambda(x_1)\ldots\Lambda(x_m) \leq \Lambda^n(x_1+\log n,\ldots,x_m + \log n) = \Lambda(x_1,\ldots,x_m)$.

Gumbel's inequality leads to

$$1 - \frac{\sum_{i\neq j} (1-\Lambda_{ij}(x_i,x_j))}{2(m-1)} \leq \Lambda(x_1,\ldots,x_m)$$

and Bonferroni's inequality yields

$$\Lambda(x_1,\ldots,x_m) \leq 1/2 \sum_{i\neq j} \Lambda_{ij}\,(x_i,x_j)-(m-2)\,\Sigma\,\Lambda(x_i) + \frac{(m-1)(m-2)}{2}$$

[see e.g. Fréchet (1940) concerning these inequalities]. Using the stability we get the second double inequality.

Using the second double inequality we obtain the basic result concerning independence:

The multivariate extreme distribution functions split in the product of the univariate marginals $[\Lambda(x_1,\ldots,x_m) = \Lambda(x_1)\ldots\Lambda(x_m)]$ if and only if all the bivariate extreme distribution marginals split into the product of univariate distributions.

REFERENCES

[1] Dodd, E. L. (1923). The greatest and the least variate under general laws of error. Trans. Amer. Math. Soc. 25.

[2] Feller, W. (1960). An Introduction to Probability Theory and Its Applications, II. John Wiley and Sons, Inc., New York.

[3] Finkelshteyn, B. V. (1953). Limiting distributions of extreme values of the variational series of two dimensional random variable. Dokl. Ak. Nauk. SSSR 91, 2.

[4] Fisher, R. A. and Tippett, L. H. C. (1928). Limiting forms of frequency distributions of the largest and smallest member of a sample. Proc. Camb. Phil. Soc. 24, part 2.

[5] Fréchet, M. (1927). Sur la loi de probabilité de l'cart maximum. Ann. Soc. Polon. Math. (Krakov) 6.

[6] Fréchet, M. (1940). Les probabilitiés associées a un systéme d'evenements compatibles et dependents. Act. Sc. et Ind. 859, Paris.

[7] Fréchet, M. (1951). Sur les tableaux de correlation dont les marges sont données. Ann. Univ. Lyon, ser. III, XIV A, Math. & Astr.
[8] Geffroy, J. (1958/1959). Contribution a la théorie des valeurs extrêmes. Publ. Inst. Statist. Univ. Paris, 718.
[9] Gnedenko, B. (1943). Sur la distribution limite du terme maximum d'une série aléatoire. Ann. Math. 44.
[10] Gumbel, E. J. (1935). Les valeurs extrêmes des distributions statistiques. Ann. Inst. Henri Poincaré, V.
[11] von Mises, R. (1936). La distribution de la plus grande de n valeurs. Rev. Math. Union Interbalkanique, I.
[12] Sibuya, M. (1960). Bivariate extremal statistics. Ann. Inst. Statist. Math. 11.
[13] Tiago de Oliveira, J. (1958). Extremal distributions. Rev. Fac. Cienc. Univ. Lisboa, 2 ser., A, Mat.
[14] Tiago de Oliveira, J. (1962/1963). Structure theory of bivariate extremes, extensions. Est. Mat. Estat e Econom. 7.

ON THE DISTRIBUTION OF THE MINIMUM AND OF THE MAXIMUM OF A RANDOM NUMBER OF I.I.D. RANDOM VARIABLES*

Moshe Shaked

Department of Statistics, University of Rochester, Rochester, New York, U.S.A.

SUMMARY. Let X_i, $i = 1,2,\ldots$ be i.i.d. random variables with a common distribution function $F(x)$. Let N be a positive integer-valued random variable, independent of the X's, with a probability generating function $\psi(u)$. Then $G(x) = \psi(F(x))$ and $H(x) = 1-\psi(1-F(x))$ are, respectively, the distributions of $\max(X_1,\ldots,X_N)$ and of $\min(X_1,\ldots,X_N)$. In sections 2 and 3 we derive some inequalities concerning F, G and H, their densities and their hazard rates. We also discuss there some cases in which the monotonicity of the hazard rate of F is preserved by G and H. In section 4 we give some solutions of a functional equation that appears as a condition in some of the theorems of section 3. We also characterize the geometric distributions by requiring the solutions to satisfy an additional assumption. By taking X_i and/or N as random vectors, we introduce in section 5 some methods of construction of multivariate distributions with desired marginals.

KEY WORDS. Series and parallel systems, hazard rate, random walks, geometric distribution, Jensen's inequality, functional equation, multivariate distributions with desired marginals, log-convex functions.

---

*This work was partially supported by National Science Foundation Grant GP-30707X1.

G. P. Patil et al. (eds.), Statistical Distributions in Scientific Work, Vol. 1, 363-380.

1. INTRODUCTION. Let X_i, i = 1,2,... be i.i.d. random variables having a common distribution function F(x). Let N be a positive integer-valued random variable, independent of the X_i's, having a probability generating function (p.g.f.) $\psi(u)$. Then

$$G(x) \equiv \psi(F(x)) \tag{1.1}$$

and

$$H(x) \equiv 1 - \psi(1-F(x)) \tag{1.2}$$

are, respectively, the distribution functions of $\max(X_1,X_2,\ldots,X_N)$ and $\min(X_1,X_2,\ldots,X_N)$. The relation (1.2) may alternatively be expressed as

$$\overline{H}(x) = (\overline{F}(x)) \ , \tag{1.2'}$$

where $\overline{H}(x) = 1-H(x)$ and $\overline{F}(x) = 1-F(x)$ (we will use this notation henceforth). Raghunandanan and Patil (1972) obtained formulae for the distribution of the ith order statistic when the sample size is random [see also Buhrman (1973)]. Formulas (1.1) and (1.2) are in fact, special cases of theirs. In section 2 we derive some inequalities concerning F, G, H and their densities.

Distributions of the forms (1.1) or (1.2) arise in reliability theory. If F(0-) = 0 then each of the X's can be thought of as a life length of an object which we will call a component. Consider now a system, consisting of n independent components, which fails when the first component fails. Such a system is called a series system. The life length of a series system is $\min(X_1,\ldots,X_n)$; its distribution function is $1-(\overline{F}(x))^n$. A system, consisting of n independent components, which functions as long as at least one of the components functions is called a parallel system. The life length of a parallel system is $\max(X_1,\ldots,X_n)$; its distribution function is $(F(x))^n$. The distributions of the life lengths of a parallel system and of a series system with a random number of components are given, respectively, by (1.1) and (1.2).

The hazard rate of a distribution F which has a density f is defined by the function

$$r(x) = f(x)/\overline{F}(x)$$

for every x for which $\overline{F}(x) > 0$. If F is a distribution of a life length of an object then r(x)dx represent the probability that an object of age x will fail in the interval [x,x+dx] [for a complete discussion see e.g. Barlow and Proschan (1965)]. In section 3 we compare the hazard rates of F, G and H (of (1.1) and (1.2)) and discuss some cases in which the monotonicity of the hazard rate of F is preserved by G or by H.

Consider now the following situation. Let X_{ij}, i,j=1,2,... be i.i.d. random variables having a common distribution F(x) and let N_i, i=1,2,... and N be i.i.d. positive inter-valued random variables having a common p.g.f. $\psi(u)$ and assume that the X's and the N's are independent. Denote $Y_i = \max(X_{i1}, X_{i2}, \ldots, X_{iN_i})$, i=1,2,... . Then the distribution of $\min(Y_1, Y_2, \ldots, Y_N)$ is $1-\psi(1-\psi(F(x)))$. Of special interest are p.g.f.'s for which the equality

$$1-\psi(1-\psi(F(x))) = F(x) \qquad -\infty < x < \infty \tag{1.3}$$

holds for every F. These p.g.f.'s have intuitive meaning and also analytical importance which will be indicated in Theorem 3.5. They will be discussed in section 4.

Generalizing (1.1) and (1.2) we introduce in section 5 methods of construction of multivariate distributions with desired marginals.

2. SOME INEQUALITIES INVOLVING F, G, H AND THEIR DENSITIES.

Theorem 2.1. Let X_i, i=1,2,... be i.i.d. random variables will a common distribution function F. If F has all its mass on the interval (a,b) ($-\infty \leq a < b \leq \infty$) and is invertible there, then G[H] is a distribution function of a maximum[minimum] of a random number of X_i' s if and only if $G(F^{-1}(u))$ $[\bar{H}(\bar{F}^{-1}(u))]$ is absolutely monotone on (0,1).

Proof. If F(x) is invertible on (a,b) then (1.1) can be written as

$$G(F^{-1}(u)) = \psi(u) \qquad 0 \leq u \leq 1,$$

and (1.2) can be written as

$$\bar{H}(\bar{F}^{-1}(u)) = \psi(u) \qquad 0 \leq u \leq 1.$$

Applying a well known result about absolutely monotone functions [see Widder (1941), p. 147] the statement of the theorem follows.

As an illustration let $F(x) = 1-e^{-\lambda x}$ and $G(x) = 1-e^{-\mu x}$, $0 < x < \infty$, $\lambda > 0$, $\mu > 0$. Then $G(F^{-1}(u)) = 1 - (1-u)^{\mu/\lambda}$ which is absolutely monotone if and only if $\mu \leq \lambda$. That is, if X_i are i.i.d. exponential random variables with mean λ^{-1} then every exponential distribution with mean $\mu^{-1} \geq \lambda^{-1}$ can be a distribution of a maximum of a random number of X_i's. On the other hand $\overline{G}(\overline{F}^{-1}(u)) = u^{\mu/\lambda}$ is absolutely monotone if and only if μ/λ is a non-negative integer. Thus G is a distribution of a minimum of a random number of X_i's if and only if it is the distribution of a minimum of a fixed number of X_i's. P.g.f.'s of the form $1-(1-u)^q$, $0 < q \leq 1$, arise in the analysis of ladder epochs [Feller (1971), Ch. 12]; they will be further discussed in Example 4.1.

Another application of Theorem 2.1 concerns the logistic distribution. We denote $F_\mu(x) = [1 + \exp(-(x-\mu))]^{-1}$, $-\infty < x < \infty$, $-\infty < \mu < \infty$. Let $\alpha \geq 0$ and μ be two constants. Can $F_{\mu+\alpha}(x)$ [respectively $F_{\mu-\alpha}(x)$] be a distribution of a maximum [minimum] of a random number of i.i.d. random variables with a common distribution $F_\mu(x)$? A simple computation shows that $F_{\mu+\alpha}(F_\mu^{-1}(u)) = \overline{F}_{\mu-\alpha}(\overline{F}_\mu^{-1}(u)) = e^{-\alpha}u/(1-(1-e^{-\alpha})u) \equiv \psi(u)$. Clearly $\psi(u)$ is a geometric p.g.f., hence the answer is in the affirmative. The geometric p.g.f.'s will be further discussed in section 4.

From Theorem 2.1 we see that a necessary condition for G to be a distribution of a maximum[minimum] of a random number of i.i.d. X_i's with a common distribution F is that $G(F^{-1}(u))$ $[\overline{G}(\overline{F}^{-1}(u))]$ is convex on (0,1), that is, that $G[\overline{G}]$ is convex with respect to $F[\overline{F}]$, [see Hardy, Littlewood and Polya (1952), p. 75]. Hence, by Jensen's inequality*

---

* We will discuss here only inequalities that are implied by the convexity of $G(F^{-1}(u))$, similar inequalities can be obtained when $\overline{G}(\overline{F}^{-1}(u))$ is convex.

$$\int_{0+}^{1-} G(F^{-1}(u))d\mu(u) \geq G(F^{-1}(\int_{0+}^{1-} ud\mu(u))) \tag{2.1}$$

for every probability measure μ which is defined on subsets of (0,1).

By choosing $d\mu(u) = du$ we obtain

$$E\ G(X_1) \geq G(M) \tag{2.2}$$

where M is the median of F.

If F has a bounded support with $\underline{L}$ and $\overline{L}$ being, respectively, the left and the right end points of the support then by taking $d\mu(u) = d(F^{-1}(u))/(\overline{L}-\underline{L})$ in (2.1) one obtains

$$G^{-1}[(\overline{L}-EY)/(\overline{L}-\underline{L})] \geq F^{-1}[(\overline{L}-EX_1)/(\overline{L}-\underline{L})] , \tag{2.3}$$

where Y is a random variable with distribution function G.

In fact, $F_1(u) \equiv F^{-1}(u)/(\overline{L}-\underline{L})$ and $G_1(u) \equiv G^{-1}(u)/(\overline{L}-\underline{L})$ are distribution functions on (0,1) and $G_1^{-1}(F_1(u))$ is convex if and only if $G(F^{-1}(u))$ is convex. Thus, the convexity of $G(F^{-1}(u))$ implies that F_1 c-precedes G_1 according to the definition of van Zwet (1964), p. 48. In his monograph, van Zwet obtained inequalities concerning the moments of F_1 and G_1. These moments are in fact

$$\gamma_n \equiv \int_{0+}^{1-} u^n dF_1(u) = \int_{\underline{L}}^{\overline{L}} (F(u))^n du/(\overline{L}-\underline{L})$$

and

$$\delta_n \equiv \int_{0+}^{1-} u^n dG_1(u) = \int_{\underline{L}}^{\overline{L}} (G(u))^n du/(\overline{L}-\underline{L}).$$

Thus van Zwet's inequalities can be applied here. Noting, for example, that $\gamma_1 = (\overline{L}-EX_1)/(\overline{L}-\underline{L})$ and $\delta_1 = (\overline{L}-EY)/(\overline{L}-\underline{L})$ one sees that inequality (2.3) is a special case of Theorem 4.2.1 of van Zwet. Further inequalities concerning γ_n and δ_n can be obtained from Theorems 4.2.1 and 4.2.2 of van Zwet.

More bounds on G and $\overline{H}$ are given by (3.11) and (3.12).

In obtaining the inequalities (2.1)-(2.3) the only fact we used is that the first two derivatives of $GF^{-1}(u)$ are non-negative. Taking into account the fact that also the third derivative is non-negative we can obtain further inequalities that involve the densities of F and G. For this purpose we assume that F has a density f and this assumption implies the existence of a density g of G. From the fact that $G(F^{-1}(u))$ is absolutely monotone, one can easily see that $g(F^{-1}(u))/f(F^{-1}(u))$ is absolutely monotone and thus, by Jensen's inequality

$$\int_{0+}^{1-} \frac{g(F^{-1}(u))}{f(F^{-1}(u))} d\mu(u) \geq \frac{g(F^{-1}(\int_{0+}^{1-} u d\mu(u)))}{f(F^{-1}(\int_{0+}^{1-} u d\mu(u)))} \tag{2.4}$$

for every probability measure μ which is defined on subsets of (0,1).

Substituting in (2.4)

$$d\mu(u) = \begin{cases} du/p & 0 < u \leq p \\ 0 & p < u < 1 \end{cases} \qquad 0 < p \leq 1$$

one obtains for every $0 < p \leq 1$

$$(1/p)G(F^{-1}(p)) \geq g(F^{-1}(p/2))/f(F^{-1}(p/2)) , \tag{2.5}$$

which for $p = 1$ reduces to

$$f(M) \geq g(M) \tag{2.6}$$

where M is the median of F. If F has bounded support with $\underline{L}$ and $\overline{L}$ being its endpoints, then by substituting $d\mu(u) = dF^{-1}(u)/(\overline{L}-\underline{L})$ in (2.4) we get the inequality

$$(\overline{L}-\underline{L})^{-1}E(1/f(Y)) \geq g\left(F^{-1}\left(\frac{EX-\underline{L}}{\overline{L}-\underline{L}}\right)\right)/f\left(F^{-1}\left(\frac{EX-\underline{L}}{\overline{L}-\underline{L}}\right)\right) \tag{2.7}$$

where Y is a random variable with a distribution function G.

Note that all the inequalities (2.1)-(2.7) are sharp in the sense that there exists a generating function $\psi(u)$ such that for $G = \psi(F)$ equalities hold in (2.1)-(2.7). This generating function is obviously $\psi(u) = u$.

3. SOME INEQUALITIES INVOLVING THE HAZARD RATES OF F, G, AND H.

In this section we compare the hazard rate of F with the hazard rates of G and of H. To see the intuitive meaning of the inequalities that are obtained below it is advisable to recall that if $F(0-) = 0$ then G[H] is a distribution of a life length of a parallel [series] system with a random number N of i.i.d. components. It is easily verified that when N is a constant then the hazard rate of H is $Nr(x)$ where $r(x)$ is the hazard rate of F. The bounds for the hazard rate of H of Theorem 3.2 below reduce to this relation when N is degenerate. A simple relation between the hazard rate of a parallel system with a fixed number of components and the hazard rate of its components is not available as in the case of a series system. This fact will be reflected later in Theorem 3.3.

Let F be an absolutely continuous distribution function and denote its hazard rate by $r(x)$. Let ψ be a p.g.f. of a positive integer-valued random variable. Define H as in (1.2) and denote its hazard rate by $\rho(x)$. Then, as is easily verified,

$$\rho(x) = r(x) \cdot u\psi'(u)/\psi(u)\Big|_{u=\bar{F}(x)}. \tag{3.1}$$

Formula (3.1) points out the importance of the transform $u\psi'(u)/\psi(u)$. The next lemma shows that this transform is monotone. The simple proof was suggested by J.H.B. Kemperman.

Lemma 3.1. If $\phi(x)$ is a distribution function of a non-negative random variable N and if

$$\psi(u) \equiv \int_0^\infty u^x d\phi(x) \tag{3.2}$$

converges on $[0,A]$ then $u\psi'(u)/\psi(u)$ is non-decreasing on $[0,A]$.

Proof. Write $\psi(u) = \hat{\omega}(\ell n\, u)$ where $\hat{\omega}$ is the moment generating function of ϕ. Then $u\psi'(u)/\psi(u) = \hat{\omega}'(\ell n\, u)/\hat{\omega}(\ell n\, u)$. It is well known that $\ell n\, \hat{\omega}$ is convex hence $\hat{\omega}'(\ell n\, u)/\hat{\omega}(\ell n\, u)$ is non-decreasing.

It is clear that $\psi(u)$ of (3.2) converges on $[0,1]$ and thus $u\psi'(u)/\psi(u)$ is monotonic there. If N of Lemma 3.1 is a non-negative integer valued random variable then $\psi(u)$ is its p.g.f.. In this case it is easy to see that $u\psi'(u)/\psi(u)$ is either strictly increasing or a constant according to whether N is a non degenerate or a degenerate random variable. If the Laplace transform of ϕ is $\omega(\cdot)$ then $\psi(u) = \omega(-\ell n\ u)$.

Using Lemma 3.1 we prove

Theorem 3.2. Let $\{X_i\}_{i=1}^{\infty}$ be i.i.d. random variables with a common distribution F on (a,b), $-\infty \leq a < b \leq \infty$ and let N be a positive integer valued random variable independent of $\{X_i\}_{i=1}^{\infty}$. Assuming F has a density, then the distribution H of $Y \equiv \min(X_1,\ldots,X_N)$ has a density and

$$Kr(x) \leq \rho(x) \leq r(x)EN \qquad a \leq x \leq b \tag{3.3}$$

where $r(x)$ and $\rho(x)$ are, respectively, the hazard rates of F and H and $K = \min\{k:P(N=k) > 0\}$. (EN may be infinity).

Proof. Let ψ be the p.g.f. of N. Then $\bar{H}(x) = \psi(\bar{F}(x))$ and if F has a density then H has. The relation between $r(x)$ and $\rho(x)$ is given by (3.1). Hence by Lemma 3.1

$$r(x)\lim_{x\to b} u\psi'(u)/\psi(u)\big|_{u=\bar{F}(x)} \leq \rho(x) \leq r(x)\lim_{x\to a} u\psi'(u)/\psi(u)\big|_{u=\bar{F}(x)}, \tag{3.4}$$

but

$$\lim_{x\to a} u\psi'(u)/\psi(u)\big|_{u=F(x)} = \lim_{u\to 1} u\psi'(u)/\psi(u) = EN$$

and

$$\lim_{x\to b} u\psi'(u)/\psi(u)\big|_{u=\bar{F}(x)} = \lim_{u\to 0} u\psi'(u)/\psi(u) = K, \text{ (by l'Hôpital's rule)}$$

thus (3.4) implies (3.3).

Consider now the distribution of $\max(X_1,\ldots,X_N)$ which is $G(x) = \psi(F(x))$. The relationship between the hazard rates $r(x)$ and $\eta(x)$ of F and G is

$$\eta(x) = r(x)\{(1-u)\psi'(u)/(1-\psi(u))\}\big|_{u=F(x)}. \tag{3.5}$$

The convexity of $\psi(u)$ on $[0,1]$ implies $(1-\psi(u))/(1-u) \geq \psi'(u)$ for $0 \leq u \leq 1$ hence $\eta(x) \leq r(x)$. Also $(1-\psi(u))/(1-u)$ $\leq \lim_{u\uparrow 1} 1-\psi(u))/(1-u) = \psi'(1) = EN$ and $\psi'(u) \geq \psi'(0) = P(N=0)$, thus, from (3.5) we get $\eta(x) \geq r(x)(P(N=1)/EN)$ and thus obtain

Theorem 3.3. Let $\{X_i\}_{i=1}^{\infty}$ be i.i.d. random variables with a common distribution F on (a,b), $-\infty \leq a < b \leq \infty$, and let N be a positive integer valued random variable independent of $\{X_i\}_{i=1}^{\infty}$. Assuming F has a density then the distribution G of $Y \equiv \max(X_1,\ldots,X_N)$ has a density and

$$r(x)(P(N=1)/EN) \leq \eta(x) \leq r(x), \quad a \leq x \leq b \tag{3.6}$$

where $r(x)$ and $\eta(x)$ are, respectively, the hazard rates of F and G. (EN may be infinity).

With an additional assumption in Theorem 3.3 we can obtain inequalities analogous to (3.3) which are sharper than (3.6). The additional assumption is that ψ, the generating function of N, satisfies the functional equaltion

$$1 - \psi(1-\psi(u)) = u \quad 0 \leq u \leq 1 \quad . \tag{3.7}$$

This equation is identical to the equation (1.3). Geometrically (3.7) means that the graph $v = \psi(u)$ is symmetric about the line $u+v = 1$ in the unit square $\{(u,v)\colon 0 \leq u, v \leq 1\}$. Examples of generating functions that satisfy (3.7) will be given in section 4. Having the geometrical meaning in mind we can replace in Lemma 3.1 u by $1-\psi(u)$ and $\psi'(u)$ by $(\psi'(u))^{-1}$ and obtain the following lemma.

Lemma 3.4. Let $\psi(u)$ be as in Lemma 3.1 and assume it satisfies (3.7). Then $(1-u)\psi'(u)/(1-\psi(u))$ is non-decreasing on $[0,A]$.

Using this lemma we prove a theorem analogous to Theorem 3.2:

Theorem 3.5. Let F, G, η, r and N be as in Theorem 3.3 and assume that the generating function of N satisfies (3.7). Then

$$r(x)(EN)^{-1} \leq \eta(x) \leq K^{-1}r(x), \quad a \leq x \leq b \tag{3.8}$$

where $K = \min\{k: P(N=k)>0\}$.

Proof. Using relation (3.5) and Lemma 3.4 we see, as in the proof of Theorem 3.2, that it is enough to prove that $\lim_{u\to 0}(1-u)\psi'(u)/(1-\psi(u)) = (EN)^{-1}$ and that

$\lim_{u\to 1}(1-u)\psi'(u)/(1-\psi(u)) = K^{-1}$. The first limit is obtained by noting that if ψ satisfies (3.7) then $\psi'(0) = (EN)^{-1}$. The second limit equals by assumption (3.7) to $\lim_{u\to 0}\psi(u)/u\psi'(u) = K^{-1}$, thus completing the proof.

Note that if ψ satisfies (3.7) then $(EN)^{-1} = P(N=1)$ and hence $EN < \infty$ if and only if $P(N=1) > 0$.

The inequality $Kr(x) \leq \rho(x)$ of (3.3) is sharp in the following sense. Given $EN = C \geq K$ there exist a sequence $\{N_m\}_{m=[C]+1}^{\infty}$ of positive integer-valued random variables such that

$$EN_m = C \tag{3.9}$$

and

$$\lim_{m\to\infty} \rho_m(x) = Kr(x) \tag{3.10}$$

for every x. Here $\rho_m(x)$ denotes the hazard rate of $Y = \min(X_1,\ldots,X_{N_m})$ and $r(x)$ is the hazard rate of X_1; (the assumptions of independence are the same as in Theorem 3.2). Let N_m be a random variable such that

$P(N_m=K) = (m-C)/(m-K)$ and $P(N_m=m) = (C-K)/(m-K), m = [C],[C]+1,\ldots$.
It is easy to verify that $\{N_m\}_{m=[C]+1}^{\infty}$ satisfies (3.9) and (3.10).

Similar sharpness property is enjoyed by the inequality (3.6).

A discussion with R. Srinivasan in Calgary drew my attention to the following bounds on $\bar{H}$ and G:

$$[\bar{F}(x)]^{EN} \leq \bar{H}(x) \leq [\bar{F}(x)]^{K} \quad a \leq x \leq b, \tag{3.11}$$

and

$$[F(x)]^{EN} \leq G(x) \leq [F(x)]^{K} \quad a \leq x \leq b, \tag{3.12}$$

where F, G, H, K and N are as in Theorems 3.2 and 3.3. The right hand side inequalities (3.11) and (3.12) are trivial. The left hand side inequalities can be proved using Jensen's inequality and the fact that $\bar{H}(x) = E[\bar{F}(x)]^N$ and $G(x) = E[F(x)]^N$, $a \leq x \leq b$. Note that (3.11) can be obtained also from (3.3).

Distributions with monotone hazard rate are called DHR (decreasing hazard rate) if the hazard rate is non-increasing and IHR (increasing hazard rate) if the hazard rate is non-decreasing. From equation (3.1) and Lemma 3.1 one immediately sees that if F (of Theorem 3.2) is DHR then H (of the same theorem) is DHR. This is in fact a special case of Theorem 5.2 of Barlow and Proschan (1965, p. 37) which states that the mixture of DHR distributions is DHR. Applying the same arguments to Lemma 3.4 and relation (3.5) we get the following:

Corollary 3.6. Under the assumptions of Theorem 3.5 if F is IHR then G is IHR.

Barlow and Proschan (1965) (p. 220) used the same method of proof to show that the distributions of the life length of some coherent structures are IHR if their components have IHR life length distributions.

Some of the results that were obtained under the assumption that ψ of (1.1) and (1.2) is a p.g.f. remain valid under the weaker assumption that ψ is of the form (3.2). Under this assumption ψ is not necessarily absolutely monotone, yet, Theorems 3.2 and 3.5 remain true under quite general conditions. Using the monotonacity of $u\psi'(u)/\psi(u)$ (which is established in Lemma 3.1) one proves immediately the right side inequality of (3.3) where $EN = \int_0^\infty x d\phi(x)$. To obtain an anologue of the left side inequality we need a further assumption. Let a be the left end point of the support of ϕ and let μ be the distribution of N-a. We assume that μ varies regularly at the origin [for exact definition see e.g. Feller (1971) p. 276]. Note that $\psi(u) = \omega(-\ln u)$ where ω is the Laplace transform of ϕ. Hence $u\psi'(u)/\psi(u) = -\omega(-\ln u)/\omega(-\ln u)$ and

$$\lim_{u\to 0} u\psi'(u)/\psi(u) = \lim_{S\to\infty} -\omega'(s)/\omega(s)$$

$$= a - \lim_{S\to\infty} \tilde{\omega}'(s)/\tilde{\omega}(s)$$

where $\tilde{\omega}$ is the Laplace transform of μ. Using Theorem 1 on page 281 of Feller (1971) it may be proved that $\lim_{s\to\infty} \tilde{\omega}'(s)/\tilde{\omega}(s) = 0$ which yields an analogue of (3.3), namely, $ar(x) \leq \rho(x) \leq r(x)EN$. An analogue of (3.8) can be obtained similarly. It is obvious that inequalities (3.11), (3.12) and Corollary 3.6 remain valid when N is any positive random variable.

As an illustration substitute in (3.2) $d\phi(x) = e^{-x}dx\ x \geq 0$ and substitute the resulting ψ and $F(x) = \exp(-e^{-x}) - \infty < x < \infty$ in (1.1). With this choice one obtains the logistic distribution. This derivation is due to Dubey (1969).

Before closing this section we mention an application of Lemma 3.1. Let A and B be two positive constants, let ψ: (0,A) $\to$ (0,B) be a differentiable strictly increasing function and let f: (0,A) $\to$ (0,A) be a non-increasing function. Marshall and Proschan (1972) proved that if in addition to these assumptions f is log convex on (0,A), ψ is convex and $u\psi'(u)/\psi(u)$ is non-increasing on (0,A), then $\psi f \psi^{-1}$ is log convex on (0,B). Thus, by Lemma 3.1, if ψ is of the form (3.2) and if f is log convex on (0,A) then $\psi f \psi^{-1}$ is log convex on $(0,\infty)$. Marshall and Proschan also proved that if in addition to the assumptions in the second sentence of this paragraph f is log concave on (0,A , ψ is concave and $u\psi'(u)/\psi(u)$ is non-increasing on (0,A), then $\psi f \psi^{-1}$ is log concave on (0,B). Noting that $u\psi'(u)/\psi(u)$ is non-decreasing in (0,A) if and only if $u(\psi^{-1}(u))'/\psi^{-1}(u)$ is non-increasing in (0,B) and that $\psi(u)$ is convex on (0,A) if and only if $\psi^{-1}(u)$ is concave on (0,B), we obtain from Lemma 3.1 that if ψ is of the form (3.2) and if f is log concave on $(0,\infty)$ then $\psi^{-1} f \psi$ is log concave on (0,A).

4. THE FUNCTIONAL EQUATION $\psi(1-\psi(1-u)) = u$ AND A CHARACTERIZATION OF THE GEOMETRIC DISTRIBUTION. In this section we will introduce a method of constructing p.g.f.'s that satisfy equation (3.7) or, equivalently, are solutions of

$$\psi(1-\psi(1-u)) = u, \quad 0 \leq u \leq 1 \quad . \tag{4.1}$$

From (4.1) it is clear that ψ must satisfy $\psi(0) = 0$ and $\psi(1) = 1$. Denote $\phi(u) = \psi(1-u)$ then (4.1) can be written as

$$\phi(\phi(u)) = u, \quad 0 \leq u \leq 1 \tag{4.2}$$

and our task is to look for solutions of (4.2) which satisfy $\phi(0) = 1$, $\phi(1) = 0$ and are completely monotone on (0,1). The first theorem characterizes the geometric distributions. It is tacitly assumed in the theorem that $\psi(u)$ as a function of a complex variable is single-valued.

Theorem 4.1. If $\psi(u)$ is a p.g.f. that, as a function of a complex variable, is meromorphic and ψ satisfies (4.1) then $\psi(u) = qu/(1-(1-q)u)$ for some $0 \leq q < 1$.

Proof. By Lemma 15.4 of Kuczma (1968) the only meromorphic solution of (4.2) is a linear fractional function. The only such functions which satisfy $\phi(0) = 1$ and $\phi(1) = 0$ are of the form

$$\phi(u) = q(1-u)/(1-(1-q)(1-u)) \tag{4.3}$$

for some q. Now $\psi(u) = \phi(1-u)$ is absolutely monotone and convergent on [0,1] if and only if $0 \leq q < 1$.

If $\psi(u)$ as a function of a complex variable is not single-valued then ψ cannot be defined uniquely on the complex plane and Theorem 4.1 cannot be applied. In this case we can have many absolutely monotone solutions for equation (4.1) as will be shown below. The construction that we show is inspired by the works of Karlin and McGregor (1968a, 1968b).

One can easily verify that if $\psi(u)$ and $\chi(u)$ are two solutions of (4.1) and if $\psi(\chi(u)) = \chi(\psi(u))$, $0 \leq u \leq 1$ then $\psi(\chi(u))$ is a solution of (4.1). Hence if $\psi(u)$ a solution of (4.1) then all its integer iterates $\psi_n(u) = \underbrace{\psi(\psi(\ldots\psi(u)\ldots))}_{n \text{ times}}$ are solutions of (4.1).

This observation supplies us with many p.g.f.'s which are solutions of (4.1) once we have one solution.

Let $\chi(u)$ be a solution of (4.1) (not necessarily absolutely monotone) and let $\alpha(u)$ be an invertible function from [0,1] onto [0,1] then it is easy to verify that

$$\psi(u) = \alpha(\chi(1-\alpha^{-1}(1-u))) \ , \ 0 \leq u \leq 1 \tag{4.4}$$

is a solution of (4.1). If $\psi(u)$ in (4.4) is absolutely monotone then (4.4) gives us a p.g.f. which satisfies (4.1). Note that (4.4) gives us solutions of (4.1) without any apriori known solution. Many interesting solutions are obtained by taking $\chi(u) = u$.

Example 4.1. If in (4.4) $\alpha(u) = 1 - u^{1/m}$ where m is a positive integer and $\chi(u) = u$ we get

$$\psi(u) = 1 - (1-u^m)^{1/m}, \quad 0 \leq u \leq 1 . \tag{4.5}$$

This generating function arises in the following manner. Let Y_i, $i=1,2,\ldots$ be i.i.d. Cauchy random variables located such that $P(Y_1>0) = 1/m$. Let $S_n = \sum_{i=1}^{n} Y_i$, then $P(S_n>0) = \frac{1}{m}$ $(n\geq 1)$. Define N as the random integer that satisfies $S_i \leq 0$, $i=1,2,\ldots,N-1$ and $S_N > 0$, that is, N is the epoch of the first visit of this random walk in $(0,\infty)$. Then [see Feller (1971) p. 415] the generating function of mN is (4.5).

To see the applicability of this generating function in our setting let $\{X_i\}_{i=1}^{\infty}$ be i.i.d. random variables independent of N with distribution function $F(x)$. If in every epoch till and including the epoch of the first visit in $(0,\infty)$ m of the X_i's are observed then the distribution of $\max(X_1,X_2,\ldots,X_{mN})$ is $1 - (1-F^m(x))^{1/m}$.

Example 4.2. A special case of (4.5) can be obtained from (4.4) by choosing $\alpha(u) = 1 - \sin(\pi u/2)$ and $\chi(u) = u$. Then $\psi(u) = 1 - (1-u^2)^{1/2}$. This is the generating function of the epoch of the first visit to the origin of a discrete symmetric random walk [Feller (1968 p. 273].

Example 4.3. If $\chi(u) = u/(q+(1-q)u)$, $0 < q \leq 1$ and $\alpha(u) = (1-u)^m$ then $\psi(u)$ of (4.4) is

$$\psi(u) = q^m(1-(1-u)^{1/m})^m/(1-(1-q)(1-(1-u)^{1/m}))^m$$

$$= \psi_1(\psi_2(u)) ,$$

where $\psi_1(u) = [qu/(1-(1-q)u))]^m$ and $\psi_2(u) = 1 - (1-u)^{1/m}$. Clearly ψ_1 and ψ_2 are p.g.f.'s, hence $\psi(u)$ is a p.g.f. [compare Karlin and McGregor (1968b), p. 145].

The lower bound of Theorem 3.5 is non-trivial if $EN < \infty$. The p.g.f.'s of Theorem 4.1 are associated with finite expectations but none of the p.g.f.'s of Examples 4.1-4.3 has this property. To show that there exist non-geometric p.g.f's that satisfy (4.1) and are associated with finite expectations we have the following:

Example 4.4. Consider the function

$$\psi(u) = 1 - (1/\theta)\log(1+e^{\theta}-e^{\theta u}) \; , \; 0 \le u \le 1 \tag{4.6}$$

where $\theta > 0$. This function is a solution of (4.1) for every $\theta > 0$ as can be verified directly or by constructing it from (4.4) with $\chi(u) = u$ and $\alpha^{-1}(u) = (e^{\theta(1-u)}-1)/(e^{\theta}-1)$. Its derivative is easily seen to be a composition of two absolutely monotone functions. Hence $\psi(u)$ is a generating function.

Note that ψ of (4.6) satisfies $\psi'(1) = e^{\theta} < \infty$ hence it is associated with a finite expectation.

5. SOME MULTIVARIATE GENERALIZATIONS. In this section we indicate shortly some generalizations of (1.1) and (1.2). Let $(X_{i1},X_{i2},\ldots,X_{ik})$, $i=1,2,\ldots$ be a sequence of i.i.d. random vectors with a common distribution $F(x_1,\ldots,x_k)$ and let N be a positive integer-valued random variable independent of $\underset{\sim}{X}_i$, $i=1,2,\ldots$ with a p.g.f. $\psi(u)$, then

$$G(x_1,x_2,\ldots,x_k) \equiv \psi(F(x_1,\ldots,x_k)) \tag{5.1}$$

is the joint distribution of $(Y_1,Y_2,\ldots,Y_k) \equiv \left(\max_{1\le i\le N}\{X_{i1}\},\ldots,\max_{1\le i\le N}\{X_{ik}\}\right)$. The univariate marginals of G, that is, the distributions $F_j(x)$ of Y_j are $\psi(F_j(x))$ where $F_j(x)$ is the distribution of X_{1j}, $j=1,2,\ldots,k$. Similarly, if $\psi(u)$ is of the form (3.2) then (5.1) is a multivariate distribution. Of special interest is the case where for every i, X_{ij}, $j=1,2,\ldots,k$ are i.i.d. with a common (univariate) distribution F(x). Then G of (5.1) has the form

$$G(x_1,x_2,\ldots,x_k) = \psi\left(\prod_{j=1}^{k} \tilde{F}(x_j)\right) . \tag{5.2}$$

Similarly the joint distribution H of $\left(\min_{1\le i\le N}\{X_{i1}\},\ldots,\min_{1\le i\le N}\{X_{ik}\}\right)$ is determined by

$$\overline{H}(x_1,\ldots,x_k) = \psi(\overline{F}(x_1,\ldots,x_k)) \tag{5.3}$$

where $\overline{H}(x_1,\ldots,x_k) = P\{\min_{1\le i\le N}\{X_{ij}\} > x_j;\ j=1,2,\ldots,k\}$ and

$\overline{F}(x_1,\ldots,x_k) = P\{X_{1j} > x_j;\ j=1,2,\ldots,k\}$. This reduces to

$$\overline{H}(x_1,\ldots,x_k) = \psi\left(\prod_{j=1}^{k} (1-\tilde{F}(x_j))\right) \tag{5.4}$$

when for every i, X_{ij}, $j=1,2,\ldots,k$ are i.i.d. Some well known distributions have the form (5.2) or (5.4).

Example 5.1. [Multivariate logistic distribution, Malik and Abraham (1973).]. If we substitute in (5.2) $F(x) = \exp\{-\exp(-x)\}$ and $\psi(u) = (1-\ln u)^{-1}$ we obtain a multivariate logistic distribution

$$G(x_1,x_2,\ldots,x_k) = \left(1 + \sum_{j=1}^{k} \exp(-x_j)\right)^{-1} . \tag{5.5}$$

Example 5.2. [Multivariate extreme value distribution, Johnson and Kotz (1972), p. 254.] If we substitute in (5.2) $\tilde{F}(x) = \{\exp -e^{-\beta x}\}$ and $\psi(u) = \exp\{-(-\ln u)^{\alpha}\}$ $0 < \alpha < 1$, [note that $\omega(s) = \exp\{-s^{\alpha}\}$, $0 < \alpha < 1$ is a Laplace transform of a non-negative distribution, Feller (1971), p. 448], we obtain

$$G(x_1,x_2,\ldots,x_k) = \exp\left\{-\left(\sum_{j=1}^{k} e^{-\beta x_j}\right)^{\alpha}\right\}, \tag{5.6}$$

Taking $\alpha = 1/m$, $\beta = m$ and $k = 2$ we get the distribution that is discussed by Johnson and Kotz.

Example 5.3. [Multivariate Burr's distribution, Takahasi (1965).]. If we substitute in (5.4) $1-\tilde{F}(x) = \exp\{-x^c\}$, $c > 0$ and $\psi(u) = (1-\ln u)^{-m}$, $m > 0$ we obtain

$$\overline{H}(x_1,\ldots,x_k) = \left(1 + \sum_{j=1}^{k} x_j^c\right)^{-m}$$

which is the multivariate Burr's distribution which was discovered by Takahasi.

A multivariate distribution which is a mixture of independent distributions with equal marginals is called in Shaked (1974) positively dependent by mixture (PDM). Some inequalities which are satisfied by PDM distributions are discussed by Dykstra, Hewett and Thompson (1973) and by Shaked (1974). In particular if the joint distribution of $(Y_1,Y_2,\ldots,Y_k)$ is PDM then $\mathrm{Cov}(h(Y_j),h(Y_\ell)) \geq 0$ for every Borel measurable function $h(1 \leq j, \ell \leq k)$. It is easy to see that distributions of the forms (5.2) and (5.4) are PDM.

Another generalization of (1.1) is the following. Let $\{X_{ij}, i=1,2,\ldots,k, j=1,2,\ldots\}$ be an array of independent random variables such that for every i $(1 \leq i \leq k)$, $X_{ij} (j=1,2,\ldots)$ have the distribution F_i. Let $(N_1,N_2,\ldots,N_k)$ be a random vector of integers, independent of the X's and having the p.g.f. $\psi(U_1,U_2,\ldots,U_k)$. Then

$$G(x_1,\ldots,x_k) = \psi(F_1(x_1),\ldots,F_k(x_k)) \tag{5.7}$$

is the joint distribution of $(\max_{1\leq j\leq N_1} \{X_{1j}\},\ldots, \max_{1\leq j\leq N_k} \{X_{kj}\})$.
The univariate marginals of G are $\psi_i(F_i(x))$ $(1 \leq i \leq k)$ where $\psi_i(u)$ is the p.g.f. of N_i. A similar formula gives the joint distribution of the minimums.

The marginal distributions of (5.1), (5.3) or (5.7) can be well known univariate distributions and thus, these formulae give us multivariate distributions with desired marginals which have, usually, very simple analytical forms.

A further discussion about these distributions, their properties, and applications will appear in the forthcoming report.

ACKNOWLEDGEMENT. I am grateful to A. W. Marshall for fruitful conversations and encouragement. I also thank J.H.B. Kemperman for pointing out the simple proof of Lemma 3.1 and making other useful comments on an earlier draft of this paper.

REFERENCES

[1] Barlow, R. E. and Proschan, E. (1965). Mathematical Theory of Reliability. Wiley and Sons, New York.

[2] Buhrman, J. M. (1973). Statistica Neerlandica 27, 125-126.
[3] Dubey, D. S. (1969). Naval Res. Logist. Quart. 16, 37-40.
[4] Dykstra, R. L., Hewett, J. E. and Thompson, W. A. (1973). Ann. Statist. 1, 674-681.
[5] Feller, W. (1968). An Introduction to Probability Theory and Its Applications, Vol. 1 (3rd ed.). John Wiley and Sons, Inc., New York.
[6] Feller, W. (1971). An Introduction to Probability Theory and Its Applications, Vol. 2 (2nd ed.). John Wiley and Sons, Inc., New York.
[7] Hardy, G. H., Littlewood, J. E. and Polya, G. (1952). Inequalities, (2nd ed.). Cambridge University Press.
[8] Johnson, N. L. and Kotz, S. (1972). Distributions in Statistics, Continuous Multivariate Distributions. John Wiley and Sons, Inc., New York.
[9] Karlin, S. and McGregor, J. (1968a). Amer. Math. Soc. Trans. 132, 115-136.
[10] Karlin, S. and McGregor, J. (1968b). Amer. Math. Soc. Trans. 132, 137-145.
[11] Kuczma, M. (1968). Functional Equations in a Single Variable. Polish Scientific Publishers.
[12] Malik, H. J. and Abraham, B. (1973). Ann. Statist. 1, 588-590.
[13] Marshall, A. W. and Proschan, F. (1972). In Inequalities, O. Shisha (ed.), 3, 225-234.
[14] Raghunandanan, K. and Patil, S. A. (1972). Statistica Neerlandica 26, 121-126.
[15] Shaked, M. (1974). A concept of positive dependence for exchangeable random variables. Submitted for publication. (Abstracted in Bull. Inst. Math. Statist. 2, 218-219.)
[16] Takahasi, K. (1965). Ann. Inst. Statist. Math. 17, 257-260.
[17] Widder, D. V. (1941). The Laplace Transform. Princeton University Press.
[18] van Zwet, W. R. (1964). Convex Transformation of Random Variables. Mathematical Center, Amsterdam.

TRANSFORMATION OF THE PEARSON SYSTEM WITH SPECIAL REFERENCE TO TYPE IV

K. O. Bowman and W. E. Dusenberry

Computer Sciences Division,* Union Carbide Corporation, Oak Ridge, Tennessee and Lilly Research Laboratories, Indianapolis, Indiana, U.S.A.

SUMMARY. The Pearson system of distributions consists of several types, including beta, gamma, and F-ratio families. There are interrelations between types, resulting from non-linear transformation of the variates. A description of these interrelations is given along with a new transformation which carries Type IV family into distributions for which $(\sqrt{\beta_1}, \beta_2)$ points lie in Type I region. The usefulness of the transformation is illustrated in an application where the percentage points of the moment estimator for the shape parameter of the gamma distribution are derived.

KEY WORDS. Transformation, Pearson system of curves.

1. INTRODUCTION. The Pearson system of distributions, arising from solutions to the differential equation $\frac{1}{y}\frac{dy}{dx} = \frac{(b+x)}{a_0+a_1x+a_2x^2}$,
and introduced by Karl Pearson (1895), have proved to be of wide application in data analysis, and in the approximation of one distribution by a four-moment equivalent distribution. It includes the normal, the uniform, as points; also the beta, F-ratio, Student's t, and gamma families. A description of the various forms of the Pearson system is to be found in Tables for Statisticians, Biometrika, Part 2, 1914. It may be true to say that the system has few rivals; the Gram-Charlier and Edgeworth

---

*Operated for the U. S. Atomic Energy Commission under U. S. Government Contract W-7405-eng-26.

G. P. Patil et al. (eds.), Statistical Distributions in Scientific Work, Vol. 1, 381-390.

expansion of basic kernels frequently run into the problem of negative densities whereas the Burr (1968) system of cumulative functions is difficult to associate meaningfully with physical concepts. However, we must remark that the Johnson (see for example Johnson and Kotz, 1970) system (univariate and bivariate) of transformed normal distribution is proving to have many applications in recent years (see for example, Pearson (1963, 1965), D'Agostino (1970), D'Agostino and Pearson (1973), Bowman and Shenton (1973), Bowman (1973), Shenton and Bowman (1975)), the advantage here being that near-normality is induced which in turn simplifies (a) problems of evaluating approximate percentage points, and (b) simulation studies.

It happens from time to time that sampling distributions of statistics either have a semi-infinite or doubly infinite range; for example, the kurtosis statistic b_2 must be greater than unity, whereas for Student's t, $-\infty < t < \infty$. It is not surprising then to encounter four-moment Pearson approximations to sampling distributions which have $2\beta_2 - 3\beta_1 - 6 > 0$ (β_1, β_2 referring to skewness and kurtosis), so that the Pearson Type IV distribution (density $y_0 \exp(-k \arctan x)/(x^2 + 1)^m$) may be the appropriate choice. This distribution is the maverick of the family, and is mathematically intractable and rarely encountered as a theoretical model. It is a transformation of this case which we now describe and which may have uses in applications to distribution approximation theory.

2. TRANSFORMATION OF THE PEARSON TYPE IV CURVE.

The Pearson Type IV has a density proportional to $\exp(-\nu \arctan(x/a))/(x^2 + a^2)^m$ and includes Student's t as a special case ($\nu = 0$). Clearly it has few mathematically attractive characteristics. Johnson and Kotz (1970) remark that

> "On account of technical difficulties associated with the use of Type IV distributions, efforts have been made to find other distributions with simpler mathematical form, which, according to circumstances, are close enough to Type IV distributions to replace them."

Figure A gives a graphic picture of the area in the (β_1, β_2) region which corresponds to Pearson Type IV curves, i.e., the area below the Type V line. One of the reasons that Type IV is difficult to work with is that, in various parts of the Type IV region, one or more of the moments of the distribution (from the fifth on up) become infinite.

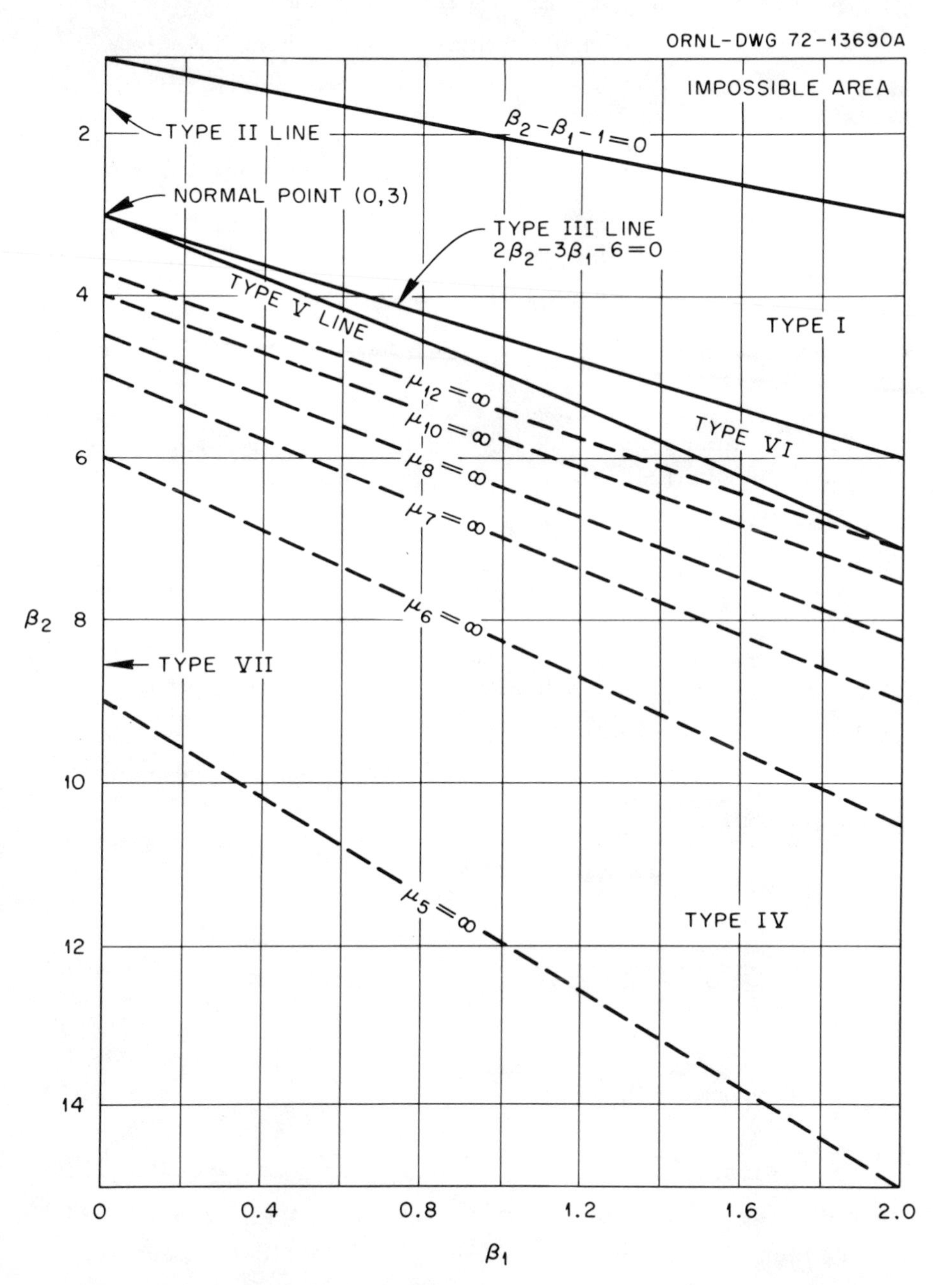

Figure A. Moment Ratio Plane Showing Pearson Types I-VII and the Area of Infinite Moments in the Type IV Region.

The fifth and higher moments for β_1, β_2 points below the $\mu_5 = \infty$ line in Figure A are infinite. Thus, there is a fairly large region where no moments higher than the fourth exist, and there is a very small region where moments higher than the twelfth exist. In the results that follow, a new distribution denoted by TR_4 (and an associated transformation) will be described which does not have these higher moment deficiencies.

2.1 The Pearson Type IV Distribution

The density function for the Pearson Type IV distribution as given by Elderton and Johnson (1969) is:

$$f(x) = \frac{1}{aF(r,\nu)} \left(1 + \frac{x^2}{a^2}\right)^{-m} e^{-\nu \tan^{-1}(x/a)}, \quad -\infty < x < \infty \tag{2.1}$$

where

$$F(r,\nu) = e^{\frac{-\nu\pi}{2}} \int_0^{\pi} \sin^r \phi e^{\nu\phi}\, d\phi, \quad r = 2m-2. \tag{2.2}$$

Using relationships among the moments of the Type IV distribution, the parameters r, ν, and a can be expressed in terms of $\beta_1(x)$, $\beta_2(x)$, and $\mu_2(x)$ as follows:

$$r = \frac{6(\beta_2 - \beta_1 - 1)}{2\beta_2 - 3\beta_1 - 6}$$

$$\nu = \frac{-r(r-2)\sqrt{\beta_1}}{\sqrt{16(r-1) - \beta_1(r-2)^2}}$$

and

$$a = \sqrt{\mu_2\{r-1 - \beta_1(r-2)^2/16)\}}$$

Each point in the (β_1, β_2) region represents two distributions; one for $\mu_3(x) > 0$ $(\sqrt{\beta_1}(x) > 0)$ and another for $\mu_3(x) < 0$ $(\sqrt{\beta_1}(x) < 0)$. The two distributions are identical except for being skewed in opposite directions. The transformation will be discussed as two separate cases corresponding to these two distributions.

2.2 Transformation Case 1: $\mu_3(x) < 0$

Consider the transformation

$$y = e^{-\tan^{-1}(x/a)} \quad . \tag{2.4}$$

The noncentral moments of y are easily found as

$$\mu_k'(y) = E(y^k) = \frac{F(r,\nu+k)}{F(r,\nu)} \quad . \tag{2.5}$$

The density function for the transformed variable y, denoted as TR_4, is given as

$$g(y) = \frac{1}{F(r,\nu)} [1 + \tan^2(-\ell n\ y)]^{-(m-1)} y^{\nu-1}$$

$$\exp(-\pi/2) < y < \exp(\pi/2) \quad . \tag{2.6}$$

The moments and the values of $(\beta_1(y), \beta_2(y))$ for the transformed distribution are easily computed using (2.5).

To determine the Pearson curve type to which a particular value of (β_1, β_2) belongs, the value

$$\text{Kappa} = \frac{\beta_1(\beta_2 + 3)^2}{4(4\beta_2 - 3\beta_1)(2\beta_2 - 3\beta_1 - 6)} \tag{2.7}$$

is computed, where Kappa refers to the x-variable. If Kappa(x) is such that 0 < Kappa(x) < 1, then the original distribution is Pearson Type IV. If the first four moments of the TR_4 transformation give Kappa(y) < 0, then the transformed distribution is Pearson Type I according to the information contained in its first four moments. Therefore, the region in the (β_1, β_2) plane that transforms from Type IV to the Type I region by use of the first four moments of the transformation given in (2.4) is the space containing all points $(\beta_1(x), \beta_2(x))$ which result in Kappa(y) < 0.

Figure B depicts 290 values of $(\beta_1(x), \beta_2(x))$ selected to cover uniformly the Type IV region that is most often considered in the literature. The 290 points (represented by circles) transform (according to their first four moments) into that portion of the Type I region bounded by the dashed line and the Type III line. Six points, indicated by numbers in parentheses, were selected from the Type IV region of Figure B. For each of these points, the first ten central moments of the standardized Pearson

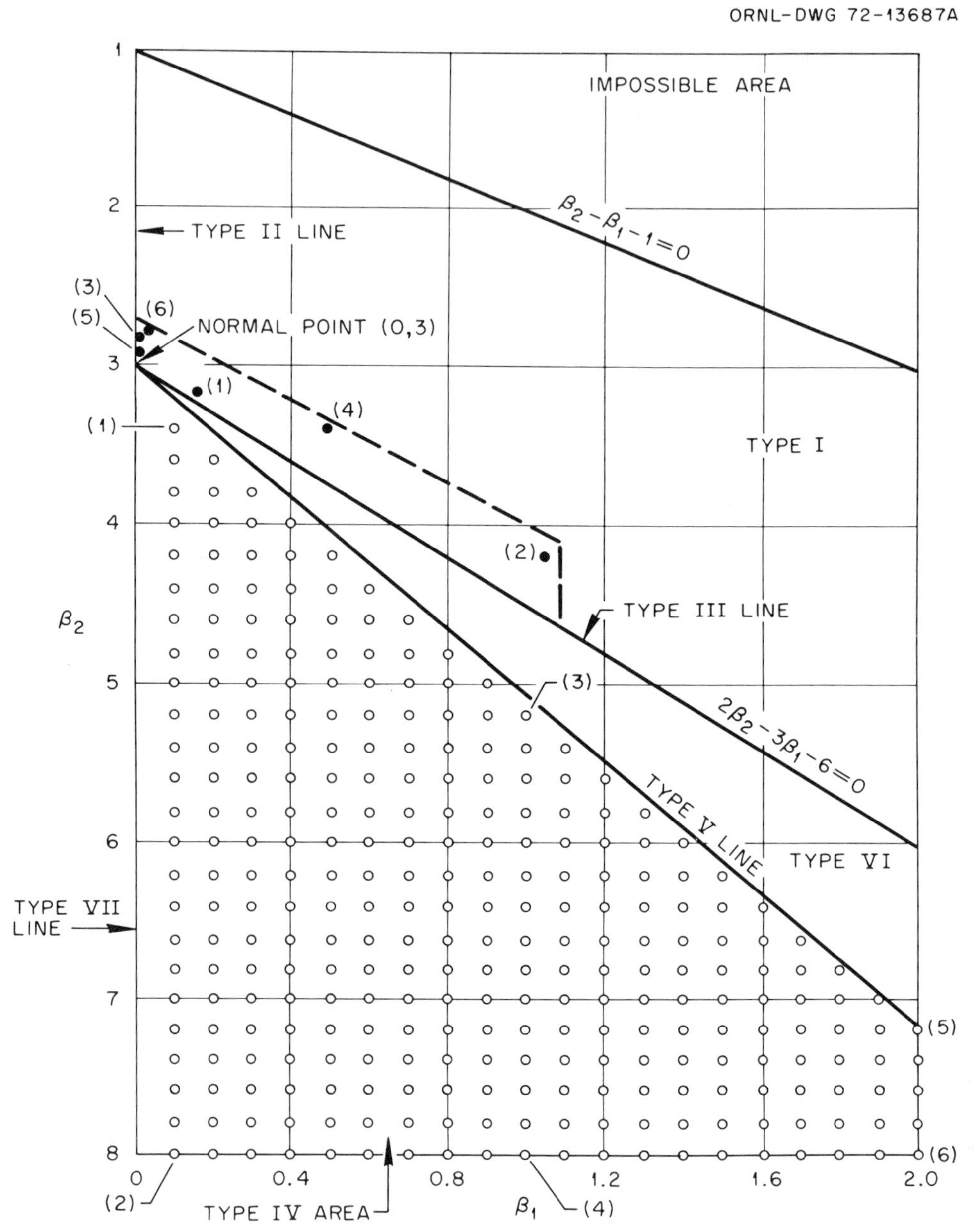

Figure B. TR_4 Transformation of 290 Pearson Type IV Points.

curve, with the corresponding values of (β_1, β_2), were computed for the original Type IV point and for the TR_4 transformed distribution. The results are given in Table 1 and show that the higher moments in the Type IV region often become infinite, as is shown graphically in Figure A. By contrast, the higher moments of the TR_4 transformation are nearly uniformly smaller (in absolute value) than their Type IV counterparts and they never become infinite. Thus, in terms of the higher-order moments, the TR_4 transformation takes one from ill-behaved points in the Type IV region to well-behaved points in the transformed distribution.

As an illustration, consider the Type IV case $\beta_1 = 1$, $\beta_2 = 8$, $\mu_2 = 1$ (Table 1, No. 4). The parameters are $r = 5.142857$, $\nu = -2.152079$, $a = 1.877634$, $F(r,\nu) = 1.520091$, and $m = (r+2)/2 = 3.571629$ and these define $g(y)$ given in (2.6).

2.3 Transformation Case 2: $\mu_3(x) > 0$

It is easily shown that Case 2 can be handled using the results of Case 1. The sign of ν is opposite to that of $\sqrt{\beta_1}(x)$ so Case 2 is the case for $\nu < 0$. The argument follows simply from the facts that (i) if U has a Type IV distribution, then $V = -U$ has a Type IV distribution and (ii) that $F(r,\nu) = F(r,-\nu)$. Thus the TR_4 transformation to use in Case 2, $\mu_3(x) > 0$, is

$$y = e^{\tan^{-1}(x/a)} \quad . \tag{2.8}$$

Using this transformation, the non-central moments of y become

$$\mu'_k(y) = \frac{F(r, \nu-k)}{F(r,\nu)} \quad , \text{ where } \nu < 0. \tag{2.9}$$

However, from the fact that $F(r,\nu) = F(r,-\nu)$ it can be seen that $\mu'_k(y)$ for $\nu < 0$, using the Case 2 transformation in (2.8), will be exactly the same as $\mu'_k(y)$ for $\nu > 0$ in Case 1, using the transformation in (2.4). Thus, attention can be restricted to either Case 1 or Case 2, and the other case is then easily derived from the one already done. Case 1 is used in the sequel.

3. APPLICATION OF THE TR_4 TRANSFORMATION IN A PERCENTAGE POINTS PROBLEM.

Dusenberry and Bowman (1973) computed percentage points

Table 1. Comparison of Moments[a] for Six Selected Type IV Points[b] and Their TR_4 Transforms

	Original Type IV	TR_4 Transformed Point		Original Type IV	TR_4 Transformed Point
(1)	$\beta_1 = 0.1$	$\beta_1 = 0.12714$	(4)	$\beta_1 = 1.0$	$\beta_1 = 0.46899$
	$\beta_2 = 3.4$	$\beta_2 = 3.12271$		$\beta_2 = 8.0$	$\beta_2 = 3.32856$
	-0.316228	0.356564		-1.0	0.684825
	3.4	3.12271		8.0	3.32856
	-3.75829	3.5354		-58.5	6.24967
	23.3744	17.9396		4377.5	21.9368
	-54.051	38.0479		Infinite	62.5248
	285.621	161.237		Infinite	217.289
	-1058.72	490.115		Infinite	742.563
	5749.81	2028.85		Infinite	2732.10
(2)	$\beta_1 = 0.1$	$\beta_1 = 1.03113$	(5)	$\beta_1 = 2.0$	$\beta_1 = 0.017303$
	$\beta_2 = 8.0$	$\beta_2 = 4.17731$		$\beta_2 = 7.4$	$\beta_2 = 2.826$
	-0.316228	1.01544		-1.41421	-0.131542
	8.0	4.17731		7.4	2.826
	-58.2346	10.9033		-37.4246	-1.17950
	Infinite	39.6849		292.368	12.7312
	Infinite	143.129		-3239.58	-10.5300
	Infinite	569.609		56151.3	77.7526
	Infinite	2363.80		-1804640.	-102.561
	Infinite	10298.0		209006000.	596.292
(3)	$\beta_1 = 1.0$	$\beta_1 = 0.00085$	(6)	$\beta_1 = 2.0$	$\beta_1 = 0.02326$
	$\beta_2 = 5.2$	$\beta_2 = 2.8335$		$\beta_2 = 8.0$	$\beta_2 = 2.74671$
	-1.0	0.0291521		-1.41421	0.152498
	5.2	2.8335		8.0	2.74671
	-17.7765	0.251576		-46.063	1.25853
	97.675	12.6838		462.286	11.7369
	-620.754	2.16735		-8389.12	10.1279
	5085.27	75.5761		498832.	66.3717
	-52748.6	20.3408		Infinite	87.1988
	713259.	551.943		Infinite	459.519

[a] Tabled values are central moments $\mu_3, \mu_4, \ldots, \mu_{10}$; $\mu_1 = 0$, $\mu_2 = 1$.

[b] Selected Points from Figure B. Parenthetic numbers correspond to those in Figure B.

of the moment estimator (ρ^*) for the shape parameter (ρ) of the gamma distribution with probability density function

$$\begin{cases} f(x) = e^{-x/a}(x/a)^{\rho-1}/a\Gamma(\rho) \quad x \geq 0,\ a, \rho \quad 0 \quad ; \\ \quad\;\; = 0,\ x < 0. \end{cases} \tag{3.1}$$

The moment estimator is computed as $\rho^* = m_1'^2/m_2$ and the $100\alpha\%$ point, τ, of ρ^* is defined by the probability statement $\Pr(\rho^* < \tau) = \alpha$. This probability statement is rearranged as (see David and Johnson (1951))

$$\Pr(\rho^* < \tau) = \Pr(y < 0) = \alpha \tag{3.2}$$

where $y = m_1'^2 - \tau m_2$ is referred to as the linearized statistic y. The exact sampling moments of the linearized statistic y were found, and then Pearson curves were fitted to the first four moments of y, for various values of a, ρ, τ, and n, to give an approximation to the sampling distribution of y. More than 10,000 combinations of a, ρ, τ, and n were studied, with the primary set of parameter combinations being

$$\begin{aligned} a &= 1; \\ \rho &= 0.5, 1.0, 1.5, 2.0, 2.5, 3.0, 4.0, 5.0, 10.0, \\ &\quad 15.0, 20.0; \\ n &= 50, 75, 100, 150, 200, 300, 500, 750, 1000. \end{aligned} \tag{3.3}$$

One hundred or more values of τ were considered for each (ρ,n) combination, where τ took on values in a range surrounding each value of ρ. The results showed that the points were best approximated by primarily Pearson Type IV curves and very occasionally by Pearson Type VI curves. It was also shown that known percentage points of Pearson Type IV curves in available tables could be reproduced quite accurately by transforming the Type IV curve according to (2.4) and using the TR_4 transformed moments in a Cornish-Fisher (C-F) series expansion. This points up the fact that the exact percentage points of the TR_4 transformed distribution $g(y)$ in (2.6) can be computed quite accurately by utilizing the moments of y in a C-F series expansion.

The sampling moments of the linearized statistic y were used as input to a Cornish-Fisher series expansion with normal kernel in computing percentage points for ρ^*, $\Pr(\rho^* < \tau) = \alpha$. Plots of τ

versus α, for the same values of a, ρ, and n as given in (3.3), produced graphs which represent approximations to the cumulative distribution function of ρ^*. The computation of these curves involved three separate steps or stages. Details and graphs can be obtained from the authors.

These graphs of the cumulative distribution function of ρ^* can be of use in problems concerning inferences about the shape parameter (ρ) of the gamma distribution, either for finding confidence intervals for ρ or for testing hypotheses concerning ρ as shown by the examples in Dusenberry and Bowman (1973).

ACKNOWLEDGMENT. The authors wish to thank Dr. L. R. Shenton for his many helpful comments.

REFERENCES.

[1] Bowman, K. O. and Shenton, L. R. (1973). Biometrika 60, 155-167.
[2] Bowman, K. O. (1973). Biometrika 60, 623-628.
[3] Box, G. E. P. and Cox, D. R. (1964). J. R. Statist. Soc. B 26, 211-252.
[4] Burr, I. W. (1942). Ann. Math. Statist. 13, 215-232.
[5] D'Agostino, R. (1970). Biometrika 57, 679-81.
[6] D'Agostino, R. and Pearson, E. S. (1973). Biometrika 60, 613-622.
[7] David, F. N. and Johnson, N. L. (1951). Biometrika 38, 43-57.
[8] Dusenberry, W. E. and Bowman, K. O. (1973). ORNL Report No. 4876, Oak Ridge National Laboratory, Oak Ridge, Tennessee.
[9] Elderton, W. P. and Johnson, N. L. (1969). Systems of Frequency Curves. Cambridge University Press, London.
[10] Johnson, N. L. and Kotz, S. (1970). Distributions in Statistics, Continuous Univariate Distributions - 1. Houghton Mifflin, Boston, and John Wiley.
[11] Pearson, E. S. (1963). Biometrika 50, 1 and 2, 95-112.
[12] Pearson, E. S. (1965). Biometrika 52, 282-285.
[13] Pearson, K. (1895). Phil. Trans. Roy. Soc. London Ser. A 186, 343-414.
[14] Shenton, L. R. (1965). Georgia Ag. Exp. Sta. Tech. Bull. N.S. 50, pp. 1-48.
[15] Shenton, L. R. and Bowman, K. O. (1975). Johnson's S_U and the skewness and kurtosis statistics. Jour. Amer. Statist. Assoc. (to appear).

DISTRIBUTIONS OF CHARACTERISTIC ROOTS OF RANDOM MATRICES

V. B. Waikar

Department of Mathematics and Statistics, Miami University, Oxford, Ohio, U.S.A.

SUMMARY. This is a review of the author's work in the area of distributions of characteristic roots of random matrices done in the last four to five years.

KEY WORDS. Multivariate distributions, characteristic roots, Wishart matrix, MANOVA matrix.

1. INTRODUCTION. The distribution problems asoociated with the characteristic roots of random matrices are of interest not only to statisticians but also to physicists. For example, the distributions of intermediate roots can be used for reduction of dimensionality in pattern recognition problems [see Cooper and Cooper (1964)] and principal component analysis [see Anderson (1958)]. In nuclear physics, the distributions of any few consecutive ordered roots are useful for finding the distributions of the spacings between the energy levels of certain complicated systems [see Carmeli (1973) and Mehta (1960)]. There are also practical applications in the areas of statistical signal analysis [see Cooper and Cooper (1964) and Liggett (1972)], target detection, identification procedure [Walter (1971)] and multiple time series [see Wahba (1968) and Hannan (1970)]. The author has made several contributions in this area in the last four to five years and has authored or co-authored several papers published in various journals. This paper is a brief review of this work.

2. REAL RANDOM MATRICES. Let $\ell_1<\ell_2<\ldots<\ell_p$ denote the characteristic roots of a real random matrix L and let their joint

G. P. Patil et al. (eds.), Statistical Distributions in Scientific Work, Vol. 1, 391-400.

probability density be given by $h_1(\ell_1,\ldots,\ell_p) =$

$$C \prod_{i=1}^{p} \psi(\ell_i) \prod_{i>j} (\ell_i-\ell_j) \qquad a<\ell_1<\ldots<\ell_p<b \tag{1}$$

where C is a known constant and $\psi(\cdot)$ is a certain function. The class of random matrices [defined by (1)] to which L belongs includes some important (central) random matrices like the Wishart matrix, MANOVA matrix, canonical correlation matrix, and also a random matrix which received special attention of the nuclear physicists.

Krishnaiah and Waikar (1971) derived an exact expression for the joint marginal density of any few consecutive ordered roots of L, i.e., of $\ell_{r+1}, \ell_{r+2}, \ldots, \ell_{r+s}$ $(0\leq r\leq p-1,\ 1\leq s\leq p-r)$. In the same paper, we also derived an expression for the cumulative distribution function of an intermediate root ℓ_r $(1\leq r<p)$. In deriving these densities and c.d.f. we used the following lemma. Let $\rho(\psi;q,\{k_1,\ldots,k_q\}, a,b)$

$$= \int_{a\leq x_1\leq\cdots\leq x_q\leq b} \cdots \int V(x_1,\ldots,x_q;\ k_1,\ldots,k_q) \prod_{i=1}^{q} \{\psi(x_i)dx_i\} \tag{2}$$

where V is the determinant $V(x_1,\ldots,x_q;\ k_1,\ldots,k_q) = |(x_j^{k_i})\ i,j=1,\ldots,q|$ and $k_1,\ldots,k_q$ is a set of non-negative integers. Further, let $F_s^t(a,b) = \int_a^b F_s(a,\theta)\theta^t\psi(\theta)d\theta$ where $F_s(a,\theta) = \int_a^\theta x^s\psi(x)dx$ and further let $f_s^{\,t}(a,b) = F_s^{\,t}(a,b) - F_t^{\,s}(a,b)$. In addition, let $\Delta(\psi;2m,\{k_1,\ldots,k_{2m}\},a,b) = |(f_{k_i}^{k_j}(a,b))\ i,j=1,\ldots,2m|^{1/2}$ and $G_t(\psi;2m+1,\{k_1,\ldots,k_{2m+1}\},a,b) = |(f_{k_i}^{k_j}(a,b))\ i,j=1,\ldots,t-1,t+1,\ldots,2m+1|^{1/2}$ with the understanding that $G_1(\psi;1,k_1,a,b) = 1$.

<u>Lemma 1.</u> Let $\psi(\cdot)$ be a function such that the integral given in (2) exists and let a<b be real constants. Then $\rho(\psi;q,\{k_1,\ldots,k_q\},a,b) = \Delta(\psi;2m,\{k_1,\ldots,k_{2m}\},a,b)$ when q = 2m and $\rho(\psi;q,\{k_1,\ldots,k_q\},a,b)$ (3)

$$= \sum_{i=1}^{2m+1} (-1)^{i+1} F_{k_i}(a,b) G_i(\ ;2m+1, k_1,\ldots,k_{2m+1} ,a,b) \qquad (4)$$

when $q = 2m+1$. Mehta (1960) proved equation (3) in the special case when $K_i=i-1 (i=1,\ldots,q)$ and $(x)=x^r\exp(-x^n)$. Using the method in Mehta (1960), Krishnaiah and Chang (1971) proved the above lemma for the case when $k = i-1$ $(i=1,\ldots,q)$. Our proof of this lemma follows on the same lines.

Using the above lemma we obtained the joint marginal density of $\ell_{r+1}, \ell_{r+2},\ldots,\ell_{r+s}$ as $h_2(\ell_{r+1},\ldots,\ell_{r+s})$

$$= C \Sigma_1 \Sigma_2 (-1)^{\gamma(r,s)+\Sigma_1^r k_i+\Sigma_1^s \alpha_i} \prod_{i=r+1}^{r+s} \psi(\ _i)$$

$$x V(\ell_{r+1},\ldots,\ell_{r+s};\alpha_1,\ldots,\alpha_s) \ x \ \rho(\psi;r,\{k_1,\ldots,k_r\},a,\ell_{r+1})$$

$$x\rho(\psi;p-r-s,\{\beta_1,\ldots,\beta_{p-r-s}\},\ell_{r+s},b) \quad a<\ell_{r+1}<\cdots<\ell_{r+s}<b \qquad (5)$$

where $k_1<\cdots<k_r$ is a subset of the set of integers $\{0,1,\ldots,p-1\}$, $t_1<\cdots<t_{p-r}$ is the subset complementary to $k_1<\cdots<k_r$ and Σ_1 denotes summation over all $\binom{p}{r}$ possible choices of $k_1<\cdots<k_r$. Further, $\alpha_1<\cdots<\alpha_s$ is a subset of the set $\{t_1,\ldots t_{p-r}\}$, $\beta_1<\cdots<\beta_{p-r-s}$ is the subset complementary to $\alpha_1<\cdots<\alpha_s$ and Σ_2 denotes the summation over all $\binom{p-r}{s}$ possible choices for $\alpha_1,\ldots,\alpha_s$. Finally, $\gamma(r,s) = (1/2)[r(r+3)+s(s+3)]$ and the ρ functions are given by equations (3) or (4) according as the second argument is even or odd.

When r=o or s=p-r, the expression in (5) is simplified considerably.

The c.d.f. of an intermediate root ℓ_r $(1\leq r<p)$ was obtained in the form of a recurrence formula as $P[\ell_r<x] = P[\ell_{r+1}<x]$ $+ C \Sigma_1 \pm \rho(\psi;r,\{k_1,\ldots,k_r\},a,x) \cdot \rho(\psi;p-r,\{t_1,\ldots,t_{p-r}\},x,b)$ (6)

where ρ's are given by (3) or (4), Σ_1, k_i's, t_i's are defined as before and the sign after Σ_1 is positive or negative according as $(1/2)r(r+3)+\Sigma k_i$ is even or odd. Now, Krishnaiah and Chang

(1971) gave the c.d.f. of the largest root ℓ_p as $P[\ell_p<x] =$
$C\,\rho(\psi;p,\{0,1,\ldots,p-1\},a,x)$ (7)
and thus formula (6) along with (7) can be considered as the c.d.f. of ℓ_r.

All the above expressions, i.e., (5) and (6), can be computed since they involve only finite linear combinations of products of double integrals. In fact, we used formula (6) (with appropriate C, ψ, a and b) to compute the tables of upper five percent and one percent points of the intermediate root ℓ_r $(r=2,3,\ldots,p-1)$ of the MANOVA matrix for p=4,5,6,7,8. These are given in Krishnaiah, Schuurmann and Waikar (1973). Further, Schuurmann and Waikar (1974) constructed the tables of the upper 100α percentage points of the smallest root ℓ_1 of the MANOVA matrix for α=0.1, 0.05, 0.025, 0.01 and p=6,7,8,9,10 using the exact expression for the c.d.f. of ℓ_1 given by Krishnaiah and Chang (1971). Clemm, Chattopadhyay and Krishnaiah (1973) used formula (6) to construct the upper percentage points of the individual roots of the Wishart matrix for p=2(1)10.

For testing hypothesis that the covariance matrix of a p-variate normal distribution is equal to a specified matrix, Roy (1957) proposed a test with acceptance region $U_1 \leq \ell_1 < \ell_p \leq U_2$ where ℓ_1 and ℓ_p are the extreme roots of the sample covariance matrix which has Wishart distribution under null hypothesis. One simple choice of U_1 and U_2 is $U_1 = 1/U$ and $U_2 = U$. Clemm, Krishnaiah and Waikar (1973) constructed tables of U where U satisfies

$$P[1/U \leq \ell_1 < \ell_p \leq U] = 1-2\alpha \tag{8}$$

for p=2(1)10(2)20 and α=.05, .025, .01, .005. These tables were constructed using the expression for the left hand side of (8) given by Krishnaiah and Chang (1971). Also using their expression for the left hand side of (8) when ℓ_1 and ℓ_p are the extreme roots of a MANOVA matrix Schuurmann, Waikar and Krishnaiah (1973) constructed tables of U satisfying (8) for p=2(1)10 and α= .05, .025, .01, .005. These tables can be used to apply Roy's two-sided test [Roy (1957)] for testing the hypothesis of equality of covariance matrices of two p-variate normal populations.

Waikar and Schuurmann (1973) obtained exact expression for the joint density of the largest and smallest characteristic roots of a Wishart as well as MANOVA matrix. Here, we will give the results for Wishart case: Let ℓ_1 and ℓ_p denote the smallest and largest roots respectively of a pxp Wishart matrix with n degrees of freedom and covariance matrix I. We obtained the joint density of ℓ_1 and ℓ_p as $h_3(\ell_1,\ell_p) =$

$$K(p,n)\ell_p^{(1/2)(p-1)(n-p-1)}(\ell_p-\ell_1)^{(1/2)p(p+1)-2}$$

$$\times \exp[-(1/2)\{(p-1)\ell_p+\ell_1\}]\ell_1^{(n-p-1)/2} \times\rho(\psi;p-2,\{0,1,\ldots,p-3\},0,1) \qquad 0<\ell_1<\ell_p<\infty \tag{9}$$

where $K(p,n)=\Pi^{p/2}/[2^{np/2}\prod_{i=1}^{p}\{\Gamma((n+1-i)/2)\Gamma((p+1-i)/2)\}]$,

$\psi(x) = x(1-x)(1-x+x(\ell_1/\ell_p))^{(1/2)(n-p-1)} \exp[x(\ell_p-\ell_1)/2]$ and ρ is

given by (3) or (4) according as p-2 is even or odd (p being greater than 2). Using this density (9) we also computed upper 100α percent points of the distribution of $f_p=1-\ell_1/\ell_p$ for $p=3,4,5,6,7,8$; $\alpha = .99, .975, .95, .90, .75, .50, .25, .10, .05, .025, .01$ and for certain values of n. These tables are given in the same paper and can be used in testing the sphericity hypothesis $\Sigma=I$.

3. REAL NONCENTRAL MATRICES. Though a lot of work has been done in this area as far as derivations of distributions are concerned not much is done regarding computation of powers of various tests. Our little contribution is given in Schuurmann and Waikar (1973) where we computed tables of power function of Roy's two-sided test for testing $\Sigma=I$ in the bivariate case. We summarize the results below.

Let $H_o:\Sigma=\Sigma_o$, $H_a:\Sigma \neq \Sigma_o$ where Σ is the covariance matrix of a p-variate normal distribution with mean vector 0 and Σ_o known. Further let $n^{-1}S$ be the sample covariance matrix based on a sample of size n+1. According to the procedure of Roy (1957) an acceptance region for testing H_o against H_a is given by $L\leq\ell_1<\ell_p\leq U$ where $\ell_1<\cdots<\ell_p$ are the characteristic roots of $S\Sigma_o^{-1}$ and $L<U$ are constants such that $P[L\leq\ell_1<\ell_p\leq U|H_o]= 1-2\alpha$. Note that under H_o, $S\Sigma_o^{-1}$ has a central Wishart distribution $W(I,n)$. Since the optimal choice of L and U (in the sense of maximizing power) was not known, Clemm, Krishnaiah and Waikar (1973) put $L=1/U$ and computed values of U as mentioned in the previous section. Hanumara and Thompson (1968) constructed tables for L_1 and U_1 for $\alpha= .05, .025, .01, .005$ and $p=2(1)10$ where L_1 and U_1 satisfy.

$$P[L_1 \leq \ell_1 < \ell_p \leq U_1] = 1-2\alpha \text{ and } P[\ell_1 \leq L_1] = 1-\alpha .$$

In this paper, for the bivariate case, i.e., when p=2, we computed the power for the above two-tailed test of Roy when $H_o:\Sigma=I$ and $H_a:\Sigma=\Sigma_\rho = \begin{bmatrix} 1 & \rho \\ \rho & 1 \end{bmatrix}$, $\rho>o$ using first the percentage points L and U given in Clemm, Krishnaiah and Waikar (1973) and then using the percentage points L_1 and U_1 given in Hanumara and Thompson (1968). The power has been computed for α=.05, .025, .01, .005; n= 3(1)20(3)30(5)50 and ρ= 0.1(.1)0.9. Since it was found that in most of the cases the power corresponding to L_1 and U_1 was significantly more than the power corresponding to L and U, in this paper, we gave only the tables of power based on L_1 and U_1.

In the previous section, we discussed the joint density of the extreme roots of Wishart and MANOVA matrices for the central case. In the noncentral case, Waikar (1973) obtained the joint density of the largest and smallest characteristic roots of the Wishart matrix with covariance Σ and of matrix $S_1S_2^{-1}$ where S_1 and S_2 are independently distributed Wishart matrices with covariance matrices Σ_1 and Σ_2 respectively. We will not give the expressions for the densities here but would point out that these expressions are in terms of infinite series involving zonal polynomials and hence are very difficult to use to compute power. However, there do not seem to be better expressions available in the literature.

4. COMPLEX RANDOM MATRICES. Though some distributions connected with characteristic roots of complex random matrices can be derived easily following the real case, the applications justify treating the complex case as a separate area. For the applications (for example, in multiple time series [see Wahba (1971)] it is necessary to compute percentage points of the appropriate distributions as well as power functions. Schuurmann and Waikar (1974a) computed the upper percentage points of the individual roots of the complex Wishart matrix: Let X be a complex valued random matrix of dimension p×n (n≥p) such that the columns of X are independent and have the p-variate complex normal distribution with zero mean vector and covariance matrix I_p. The distribution of the p×p matrix $X\bar{X}'$ is then complex Wishart with n degrees of freedom. Let $\theta_1<\theta_2<\cdots<\theta_p$ denote the characteristic roots of $X\bar{X}'$. In this paper we give tables of upper 100α percent points of the distribution of the individual root θ_s for s=1,2,...,p-1; p=3,4,5; α= .1, .05, .025, .01 and for several values of n.

While the statisticians have done a lot of work on the distributions problems connected with ordered roots of random matrices, Wigner (1965) and other physicists have done a considerable amount of work on the distribution problems connected with the unordered roots of certain random matrices since they are useful in studying the energy levels of certain systems in physics [see Carmeli (1973) and Wigner (1957)]. Joint density of any few unordered roots of a class of central complex random matrices is known [see Wigner (1965), Mehta (1967)]. This class includes complex analogues of the central Wishart matrix, MANOVA matrix, canonical correlation matrix and a matrix A (of interest to physicists). Waikar, Chang and Krishnaiah (1972) derived expressions for the joint densities of any few unordered roots of the above four matrices in the noncentral case. It might be interesting to discuss the A matrix which has received attention of the physicists.

Let $A = (a_{jk})$ be a p×p Hermitian random matrix such that $E(A) = M \neq 0$. Further, let $a_{jk} = r_{jk} + is_{jk}$ where r_{jk} is distributed normally with mean μ_{jk} and variance 1 for $j<k$, r_{jj} is normal with mean μ_{jj} and variance 2. Further, s_{jk} is normal with mean ν_{jk}, and variance 1 for $j<k$ and also the r_{jk}'s and s_{jk}'s are independent. Starting from the joint density of the r_{jk}'s and s_{jk}'s and making suitable transformations, we obtained the density of any r unordered roots of the random matrix A. This density is given in terms of infinite sum of zonal polynomials and multivariate complex gamma functions.

5. BASIS FOR SOME DISTRIBUTIONS. In this section, we will review some simultaneous test procedures for testing the equality of characteristic roots of (parameter) matrices against restricted alternatives when the characteristic roots of the associated random matrices have certain distributions. These were proposed by Krishnaiah and Waikar (1971a, 1972) and the random matrices considered were (i) Wishart matrix, (ii) MANOVA matrix, (iii) matrix of canonical correlations and (iv) $S_1 S_2^{-1}$ where S_1 and S_2 are independently distributed central Wishart matrices. In these papers we also derived the distributions of the associated test statistics involving characteristic roots under the null as well as under the alternate hypotheses.

Let Λ be a p×p positive definite symmetric unknown matrix and let $\lambda_1 \geq \lambda_2 \geq \cdots \geq \lambda_p > o$ be the characteristic roots of Λ. Also let the random matrix L be an estimate of Λ and let the characteristic roots of L be given by $\ell_1 > \ell_2 > \cdots > \ell_p > o$. Also let $f_{ij} = \ell_i / \ell_j$.

Now, let H: $\lambda_1=\lambda_2=\cdots=\lambda_p$, $H_{ij}:\lambda_i=\lambda_j$, $A_{ij}:\lambda_i>\lambda_j$ and $A=\bigcup_{i=1}^{p=1} A_{i\,i+1}$. Note that $H=\bigcap_{i=1}^{p-1} H_{i\,i+1}$. Consider testing H against the alternative A. If H is rejected, it is of interest to test various component hypotheses of H also. Thus, a test procedure for testing H_{12}, H_{23}, H_{34},...,H_{p-1p} and H simultaneously against the alternatives A_{12}, A_{23}, A_{34},...,A_{p-1p} and A respectively is given by: Accept or reject $H_{i\,i+1}$ according as $f_{i\,i+1} \lessgtr C_{i\alpha}$ i=1,...,p-1 where the critical constants $C_{i\alpha}$ are such that

$$P[1<f_{i\,i+1}<C_{i\alpha};\ i=1,\ldots,p-1 \mid H] = (1-\alpha). \tag{10}$$

The total hypothesis H is accepted if and only if all the component hypotheses $H_{i\,i+1}$ i=1,...,p-1 are accepted. The power of this test is given by $P_1 = 1-P[1<f_{i\,i+1}<C_{i\alpha}\ i=1,\ldots,p-1|A]$. (11)

The optimum choices (in the sense of maximizing power) of the critical constants $C_{i\alpha}$ i=1,...,p-1 are not known. For practical purposes we choose $C_{i\alpha} = C_\alpha$ i=1,...,p-1, i.e., all equal. It is obvious that we need to know the distribution of $f_{i\,i+1}$ i=1,..., p-1 under H to carry out the test (null distribution) and under A (noncentral) to compute the power given by (11). In the above papers, we have derived both of these distributions for the four cases (matrices) mentioned above. Most of the noncentral densities involve zonal polynomials and so computing power is not easy.

We have also proposed simultaneous tests for equality of characteristic roots against several other sets of alternatives and also derived the distributions of the test statistics in the above two papers. Recently Krishnaiah and Schuurmann (1974) computed some tables which are useful in carrying out some simultaneous tests discussed in this section.

ACKNOWLEDGMENTS. The author wishes to acknowledge the support from Miami University in the form of a Summer Research Fellowship for 1974. The author wishes to express his thanks to the referees for suggestions which were helpful in revising the paper.

REFERENCES

[1] Anderson, T. W. (1968). An Introduction to Multivariate Statistical Analysis, John Wiley and Sons, Inc., New York.

[2] Carmeli, M. (1973). Statistical theory of energy levels and random matrices in physics. Aerospace Res. Lab. Tech. Rept. No. 73-0044.
[3] Clemm, D. S., Chattopadhyay, A. K. and Krishnaiah, P. R. (1973). Sankhyā Ser B 35, 325-338.
[4] Clemm, D. S., Krishnaiah, P. R. and Waikar, V. B. (1971). J. Statist. Simul. 2, 65-92.
[5] Cooper, D. B. and Cooper, P. W. (1964). Information and Control 7, 416-444.
[6] Hannan, E. J. (1970). Multiple Time Series. John Wiley and Sons, Inc., New York.
[7] Hanumara, R. C. and Thompson, W. A. (1968). Biometrika 55, 505-512.
[8] Krishnaiah, P. R. and Chang, T. C. (1971). J. Multivariate Anal. 1, 108-117.
[9] Krishnaiah, P. R. and Schuurmann, F. J. (1974). On the evaluation of some distributions that arise in simultaneous tests for the equality of the latent roots of the covariance matrix. J. Multivariate Anal. (to appear).
[10] Krishnaiah, P. R. and Waikar, V. B. (1971). J. Multivariate Anal. 1, 308-315.
[11] Krishnaiah, P. R. and Waikar, V. B. (1971a). Ann. Inst. Statist. Math. 23, 451-468.
[12] Krishnaiah, P. R. and Waikar, V. B. (1972). Ann. Inst. Statist. Math. 21, 81-85.
[13] Krishnaiah, P. R., Schuurmann, F. J. and Waikar, V. B. (1973). Upper percentage points of the intermediate roots of the MANOVA matrix. Sankhyā Ser B (to appear).
[14] Liggett, W. S. (1972). Passive sonar: fitting models to multiple time series. Paper presented at the NATO Advanced Study Institute on Signal Processing held at the University of Technology, Loughborough, U.K., August 21-September 1, 1972.
[15] Mehta, M. L. (1960). Nucl. Phys. 18, 395-419.
[16] Mehta, M. L. (1967). Random Matrices. Academic Press, New York.
[17] Roy, S. N. (1957). Some Aspects of Multivariate Analysis. John Wiley and Sons, Inc., New York.
[18] Schuurmann, F. J. and Waikar, V. B. (1973). Comm. Statist. 1, 271-280.
[19] Schuurmann, F. J. and Waikar, V. B. (1974). Upper percentage points of the smallest root of the MANOVA matrix. Ann. Inst. Statist. Math. (to appear).
[20] Schuurmann, F. J. and Waikar, V. B. (1974a). Upper percentage points of individual roots of the complex Wishart matrix. Sankhyā Ser B (to appear).
[21] Schuurmann, F. J., Waikar, V. B. and Krishnaiah, P. R. (1973). J. Statist. Comp. Simul. 2, 17-38.
[22] Wahba, G. (1968). Ann. Math. Statist. 39, 1849-1862.
[23] Wahba, G. (1971). J. Roy. Statist. Soc. Ser B 33, 153-166.

[24] Waikar, V. B. (1973). S. Afr. Statist. J., 103-108.
[25] Waikar, V. B. and Schuurmann, F. J. (1973). Utilitas Mathematica 4, 253-260.
[26] Waikar, V. B., Chang, T. C. and Krishnaiah, P. R. (1972). Aust. J. Statist. 14, 84-88.
[27] Walter, C. M. (1971). Personal communication.
[28] Wigner, E. P. (1957). SIAM Review 9, 1-23.
[29] Wigner, E. P. (1965). In Statistical Theories of Spectra: Fluctuations, C. E. Porter (ed.). Academic Press, New York, 446-461.

ON THE ARITHMETIC MEANS AND VARIANCES OF PRODUCTS AND RATIOS OF RANDOM VARIABLES

Fred Frishman

George Washington University and U.S. Internal Revenue Service, Washington, D. C., U.S.A.

SUMMARY. Frequently occuring functions of random variables, that arise in the area of applied statistics, are the Product and Ratio of pairs of not necessarily independent variates. As is the case in much statistical work, in practice, attempts to understand the underlying processes usually begin with the consideration of the mean and variance. Exact expressions are given for the mean and variance of a ratio of random variables under mildly restrictive conditions. In addition approximation results are obtained which, it is argued, may be of practical use in many of the cases, including normality, where the restrictions are violated.

KEY WORDS. Ratios of random variables, products of random variables, expectations of ratios and products, variances of ratios and products.

1. INTRODUCTION. In what follows, closed form expressions are given for the mean and variance of a ratio of suitably restricted random variables. The moments are, of course, assumed to exist and certain correlation coefficients must be assumed known in order to make use of the results obtained. Included in the class of distribution for which the restrictions are not satisfied are the normal distributions and those distributions which assign positive probability to zero.

Series expressions are derived for the mean and variance of a ratio under less demanding assumptions on knowledge of the correlation coefficients but with an assumption on the support of the denominator random variable and an assumption that the mean

G. P. Patil et al. (eds.), Statistical Distributions in Scientific Work, Vol. 1, 401-406.

of the numerator random variable not be zero. Approximations obtainable by truncating the series are also discussed.

2. ON THE PRODUCT OF TWO RANDOM VARIABLES. The mean and variance of a product of random variables, as a function of the moments of the original variates, are well known [see e.g., Goodman (1960)]. Elementary calculations show that

$$E(XY) = E(X)E(Y) + \mathrm{Cov}(X,Y), \tag{2.1}$$

and that

$$V(XY) = \mathrm{Cov}(X^2,Y^2) - [\mathrm{Cov}(X,Y)]^2 - 2\,\mathrm{Cov}\,(X,Y)E(X)E(Y)$$

$$+ V(X)\,V(Y) + V(X)\,[E(Y)]^2 + V(Y)[E(X)]^2, \tag{2.2}$$

a result which appears, for example, in Frishman (1971), Goodman (1960), and Koop (1964).

If X and Y are independent or uncorrelated, we have

$$E(XY) = E(X)\,E(Y), \tag{2.3}$$

and if X and Y are independent,

$$V(XY) = V(X)V(Y)+V(X)\,[E(Y)]^2+V(Y)[E(X)]^2. \tag{2.4}$$

Further, if and only if $E(X) = E(Y) = 0$,

$$V(XY) = V(X)\,V(Y). \tag{2.5}$$

3. ONE SET OF RESULTS FOR THE MEAN AND VARIANCE OF A RATIO OF RANDOM VARIABLES. From (2.1),

$$E(\tfrac{Y}{X}) = E(Y)\,E(\tfrac{1}{X}) + \mathrm{Cov}(Y,1/X). \tag{3.1}$$

It is easy to see that $E[X^{-r}]$ exists provided that X has a density function with a zero of order at least r at the origin. For example, $E[X^{-r}]$ does not exist for the Normal Distribution.

Various results for $V(\frac{Y}{X})$ can be written down from (2.2), (2.4) and (2.5) by replacing X by 1/X.

If Y and X are uncorrelated Normals, $E(\frac{Y}{X})$ and hence $V(\frac{Y}{X})$ do not exist.

Replacing Y by Y/X in (2.1) and (2.4) and rearranging terms, we can also obtain

$$E\left(\frac{Y}{X}\right) = \frac{E(Y)}{E(X)} - \frac{\text{Cov}\left(X,\frac{Y}{X}\right)}{E(X)}, \tag{3.2}$$

provided that $E(X) \neq 0$. [See Hartley and Ross (1954) and Patil (1961)], and

$$V\left(\frac{Y}{X}\right) = \left\{ V(Y) - \text{Cov}\left(X^2, \frac{Y^2}{X^2}\right) - \left[\text{Cov}\left(X, \frac{Y}{X}\right)\right]^2 + \right. \tag{3.3}$$

$$\left. 2E(Y)\,\text{Cov}\left(X,\frac{Y}{X}\right) - \frac{V(X)}{[E(X)]^2}\left[E(Y)-\text{Cov}\left(X,\frac{Y}{X}\right)\right]^2 \right\} \frac{1}{V(X)+[E(X)]^2}.$$

If X and $\frac{Y}{X}$ are uncorrelated,

$$E\left(\frac{Y}{X}\right) = \frac{E(Y)}{E(X)} \text{ provided that } E(X) \neq 0, \tag{3.4}$$

and

$$V\left(\frac{Y}{X}\right) = \frac{V(Y)\,[E(X)]^2 - V(X)\,[E(Y)]^2}{[E(X)]^2\,\{V(X) + [E(X)]^2\}}, \tag{3.5}$$

provided that $E(X) \neq 0$.

The results in this section are highly dependent, for practical purposes, on information about relationships between the numerator random variable and the ratio of the random variables.

4. A PROCEDURE FOR DETERMINING $E\left(\frac{Y}{X}\right)$.

Rao (1952) et al determine the Expected Value of the ratio of normally distributed variates in essentially the following fashion:

Let X and Y be bivariate normal $(\mu_x,\mu_y,\sigma_x^2,\sigma_y^2,\rho)$; $-\infty<x,\ y<+\infty$.

Now, let $U = \frac{Y-\mu_y}{\sigma_y}$ and $Z = \frac{X-\mu_x}{\sigma_x}$, so that $Y=\mu_y+U\sigma_y$ and $X = \mu_x + Z\sigma_x$. From the definitions of U and Z, we have,

$E(U) = E(Z) = 0$ and $\sigma_u^2 = \sigma_z^2 = 1$.

Then,

$$\frac{Y}{X} = \frac{\mu_y}{\mu_x}\left(1 + \frac{U\sigma_y}{\mu_y}\right)\left(1 + \frac{Z\sigma_x}{\mu_x}\right)^{-1} . \quad (4.1)$$

Rao expanded the last term, assuming its validity, which can be shown to require $0 < X < 2\mu_x$ for $\mu_x > 0$, and that $2\mu_x < X < 0$ for $\mu_x < 0$. But since $-\infty < X < +\infty$ for the normal distribution, the requirement is not satisfied. The procedure requires that $\mu_x, \mu_y \neq 0$, also.

Examples for which the procedure is valid are appropriately parameterized uniform or binomial random variables. It probably is appropriate for cases where we say that the normal distribution satisfies the data, since we probably mean that a truncated Normal satisfies the data, it being rare for an item or test measurement to take on both positive and negative values. By slightly rewriting (4.1), performing the expansion, and taking expected values, we obtain

$$E\left(\frac{Y}{X}\right) = \frac{\mu_y}{\mu_x}\left\{1 + \frac{\sigma_X^2 E(Z^2)}{\mu_x^2} - \frac{\sigma_x^3 E(Z^3)}{\mu_x^3} + \cdots + \frac{(-1)^m \sigma_x^m E(Z^m)}{\mu^m} + \cdots\right\} - \frac{\sigma_y}{\mu_x}\left\{\frac{\sigma_x E(ZU)}{\mu_x} - \frac{\sigma_x^2 E(Z^2U)}{\mu_x^2} + \cdots + \frac{(-1)^{n-1}\sigma_x^n E(Z^nU)}{\mu_x^n} + \cdots\right\} . \quad (4.2)$$

$$= \frac{\mu_y}{\mu_x}\left\{1 + \sum_{i=2}^{\infty} (-1)^i \frac{\sigma_x^i E(Z^i)}{\mu_x^i}\right\} + \sigma_y \sum_{j=1}^{\infty} \frac{(-1)^j \sigma_x^j E(Z^j U)}{\mu_x^{j+1}} .$$

If, in addition to the prior conditions on the range of X and that $\mu_x \neq 0$, we specify that Z, U are bivariate central symmetrically distributed, then from [Frishman (1971), p. 23] we have that

$$E(Z^{2i+1}) = 0, \; i = 0,1,\ldots$$

and

$$E(Z^{i}U) = 0 \text{ for } i \text{ an even integer.}$$

Consequently,

$$E\left[\frac{Y}{X}\right] = \frac{\mu_y}{\mu_x}\left\{1 + \sum_{j=1}^{\infty} \frac{\sigma_x^{2j}\, E(Z^{2j})}{\mu_x^{2j}}\right\} - \sigma_y\left\{\sum_{j=1}^{\infty} \frac{{}_x^{2j-1}E(Z^{2j-1}U)}{\mu_x^{2j}}\right\}. \tag{4.3}$$

If, in addition, we add the condition of independence between Z and U or X and Y, then (4.3) reduces to

$$E\left[\frac{Y}{X}\right] = \frac{\mu_y}{\mu_x}\left\{1 + \sum_{i=1}^{\infty} \sigma_x^{2i} \frac{E(Z^{2i})}{\mu_x^{2i}}\right\} . \tag{4.4}$$

We note that (4.2), (4.3) and (4.4) presented above could have been obtained by use of a Taylor expansion of $\frac{Y}{X}$, the classical propagation of errors procedure.

We conclude this section by noting that discussions of the material presented above and in section 5 appear in various papers listed in the references.

5. THE VARIANCE OF $\frac{Y}{X}$, FOLLOWING FROM RAO'S PROCEDURE. Utilizing (4.2) and after extensive algebraic simplification and approximation, we arrive at

$$V\left(\frac{Y}{X}\right) \simeq \frac{\mu_y^2\sigma_x^2 + \mu_x^2\sigma_y^2 - 2\,\rho\mu_x\mu_y\sigma_x\sigma_y}{\mu_x^4} . \tag{5.1}$$

If we specify that X and Y (and consequently U and Z) are independent, then (5.1) reduces to

$$V\left(\frac{Y}{X}\right) \simeq \frac{\mu_y^2\sigma_x^2 + \mu_x^2\sigma_y^2}{\mu_x^4} . \tag{5.2}$$

We note that (5.1) and (5.2) provide reasonably accurate approximations as is evidenced by their common usage in sampling theory texts, for example, Hansen, Hurwitz and Madow (1953).

REFERENCES.

[1] Frishman. F. (1971). On Ratios of Random Variables and an Application to a Non-linear Regression Problem. Ph.D. dissertation, The George Washington University.
[2] Goodman, L. (1960). J. Amer. Statist. Assoc. 55, 708-713.
[3] Hansen, M. H., Hurwitz, W. N. and Madow, W. G. (1953). Sample Survey Methods and Theory. Vols. I & II. John Wiley and Sons, Inc.
[4] Hartley, H. O. and Ross, A. (1954). Nature 174, 270-271.
[5] Koop, J. C. (1973). J. Amer. Statist. Assoc. 68, 407.
[6] Koop, J. C. (1964). J. of the Roy. Statist. Soc., Part B, 26, 484-486.
[7] Patil, G. P. (1961). Sankhyā Ser A 23, 269-280.
[8] Pearson, K. (1910). Biometrika 7, 531-541.
[9] Rao, C. R. (1952). Advanced Statistical Methods in Biometrics Research. John Wiley and Sons, Inc.
[10] Sukhatme, B. V. and David, I. P. (1973). J. of the Amer. Statist. Assoc. 68, 405-406, 407-408.

EXACT AND APPROXIMATE SAMPLING DISTRIBUTION OF THE F-STATISTIC UNDER THE RANDOMIZATION PROCEDURE*

Junjiro Ogawa

The University of Calgary, Calgary, Alberta, Canada

SUMMARY. The main purpose of this paper is the justification of the traditional conclusion by sharpening the definitions of a proper blocking, randomization and the discussion of the concept of the asymptotic equivalence of probability distributions.

KEY WORDS. Randomization, blocking, F-statistic, asymptotic distribution, asymptotic equivalence.

1. INTRODUCTION. For the sake of the simplicity of the description of the problem, we consider a complete block design of b blocks of size k each and k treatments or varieties to be compared by experimentation. Suppose that the whole experimental units are labeled from 1 through $n = kb$ in the lexicographic order, in other words, the i-th unit in the j-th block bears the number $f = (j - 1)k + i$. The incidence vector of the α-th treatment $\underline{\zeta}_\alpha$ is defined as follows

$$\underline{\zeta}'_\alpha = (\zeta_{\alpha 1}\ \zeta_{\alpha 2}\ \cdots\ \zeta_{\alpha n}),$$

$$\text{where } \zeta_{\alpha f} = \begin{cases} 1, & \text{if the } \alpha\text{-th treatment is allocated to the } f\text{-th unit,} \\ 0, & \text{otherwise.} \end{cases}$$

The incidence vector of the j-th block $\underline{\eta}_j$ is defined as follows

$$\underline{\eta}'_j = (\eta_{j1}\ \eta_{j2}\ \cdots\ \eta_{jn}),$$

$$\text{where } \eta_{jf} = \begin{cases} 1, & \text{if the } f\text{-th unit belongs to the } j\text{-th block,} \\ 0, & \text{otherwise.} \end{cases}$$

*This research was supported by NRC Grant No. A7683.

G. P. Patil et al. (eds.), Statistical Distributions in Scientific Work, Vol. 1, 407-418.

As a matter of fact, in the lexicographic labeling, it turns out to be

$$\underline{\eta}'_j = (\overbrace{0 \ldots 0}^{k} \ldots\ldots \overbrace{0 \ldots 0}^{k}\ \overbrace{1 \ldots 1}^{k}\ \overbrace{0 \ldots 0}^{k} \ldots\ldots \overbrace{0 \ldots 0}^{k}).$$

$$\quad\; 1 \quad \ldots\ldots \quad j-1 \quad\; j \quad\; j+1 \quad \ldots\ldots \quad b$$

The observation obtained at the f-th unit is denoted by x_f which is assumed to have the structure

$$x_f = \gamma + \sum_{\alpha=1}^{k} \zeta_{\alpha f}\tau_\alpha + \sum_{j=1}^{b} \eta_{jf}\beta_j + \pi_f + e_f, \tag{1.1}$$

where γ is the general mean, $\tau_1,\ldots,\tau_k$ and $\beta_1,\ldots,\beta_b$ are treatment-effects and block-effects being subjected to the restrictions

$$\tau_1 + \ldots + \tau_k = 0 \quad\text{and}\quad \beta_1 + \ldots + \beta_b = 0$$

respectively, π_f is the plot-effect and by denoting $\pi_i^{(j)} = \pi_f$ for $f = (j-1)k + i$, they are subjected to the restrictions $\pi_1^{(j)} + \ldots + \pi_k^{(j)} = 0$, $j = 1, \ldots, b$, and e_f denotes the error term which is sometimes called the technical error.

The blocking was introduced in order to make the plot-effects within a block as homogeneous as possible. The meaning of this statement is rather vague as it is and it should be made more precise later on. One intuitive interpretation seems to be as follows: One can make the plot-effects $\pi_1^{(j)}, \ldots, \pi_k^{(j)}$ fairly small for each j by a proper blocking. Then after the random allocation of the treatments to plots in each block, if there is no way to distinguish the realization at hand from other possible allocations, the realization can be regarded as the average of all possible allocations. Denoting the incidence vectors of the realization by $\underline{\zeta}_\alpha^{(0)}$, one may proceed based on the model

$$x_f = \gamma + \sum_{\alpha=1}^{k} \zeta_{\alpha f}^{(0)}\tau_\alpha + \sum_{j=1}^{b} \eta_{jf}\beta_j + e_f . \tag{1.2}$$

This leads to the familiar F-distribution of the F-statistic appearing in the analysis of variance. One might notice this sort of thinking seems to underly the whole discussions made by R. A. Fisher (1926, 1959).

However if one sticks to the model (1.1) and to the randomization procedure, then one gets a different story yet obtaining the same conclusion in asymptotic sense specified later on.

The main purpose of this paper is the justification of the traditional conclusion by sharpening the definition of a proper blocking and the concept of the asymptotic equivalence of probability distributions [see Ikeda (1963, 1968), Ogawa and Ikeda (1973)].

2. DESCRIPTION OF THE ANALYSIS OF VARIANCE IN MATRIX LANGUAGE.

Let the incidence matrix for treatments be $\Phi = ||\underline{\zeta}_1 \underline{\zeta}_2 \cdots \underline{\zeta}_k||$, then it is easy to see that $\Phi'\Phi = bI_k$. Hence let the relationship matrix for treatments be $T = ||t_{fg}|| = \Phi\Phi'$, where

$$t_{fg} = \begin{cases} 1, & \text{if the f-th and g-th units receive the same treatment} \\ 0, & \text{otherwise,} \end{cases}$$

$$\left(\frac{1}{b}T\right)^2 = \frac{1}{b}T \qquad \text{(idempotency).}$$

Similarly let the incidence matrix for blocks $\psi = ||\underline{\eta}_1\ \underline{\eta}_2 \cdots \underline{\eta}_b||$, then it can be seen that $\psi'\psi = kI_b$. Hence let the relation matrix for blocks be $B = ||b_{fg}|| = \psi\psi'$, where

$$b_{fg} = \begin{cases} 1, & \text{if the f-th and g-th units belong to the same block} \\ 0, & \text{otherwise.} \end{cases}$$

As a matter of fact, B turn out to be

$$B = \left\| \begin{matrix} G_k & & & \\ & G_k & & \\ & & \ddots & \\ & & & G_k \end{matrix} \right\|$$

where G_k is a kxk matrix whose elements are all unity. Also,

$$\left(\frac{1}{k}B\right)^2 = \frac{1}{k}B \quad \text{(idempotency).}$$

The kxb matrix N defined by

$$N = ||n_{\alpha j}|| = \Phi'\psi, \quad \alpha = 1, \ldots, k;\ j = 1, \ldots, b \qquad (2.1)$$

is called the incidence matrix of the design and it can be seen that

$$n_{\alpha j} = \begin{cases} 1, & \text{if the } \alpha\text{-th treatment appears in the j-th block,} \\ 0, & \text{otherwise.} \end{cases}$$

In the case of a complete block design, N turns out to be

$$N = \left\| \begin{matrix} 1 & 1 \cdots\cdots 1 \\ 1 & 1 \cdots\cdots 1 \\ \vdots & \vdots \ddots \vdots \\ 1 & 1 \cdots\cdots 1 \end{matrix} \right\| , \quad \text{hence} \quad \begin{matrix} NN' = bG_k \\ N'N = kG_b \end{matrix} .$$

Let $\underline{1}' = (1, 1, \ldots, 1)$, then $G = \underline{1}\,\underline{1}' = G_n$.

Now if we denote the observations by $\underline{x}$, then the treatment totals and the block totals are $T_\alpha = \underline{\zeta}'_\alpha x$, $(\alpha = 1, 2, \ldots, k)$, and $B_j = \underline{\eta}'_j x$, $(j = 1, 2, \ldots, b)$, respectively and the grand total is $\underline{1}' x = n\bar{x}$.

The total sum of squares S^2 is given by $S^2 = \underline{x}'\,\underline{x} - n\bar{x}^2 = \underline{x}'\,\underline{x} - \frac{1}{n}\underline{x}'\,\underline{1}\,\underline{1}'\,\underline{x} = \underline{x}'\left(I - \frac{1}{n}G\right)\underline{x}$.

The sum of squares due to treatments is given by

$$S_t^2 = \frac{1}{b}\left(T_1^2 + \ldots + T_k^2\right) - n\,\bar{x}^2 = \underline{x}'\left(\frac{1}{b}T - \frac{1}{n}G\right)\underline{x},$$

and sum of squares due to blocks is given by

$$S_b^2 = \frac{1}{k}\left(B_1^2 + \ldots + B_b^2\right) - n\,\bar{x}^2 = \underline{x}'\left(\frac{1}{k}G - \frac{1}{n}G\right)\underline{x}.$$

Hence the sum of squares due to errors is given by

$$S_e^2 = S^2 - S_t^2 - S_b^2 = \underline{x}'\left(I - \frac{1}{b}T - \frac{1}{k}B + \frac{1}{n}G\right)\underline{x}$$

and consequently the F-statistic is given by

$$F = (b-1)\,\frac{\underline{x}'\left(\frac{1}{b}T - \frac{1}{n}G\right)\underline{x}}{\underline{x}'\left(I - \frac{1}{b} - \frac{1}{k}B + \frac{1}{b}G\right)\underline{x}} .$$

Thus our first task is to find out the conditional distribution based on the model $\underline{x} = \gamma\underline{1} + \Phi\underline{\tau} + \psi\underline{\beta} + \underline{\pi} + \underline{e}$ under the null-hypothesis H_0: $\underline{\tau} = 0$.

3. THE NULL DISTRIBUTION OF THE F-STATISTIC BEFORE THE RANDOMIZATION. Under the null-hypothesis H_0: $\underline{\tau} = 0$

$$\left(\frac{1}{b}T - \frac{1}{n}G\right) \underline{x} = \left(\frac{1}{b}T - \frac{1}{n}G\right) \underline{\pi} + \left(\frac{1}{b}T - \frac{1}{n}G\right) \underline{e}$$

and

$$\left(I - \frac{1}{b}T - \frac{1}{k}B + \frac{1}{n}G\right) \underline{x} = \left(I - \frac{1}{b}T - \frac{1}{k}B + \frac{1}{n}G\right) \underline{\pi}$$

$$+ \left(I - \frac{1}{b}T - \frac{1}{k}B + \frac{1}{b}G\right) \underline{e} \ .$$

Under the normal assumption of the error vector $\underline{e}$, i.e., $\underline{e} \sim N(0,\sigma^2 I)$, $\chi_1^2 = \underline{x}' \left(\frac{1}{b}T - \frac{1}{n}G\right) \underline{x}/\sigma^2$ is distributed as the non-central chi-square distribution with degrees of freedom (k-1) and the noncentrality parameter δ_1/σ^2, where $\delta_1 = \underline{\pi}' \left(\frac{1}{b}T - \frac{1}{b}G\right) \underline{\pi}$ $= \frac{1}{b} \underline{\Pi}' \underline{\Pi}$. The probability element of χ_1^2 is given by

$$e^{-\frac{\delta_1}{2\sigma^2}} \sum_{\mu=0}^{\infty} \frac{\left(\frac{\delta_1}{2\sigma^2}\right)^{\mu}}{\mu!} \cdot \frac{\left(\frac{\chi_1^2}{2}\right)^{\frac{k-1}{2}+\mu-1}}{\Gamma\left(\frac{k-1}{2}+\mu\right)} e^{-\frac{\chi_1^2}{2}} d\left(\frac{\chi_1^2}{2}\right) .$$

Likewise $\chi_1^2 = \underline{x}' \left(I - \frac{1}{b}T - \frac{1}{k}B + \frac{1}{n}G\right) \underline{x}/\sigma^2$ is distributed as the noncentral chi-square distribution with degrees of freedom (b-1) (k-1) and the noncentrality parameter δ_2^2/σ^2, where

$$\delta_2 = \underline{\pi}' \left(I - \frac{1}{b}T - \frac{1}{k}B + \frac{1}{n}G\right) \underline{\pi} = \underline{\pi}' \underline{\pi} - \frac{1}{b} \pi' \pi = \Delta - \delta_1 .$$

The probability element of χ_2^2 is given by

$$e^{-\frac{\delta_2}{2\sigma^2}} \sum_{\nu=0}^{\infty} \frac{\left(\frac{\delta_2}{2\sigma^2}\right)^{\nu}}{\nu!} \cdot \frac{\left(\frac{\chi_2^2}{2}\right)^{\frac{(b-1)(k-1)}{2}+\nu-1}}{\Gamma\left(\frac{(b-1)(k-1)}{2}+\nu\right)} e^{-\frac{\chi_2^2}{2}} d\left(\frac{\chi_2^2}{2}\right) .$$

Since χ_1^2 and χ_2^2 are stochastically independent, and the F-statistic is given by $F = (b - 1)\dfrac{\chi_1^2}{\chi_2^2}$, the probability element of F is

$$e^{-\frac{\Delta}{2\sigma^2}} \sum_{\ell=0}^{\infty} \frac{\left(\frac{\Delta}{2\sigma^2}\right)^{\ell}}{\ell!} \sum_{\mu+\nu=\ell} \frac{\ell!}{\mu!\nu!} \theta^{\mu}(1-\theta)^{\nu} h_{\mu\nu}(F)dF, \tag{3.1}$$

where

$$h_{\mu\nu}(F) = \frac{\Gamma\left(\frac{b(k-1)}{2} + \mu + \nu\right)}{\Gamma\left(\frac{k-1}{2} + \mu\right) \Gamma\left(\frac{(b-1)(k-1)}{2} + \nu\right)} \left(\frac{F}{b-1}\right)^{\frac{k-1}{2} + \mu - 1}$$

$$\times \left(1 + \frac{F}{b-1}\right)^{-\frac{b(k-1)}{2} - \mu - \nu} \frac{1}{b-1},$$

and $\theta = \frac{1}{b} \underline{\Pi}' \underline{\Pi}/\Delta$.

4. ASYMPTOTIC DISTRIBUTION OF θ DUE TO THE RANDOMIZATION. Let for the j-th block $\sigma_j = \begin{pmatrix} 1 & 2 & \cdots & k \\ \sigma_j(1) & \sigma_j(2) & \cdots & \sigma_j(k) \end{pmatrix}$ be a permutation of k objects, and let S_j be the permutation matrix corresponding to σ_j, i.e.,

$$S_j \begin{pmatrix} 1 \\ 2 \\ \vdots \\ k \end{pmatrix} = \begin{pmatrix} \sigma_j(1) \\ \sigma_j(2) \\ \vdots \\ \sigma_j(k) \end{pmatrix}.$$ And further let $U_\sigma = \begin{Vmatrix} S_1 & & & 0 \\ & S_2 & & \\ & & \ddots & \\ 0 & & & S_b \end{Vmatrix}$

be the permutation matrix corresponding to the randomization of the whole b blocks. Starting with any fixed incidence matrix Φ_0 of treatments, the randomization makes the incidence matrix Φ of the treatments a discrete random matrix such that $P\{\Phi = U_\sigma \Phi_0\} = \frac{1}{(k!)^b}$ for all U_σ.

Let $\Pi_j^\sigma = (\Pi_{1j}^\sigma \ldots \Pi_{kj}^\sigma)'$ where $\Pi_{\alpha j}^\sigma = \sum_{i=1}^{k} \zeta_{\alpha,i}^{(j)} \pi_{\sigma_j(i)}^{(j)}$,

$\alpha = 1, \ldots, k;\ j = 1, \ldots, b;\ \Pi^\sigma = \sum_{j=1}^{b} \Pi_j^\sigma$. It is not difficult

to see that $E\left(\underline{\Pi}_j^\sigma\right) = \underline{0}$ and $E\left(\underline{\Pi}_j^\sigma \underline{\Pi}_j^{\sigma'}\right) = \frac{\Delta_j}{k(k-1)} \Lambda_j$, where E stands

for the expectation operator with respect to the permutation distribution due to the randomization and

$$\Lambda_j = \left\| \begin{matrix} k-1 & -1 & \cdots & -1 \\ -1 & k-1 & \cdots & -1 \\ \vdots & \vdots & & \vdots \\ -1 & -1 & \cdots & k-1 \end{matrix} \right\| = kI_k - G_k \quad .$$

It should be noticed that $\sum_{j=1}^{b} \Lambda_j = b(kI_k - G_k)$.

We now state the uniformity conditions to be imposed on the plot-effects:

$$\left.\begin{aligned} &\overline{\Delta} = \frac{1}{b} \sum_{j=1}^{b} \Delta_j \to \Delta_0 \neq 0 \\ &\frac{1}{\overline{\Delta}^2} \frac{1}{b} \sum_{j=1}^{b} (\Delta_j - \overline{\Delta})^2 = 0(b) \end{aligned}\right\} \quad b \to \infty,$$

where $\Delta_j = \sum_{i=1}^{k} \pi_i^{(j)^2}$, $j = 1, \ldots, b$.

We refer to the central limit theorem due to W. Feller (1935).

<u>Theorem (W. Feller)</u>. Let $\underline{X}_1^{(k)}, \ldots, \underline{X}_n^{(k)}$ be a sequence of independent random vectors of k-dimensions such that

$$E(\underline{X}_i^{(k)}) = 0 \text{ and } E(\underline{X}_i^{(k)} \underline{X}_i^{(k)'}) = \Lambda_i^{(k)}, \quad i = 1, \ldots, n.$$

$$\frac{1}{n}\sum_{i=1}^{n}\Lambda_i^{(k)} \to \Lambda^{(k)}\ (\neq 0) \text{ and } \frac{1}{n}\sum_{i=1}^{n}\int_{||\underline{x}||>\varepsilon\sqrt{n}} ||\underline{x}||^2 dF_i^{(k)} \to 0 \text{ as } n \to \infty$$

are satisfied, where $F_i^{(k)}$ denotes the distribution function of $\underline{X}_k^{(k)}$ and $||\underline{x}||$ is the Euclidean norm of the k-dimensional vector $\underline{x}$. Then

$$L\left(\frac{1}{\sqrt{n}}\sum_{i=1}^{n}\underline{X}_i^{(k)}\right) \to N(0, \Lambda) \text{ as } n \to \infty.$$

Now take n=b and $\underline{X}_j^{(k)} = \underline{\Pi}_j^{\sigma}$, j = 1, ..., b, then

$$E(\underline{X}_j^{(k)}) = E(\underline{\Pi}_j^{\sigma}) = 0,\ E(\underline{X}_j^{(k)}\ \underline{X}_j^{(k)'}) = E(\underline{\Pi}_j^{\sigma}\ \underline{\Pi}_j^{\sigma'}) = \frac{\Delta_j}{k(k-1)}\Lambda_j.$$

Since all conditions of the theorem are satisfied and therefore

$$L\left(\frac{1}{\sqrt{b}}\underline{\Pi}^{\sigma}\right) \to N\left(0, \frac{\Delta_0}{k(k-1)}(kI_k - G_k)\right) \text{ as } b \to \infty.$$

Whence one obtains the result

$$L\left(\frac{k-1}{\Delta_0}\ \frac{1}{b}\underline{\Pi}^{\sigma'}\ \underline{\Pi}^{\sigma}\right) \to \chi^2_{k-1} \text{ in law as } b \to \infty.$$

5. ASYMPTOTIC EQUIVALENCE IN THE SENSE OF THE TYPE $(\mathcal{C})_d$ [Ikeda (1963, 1968)]. Two sequences of k_n-dimensional random variables $\{X_n^{(k_n)}\}$ and $\{Y_n^{(k_n)}\}$ where the dimensionalities of the random variables may depend on n, are said to be asymptotically equivalent in the sense of the type $(\mathcal{C})_d$ as $n \to \infty$, if

$$d\left(X_n^{(k_n)}, Y_n^{(k_n)}; \mathcal{C}\right) = \sup_{E\varepsilon\mathcal{C}}\left|P^{X_n^{(k_n)}}(E) - P^{Y_n^{(k_n)}}(E)\right| \to 0 \text{ as } n \to \infty,$$

where $\mathcal{C}$ is any given non-empty sublcass of the usual Borel field of subsets of the k_n-dimensional Euclidean space R_{k_n}. We denote this asymptotic equivalence by

$$X_n^{(k_n)} \sim Y_n^{(k_n)}, \ (C)_d, \ (n \to \infty).$$

Let M be the class of all subsets of the form

$$M = \{\underline{x} = (x_1, \ldots, x_{k_n})'; -\infty \leq x_i < a_i, \ i = 1, \ldots, k_n\},$$

where the a_i's are extended real values. $M = \phi$(empty set) if $a_i = -\infty$ for some i, and $M = R_{k_n}$ (the whole space), if $a_i = +\infty$ for all i. Let S be the class of all subsets of the form

$$S = \{\underline{x} = (x_1, \ldots, x_{k_n})'; b_i \leq x_i < a_i, \ i = 1, \ldots, k_n\},$$

where the a_i's and b_i's are extended real numbers, $S = \phi$ if $a_i = b_i$ for some i and $S = R_{k_n}$ if $b_i = -\infty$ and $a_i = +\infty$ for all i. Finally let B be the Borel field containing M. Evidently $M \leq S \leq B$.

A sequence of random variables $\left\{X_n^{(k_n)}\right\}$ is said to have the property B(S), if for any given positive number $\varepsilon > 0$, there exist a bounded subset $S \varepsilon$ and a positive integer n_0 such that

$$P^{X_n^{(k_n)}}(S) > 1 - \varepsilon \text{ for all } n \geq n_0.$$

Also it is said to have the property C(S), if for any given $\varepsilon > 0$, there can be found a positive number $\delta = \delta(\varepsilon)$ and a positive integer $n_0 = n_0(\varepsilon)$ such that

$$\mu(E) < \delta, \ E \in S \text{ implies } P^{X_n^{(k_n)}}(E) < \varepsilon \quad \text{for all } n \geq n_0,$$

where (E) stands for the Lebesque measure.
We now state three theorems which are necessary to show that the null-distribution of the F-statistic after the randomization is asymptotically equivalent to a familiar central F-distribution in the sense of the type $(M)_d$, as $b \to \infty$ if the uniformity conditions are satisfied.

<u>Theorem 1</u>. [S. Ikeda (1968)]. If $\left\{\underline{X}_n^{(k_n)}\right\}$ has the properties B(S) and C(S) simultaneously, then for any sequence of real numbers $\{c^n, \ldots, c_n^n\}$ such that $c_i^n \to 1$ as $n \to \infty$ for all i,

$$\underline{X}_n^{(k_n)} = \left(X_1^n, \ldots, X_{k_n}^n\right)' \sim \underline{Y}_n^{(k_n)} = \left(c_1^n X_1^n, \ldots, c_{k_n}^n X_{k_n}^n\right)'$$

$(\mathcal{M})_d$, $n \to \infty$.

Theorem 2. [S. Ikeda (1968)]. Suppose $\underline{X}_n^{(k_n)}$ and $\underline{Y}_n^{(k_n)}$ have the probability density functions $f_n(\underline{X})$ and $g_n(\underline{Y})$ respectively. If the Kullback-Leibler mean information converges to zero as $n \to \infty$, i.e.,

$$I(f_n; g_n) = E_{f_n}\left(\log \frac{f_n}{g_n}\right) \to 0 \text{ as } n \to \infty,$$

then

$$\underline{X}_n^{(k_n)} \sim \underline{Y}_n^{(k_n)}, \ (\mathcal{B})_d, \ n \to \infty.$$

Theorem 3. [S. Ikeda (1968)]. Let $\{(X_n, Y_n)\}$ be a sequence of 2-dimensional random variables, and let $F_n(x)$ and $p_n(y|x)$ be the cumulative distribution function of X_n and the conditional probability density function of Y_n given $X_n = x$ respectively. Further let $\{X_n^*\}$ be a sequence of random variables with cumulative distribution function $F_n^*(x)$. It is seen that Y_n has the cumulative distribution function

$$h_n(y) = \int_{R_1} p_n(y|x)d\, F_n(x).$$

Let a random variable with the probability distribution function.

$$h_n^*(y) = \int_{R_1} p_n(y|x)dF_n^*(x)$$

be Y_n^*, Then $X_n \sim X_n^*$, $(\mathcal{M})_d$, $n \to \infty$ implies $Y_n \sim Y_n^*$, $(\mathcal{M})_d$, $n \to \infty$, if the following three conditions are satisfied.

(1) There exists a sequence of real numbers $\{(c_n, d_n)\}$ such that the sequence $\{(X_n - d_n)/c_n\}$ has the property $B(\mathcal{S})$,

(2) $q_n(y|x) = p_n(y|c_n x + d_n)$ is differentiable with respect to x, and

(3) for any bounded subset $E \in S$

$$\sup_{x\ E} \int_{R_1} \left| \frac{\partial p_n(y|c_n x + d_n)}{\partial x} \right| dy$$

is bounded for all values of n which are sufficiently large.

One can show that

$$\theta \sim X_b (M)_b, \quad b \to \infty$$

where X_b has the probability density

$$f_b(x) = \frac{\Gamma\left(\frac{b(k-1)}{2}\right)}{\Gamma\left(\frac{k-1}{2}\right) \Gamma\left(\frac{(b-1)(k-1)}{2}\right)} x^{\frac{k-1}{2} - 1} (1-x)^{\frac{(b-1)(k-1)}{2} - 1}$$

$$\text{for } 0 < x < 1. \qquad (5.1)$$

One can also check that the conditions of Theorem 3 are also satisfied in our case. Thus integrating θ out from (3.1) with rexpect to the distribution (5.1), one obtains the familiar central F-distribution

$$\frac{\Gamma\left(\frac{b(k-1)}{2}\right)}{\Gamma\left(\frac{k-1}{2}\right) \Gamma\left(\frac{(b-1)(k-1)}{2}\right)} \left(\frac{F}{b-1}\right)^{\frac{k-1}{2} - 1} \left(1 + \frac{F}{b-1}\right)^{\frac{b(k-1)}{2}} d\left(\frac{F}{b-1}\right)$$

which is asymptotically equivalent to the null-distribution of the F-statistic under the randomization in the sense of the type $(M)_d$ as $b \to \infty$.

REFERENCES.

[1] Feller, W. (1935). Math. Zeitschrift 40, 521-559.
[2] Fisher, R. A. (1926). J. Min. Agr. 23, 503-513.
[3] Fisher, R. A. (1959). Statistical Methods and Scientific Inference, Oliver & Boyd, Edinburgh.
[4] Ikeda, S. (1963). Ann. Inst. Statist. Math. Tokyo 15, 87-116.

[5] Ikeda, S. (1968). Ann. Inst. Statist. Math. Tokyo 20, 339-362.

[6] Ogawa, J. and Ikeda, S. (1973). A Survey of Combinational Theory, J. N. Srivastava, et al (eds.), North-Holland Publishing Company, 335-347.

SUBJECT INDEX FOR VOLUME 1

absolutely continuous, 14, 209
absolutely monotone, 375
adaptive inference, 167
amplitude distribution, 140
analysis of variance, 408
approximate normality, 194
approximation theory, 185
association, 268, 271
asymmetric, 93
asymptotic efficiency, 59, 230, 231
asymptotic equivalence, 407, 409, 414
asymptotic expansion, 314
asymptotic normal, 95, 106, 227
asymptotic variance, 167
average noise power, 148
Bachalier's model, 98
Banach space, 188
Bayesian estimation, 110
Bennett problem, 131
Bernoullian input, 55
Bernstein polynomials, 186
Bessel, 119
best critical region, 126
beta, 14, 32, 108, 283, 381
beta function, 138
biextremal model, 358
binomial, 9, 22, 32, 43, 218, 242, 404
binomial mixture, 109
binormal, 256
bivariate Burr, 329
bivariate normal, 273, 403
bivariate t, 273
block effects, 408
blocking, 407
Bonferroni's inequality, 360
Borel field, 414
Borel-Tanner, 41
branching process, 237
breakdowns distribution, 53
Brownian motion, 190
Burr, 382
C-numbers, 19
C-type distribution, 27
Calgary Conference, 1
Calgary Course, 1
capture-recapture sampling, 33
carrier set, 214
Cauchy, 88, 165, 376
Cauchy's inequality, 351
censored family, 164
center of symmetry, 167
central limit theorem, 121, 228, 413
central location, 88
central moments, 44
Cesaro method, 188
characteristic function, 88, 115, 197, 204, 299, 325
characteristic roots, 296, 304, 346, 391
circular target, 278
complete block design, 407
complete sufficient statistic, 23, 123
complete symmetry, 173, 184
complex random matrix, 182, 396
compound Poisson process, 96
concentration parameter, 113
concordant, 274
conditional exponential family, 234
conditional unbiased test, 126
confidence cone, 126

confidence interval, 213, 289
confluent hypergeometric function, 62, 135
consistent estimator, 239
contagion, 31
contaminated independence model, 271
contingency, tables, 259
convexity, 161, 162, 278, 366
convolution, 41, 88, 104
Cornish-Fisher expansion, 261, 273, 389
correlation surface, 250
course on distributions, 4
covariance matrix, 64, 110, 115, 257, 304, 322, 394
coverage probabilities, 224
Cramer-Rao lower bound, 231
critical region, 124
Crow-Bardwell family, 60
cumulants, 41
curves of equal probability, 251
D'Alembert's functional equation, 173, 180
density free approach, 339
departure from normality, 196
dependence, 248, 274
dependent gamma rv's, 320
diagonal matrix, 181
direction vector, 116
directional data, 115
directions for dissemination, 4
directions for training, 4
directions in research, 3
Dirichlet-gamma, 283
discriminant analysis, 267
discrete limit distribution, 213
discrete rectangular, 33
distance, 193
distribution (see individual distributions)
domain of attraction, 87
double binomial, 49
double exponential, 162
double hypergeometric, 249
ecology, 56
Edgeworth expansion, 267
Edgeworth extension, 381
efficiency, 227
eigenfunctions, 259
eigenvectors, 267
elliptical probability, 93
empirical Bayes, 104
energy, 397
epidemiology, 56
equilibrium, 35
equivariant estimator, 175, 176, 184
ergodicity, 190, 229, 299
error vector, 411
Esseen's bound, 197
estimation, 106
Euclidean norm, 414
Euler method, 188
exponential, 110, 168, 267
exponential family, 161, 213, 237
exponential generating function, 21
extreme values, 356
F-distribution, 38, 381, 408
F-statistic, 411
Faa de Bruno, 47
factorial moment, 34, 64
factorization theorem, 233
Farlie-Morgenstern, 247, 262
Feller's distribution, 33
finite mixture, 103
first busy period, 52
first order autoregression, 237
Fisher ancillary principle, 126
Fisher information, 111
Fourier coefficient, 190
fractiles, 93
Frechet distribution, 262
Frechet inequality, 360
Frechet space, 190
functional equation, 208, 363
functional-integral equation, 203

Galton-Watson branching process, 235
gamma, 134, 324, 381
gamma mixing, 39
gamma-type vector, 320
Gaussian hypergeometric, 32
Gaussian noise, 132
generalized Bessel, 132
generalized Hermite polynomial, 299
generalized hypergeometric 283
generalized Laguerre polynomial, 294, 299
generalized negative binomial, 42
generalized Poisson, 42, 243
generalized power series, 84
generalized Stirling, 21
generating function, 371
geometric, 42, 83, 363
gradient projection, 97
Gumbel, 356
Hahn-Banach theorem, 189
Hajek's ordering, 166
Hankel transform, 293
Harr measure, 346
Hausdorff, matrix, 188
hazard rate, 363
heavy tail, 167
Hermite, 9
Hermite polynomial, 156, 260
Holder, 188
homogeneity, 248
homomorphism, 182
Hotelling, 295, 302
hyper-Poisson, 33, 77
hypergeometric, 31, 59, 135, 287, 313
hypergeometric with matrix argument, 281, 284
hypothesis testing, 123, 213
idempotency, 409
identifiability, 103
identification procedure, 391
incidence matrix, 409
incomplete gamma, 143
independence hypothesis, 327
infinite mixtures, 266
information matrix, 71, 87
inspection sampling, 33
intermediate root, 392
interquartile range, 162
invariant test, 125
inventory decision problem, 35
inverse Fourier transform, 326
inverse Laplace transform, 291
inverse Polya, 33
inverse sampling, 38, 110
isotropic random walk, 113
Jensen's inequality, 363
Johnson system, 382
K-matrix, 188
Katz family, 59
kernel, 185, 200
Khintchine's theorem, 356
Kullback-Leibler mean information, 416
kurtosis, 161, 198, 254, 382
Lagrange distribution, 41
Lagrange expansion, 41
Lagrangian gamma, 241
Lagrangian Poisson, 241
Laplace-Hagg, 33
large deviation, 198
largest root, 394
lattice type, 200
law of large numbers, 228
left truncated power series, 13
level of crowding, 34
liability accident model, 38
life distribution, 162
life model, 168
light tail, 167
likelihood function, 94, 108, 117, 227
limit distributions, 213, 356
limit preserving method, 188
limit theorems, 203
linear correlation, 262
linear dependence, 65, 272
linear exponential family, 268
linear fractional function, 375

linear functionals, 189
linear regression, 260
linearized statistic, 389
locally Euclidean group, 178
location family, 105
location model, 161
location scale family, 110
log characteristic function, 91
log concave, 374
log-convex, 363
log-normal, 89, 207
logarithmic, 20
logistic, 162, 359, 374
long tail, 98
loss function, 173
lost games, 38
Mandelbrot's model, 98
marched filters, 134
Marcum's distribution, 134
Markov process, 190, 227
Marlow's factorial distribution, 33
Martingale, 228
matrix-beta distribution, 340
matrix-variate hyper-geometric, 281
maximal correlation, 259
maximal invariant, 128
maximum likelihood, 17, 59, 63, 77, 94, 110, 124, 231, 239, 327
Maxwell-Boltzmann, 133
mean deviation, 167
mean direction, 127
mean distance estimator, 96
mean ergodic theorem, 190
mean squared error, 97
mean vector, 115
Meijer G-functions, 288
meromorphic, 375
meterology, 56
metric space, 193
minimal sufficient statistics, 93, 123,233
minimax unbiased estimators, 110
minimum chi-square estimator, 63, 110
minimum distance estimator, 95
minimum variance unbaised estimation, 9, 10, 13, 19
mixed distribution, 106
mixed model, 359
mixed negative-binomial, 39
mixed Poisson, 39
mixing distribution, 104, 200
mixing models, 31
mixing processes, 38
mixing proportions, 105
mixture representation, 131
mixtures, 98, 105, 110, 195, 211
modified Bessel function, 114, 132
modulo sequence, 83
moment estimators, 96, 110, 384
moment problem, 185, 203
moment sequence, 188
monotone likelihood ratio, 128, 166
monotone likelihood ratio family, 213
monotone sequence, 188
Monte Carlo techniques, 266
Moore-Penrose inverse, 342
multi-sample problems, 113
multinomial, 16, 94, 108
multiparameter Stirling, 19
multiple correlation, 16, 250
multiple normal surface, 250
multiple time series, 391
multiple truncation, 19
multiple Poisson, 16
multivariate beta, 281, 337
multivariate Burr, 329, 378
multivariate Dirichlet, 282
multivariate exponential, 268
multivariate extreme value, 355, 359, 378
multivariate folded normal, 286
multivariate gamma, 319
multivariate hazard rates, 267
multivariate hypergeometric, 268
multivariate logistic, 378
multivariate models, 266

multivariate normal, 272, 283, 319
multivariate Pearson type, 257
multivariate power series, 17
multivariate stable laws, 91
multivariate t, 286
Nakagami amplitude distribution, 131
narrowband Gaussian process, 141
near-normality, 382
negative binomial, 9, 38, 43, 153, 186, 207, 242
negative dependence, 278
negative multinomial, 16
non-central beta, 295
non-central chi, 133, 143
non-central chi-square, 15, 141, 303
non-central moments, 47, 387
non-central Wishart, 346
non-Gaussian, 137
non-symmetric, 193
non-uniform phase, 131
normal, 88, 165, 223, 248, 320, 389, 402
normal marginals, 262
normal probability paper, 90
normal random vector, 321
normal scores, 165
normalized spacing, 169
one-sided confidence limits, 213, 220
one-sided Gaussian, 133
optimality properties of tests, 115
order statistics, 94
ordered roots, 397
ordering, 161
orthogonality, 259, 296, 307, 313, 346, 353
parabolic cylinder function, 135
parallel system, 364
Pareto, 170
partial moment curves, 251
partial ordering, 274
partitions, 309
Pascal, 186
pattern recognition, 391
peakedness, 279
Pearson system, 195, 257, 381
permutation distribution, 413
permutation matrix, 412
perturbation approximation, 267
phase distributions, 140
phi-square-boundedness, 259
Pitman-Koopman, 10
planar Wishart distribution, 338
Poisson, 9, 20, 53, 62, 134, 186, 218, 242
Poisson input, 53
Poisson kernel, 185
Polya, 33
positive definite matrix, 292
positive dependence, 271
positive orthant dependence, 275
positive quadrant dependence, 272
power series, 9, 17, 236
probability contours, 258
proceedings, 3
products, 401
principal components, 391
pulse, 147
quadrant dependence, 274
quadratic form, 131, 345
queueing, 41, 245
radar, 131
radio communication, 131
Rainville, 156
random allocation, 408
random counts, 56
random matrix, 281, 300, 391
random sine wave, 136
random walk, 363
randomization, 407
Rao-Blackwell theorem, 124
ratio property, 291
ratios and products, 401
Rayleigh, 124, 133
recurrence relation, 13
regression function, 254, 272
residue, 144

Riesz theorem, 189
right tail, 163
risk function, 184
robustness, 161
sample fractiles, 94
sample kurtosis, 168
sample mean direction, 113
sample moments, 66, 246
scaling matrix, 257
Schnirelmann, 11
score function, 164
security prices, 98
series and parallel systems, 363
series representation, 345
sigmoid shape, 90
signal amplitude, 148
signal-pulse-noise, 148
similar test, 129
simultaneous tests, 398
single server queue, 52
skew surface, 249
skew correlation surface, 251
skew-symmetric matrix, 313
skewness, 193, 259, 382
spacings, 391
sphericity, 395
square-law detector, 148
stability relation, 357
stable law, 87
star-shaped function, 161
stationary process, 204, 229
stationary noise, 131
signal analysis, 391
STER, 31
stochastic difference, 275
stochastically larger, 275
strongly consistent, 96, 231
student statistic, 109, 277, 381
studentization, 299
study institute purpose, 2
sufficient statistic, 24, 233
sum symmetric power series, 9
summability, 185
superadditive function, 169
survival function, 265
Swerling, 134
symmetric convex set, 277
symmetric distribution, 161
symmetric function, 109
symmetric matrix, 302
symmetric stable laws, 94, 195
symmetric statistic, 109
symmetrized distribution, 295
symmetry parameter, 88
T matrix, 188
tail probability, 9, 93, 166
target detection, 391
Taylor expansion, 299, 405
testing goodness of fit, 60
tests of uniformity, 124
time-homogeneous Markov process, 233
trace, 296
transformation, 381
trimmed means, 167
truncated geometric, 13, 85
truncated normal, 404
truncated power series, 9
unbiased estimator, 11, 20
unbounded carrier set, 221
uncorrelated normal, 402
uniform, 104, 114, 162, 404
uniform metric, 193
uniformly most power test, 125
unit normal law, 193
unit random vector, 113
unordered roots, 397
urn model, 31, 38
variance stabilizing transformation, 123
vector measures, 190
von Mises-Fisher, 113
waiting-line models, 246
Waring, 33
weak convergence, 213
weak relative compactness, 191
Weibull, 206, 208
Weierstrass theorem, 86
weighted distributions, 31
weighted sums, 299
Whittaker function, 38
Wilcoxon scores, 165
Wilk's likelihood ratio, 295
Wishart matrix, 300, 391
zonal polynomials, 281, 306, 350, 396